IUTAM Symposium on Turbulent Mixing and Combustion

FLUID MECHANICS AND ITS APPLICATIONS
Volume 70

Aims and Scope of the Series

The purpose of this series is to focus on subjects in which fluid mechanics plays a fundamental role.

As well as the more traditional applications of aeronautics, hydraulics, heat and mass transfer etc., books will be published dealing with topics which are currently in a state of rapid development, such as turbulence, suspensions and multiphase fluids, super and hypersonic flows and numerical modelling techniques.

It is a widely held view that it is the interdisciplinary subjects that will receive intense scientific attention, bringing them to the forefront of technological advancement. Fluids have the ability to transport matter and its properties as well as transmit force, therefore fluid mechanics is a subject that is particulary open to cross fertilisation with other sciences and disciplines of engineering. The subject of fluid mechanics will be highly relevant in domains such as chemical, metallurgical, biological and ecological engineering. This series is particularly open to such new multidisciplinary domains.

The median level of presentation is the first year graduate student. Some texts are monographs defining the current state of a field; others are accessible to final year undergraduates; but essentially the emphasis is on readability and clarity.

For a list of related mechanics titles, see final pages.

IUTAM Symposium on
Turbulent Mixing and Combustion

Proceedings of the IUTAM Symposium held in Kingston, Ontario, Canada, 3–6 June 2001

Edited by

A. POLLARD
Queen's University,
Department of Mechanical Engineering,
Kingston, Ontario, Canada

and

S. CANDEL
Ecole Centrale Paris,
EM2C, CNRS, France

KLUWER ACADEMIC PUBLISHERS
DORDRECHT / BOSTON / LONDON

A C.I.P. Catalogue record for this book is available from the Library of Congress.

ISBN 1-4020-0747-7

Published by Kluwer Academic Publishers,
P.O. Box 17, 3300 AA Dordrecht, The Netherlands.

Sold and distributed in North, Central and South America
by Kluwer Academic Publishers,
101 Philip Drive, Norwell, MA 02061, U.S.A.

In all other countries, sold and distributed
by Kluwer Academic Publishers,
P.O. Box 322, 3300 AH Dordrecht, The Netherlands.

Printed on acid-free paper

Printed in the Netherlands.

IUTAM Symposium on Turbulent Mixing and Combustion

Scientific Committee:

S. Candel (France, Co-Chair)
J. Ferziger (USA)
J.C.R. Hunt (England)
A.V. Johannson (Sweden)
A.E. Perry (Australia, deceased)
N. Peters (USA)
A. Pollard (Canada, Co-Chair)
S. Pope (USA)
T. Takeno (Japan)
T. Tatsumi (Japan, IUTAM)

Local Organisation:

A. Pollard (Chair)
B. Pollard, Symposium administration
E. Cerisano, Symposium support

Session aides:

P. Lavoie (senior session aide)
M. Cunningham
C. Drevet
T. Holme
A. Johnstone
S. Waterman

Additional assistance from ITS, Queen's University: L. Brown, M. Giacomello, T. Forrest

Symposium Sponsors:

IUTAM
National Research Council of Canada
Natural Resources Canada (CANMET)
Queen's University
CAGCT, Queen's University
Canadian Gas Association

Acknowledgement

This volume could not have been assembled without the dedication and perseverance of Andrew Johnstone and Philippe Lavoie, both M.Sc. students at Queen's University. They are part of the next generation of talented researchers in turbulence, mixing and combustion.

Dedication

The Scientific Committee was saddened to learn of the death of Professor Tony Perry in January 2001. This Symposium is dedicated to his memory.

Anthony Edward Perry

˜ TONY ˜
19.2.1937 to 3.1.2001

"...he was a most distinguished researcher. Townsend said, "the best in turbulence" "

- Julian Hunt 2001

Professors Perry and Hunt are members of the Scientific Committee of the IUTAM Symposium on Turbulent Mixing and Combustion.

Preface

The goals of the Symposium were to draw together researchers in turbulence and combustion so as to highlight advances and challenge the boundaries to our understanding of turbulent mixing and combustion from both experimental and simulation perspectives; to facilitate cross-fertilization between leaders in these two fields. These goals were noted to be important given that turbulence itself is viewed as the last great problem in classical physics and the addition of chemical reaction amplifies the difficulties enormously.

The papers that have been included here reflect the richness of our subject. Turbulence is rich and complex in its own right. And, its inner structure, hidden in the morass of scales, large and small, can dominate transport. Earlier IUTAM Symposia have considered this field, Eddy Structure Identification in Free Turbulent Flows, Bonnet and Glauser (eds) 1992 and Simulation and Identification of Organized Structures in Flows, Sorensen, Hopfinger and Aubry (eds) 1997. The combustion community is well served by its specialized events, most notable is the bi-annual International Combustion Symposium, held under the auspices of the Combustion Institute. Mixing is often considered somewhere in between these two. This broad landscape was addressed in this Symposium in a somewhat temporal linear fashion of increasing complexity. The lectures considered the many challenges posed by adding one element to the base formed by others: turbulence and turbulent mixing in the absence of combustion through to turbulent mixing dominated by chemistry and combustion. The Symposium gathered experimental, theoretical and computational contributions, which were focused on either fundamental or more practical questions.

Direct numerical simulations were described in a number of papers in this Symposium. DNS have emerged as a preferred method to explore the details of turbulence and combustion. Another area of considerable interest for many practical applications is that of large eddy simulations (LES). Progress during the recent period has been made mainly for non-reactive flows. Much research is now focused on the development of large eddy simulations for combustion and particularly sub-grid scale models. This aspect was well covered by a group of papers in this Symposium. There are challenging problems in this field and many open questions; encouragingly, some clear demonstrations were presented of the potential of these novel computational approaches.

This Symposium benefited enormously from insights gained from those who work in physical experiments where initial and boundary conditions are often difficult to quantify and control and indeed, the exquisite attention to detail and error analysis that is demanded by their craft. Much can be learnt from such detailed studies when they are focused on complex flow problems (mixing of supercritical fluids, synthetic jets, mixing in micro-combustors) or when they are specifically designed to provide data for the new modelling approaches and methods.

The invited speakers' lectures gave a remarkable introduction and overview of the many facets of turbulence, turbulent mixing and combustion and provided considerable insights on the future of this field of research. Villermaux began the Symposium by providing fresh new insights into mixing mechanisms. Dimotakis followed by adding the effects of combustion to mixing. Meneveau returned to mixing, but in the context of Large Eddy Simulation and scale separation. Vervisch considered LES of partially premixed combustion and thus provided the first complete foray into combustion. Moin, the penultimate invited speaker, considered the progress made in LES of a gas turbine combustor and control of mixing and last, but definitely not least, Janicka provided a perspective on the future of turbulent mixing and combustion.

The Symposium comprised 6 invited plenary and 50 papers presented either orally or as posters. The extensive discussions that followed each presentation are included here and they added a pedagogic flavour to the symposium. Each poster session was awarded a Best Poster Award. The winners were Helie and Trouve, Johnson and Kostiuk, and Juniper, Tripathi, Scouflaire, Robon and Candel. Papers were selected for inclusion in this volume based on their impact. The editors reviewed each paper. This volume contains 38 of the 56 papers presented and are presented here grouped broadly according to:

⇒ Turbulence and mixing (Invited - Villermaux)
⇒ Turbulent Mixing with Combustion (Invited - Dimotakis)
⇒ Modeling scalars (Invited - Meneveau)
⇒ DNS/LES of flames (Invited - Vervisch)
⇒ Control (Invited - Moin)
⇒ Perspectives for Turbulent Mixing and Combustion (Invited - Janicka)

The editors hope that this volume will stimulate further interaction between the turbulence, mixing and combustion communities.

Andrew Pollard and Sebastien Candel
Editors
Kingston and Paris, February 2002.

Contents

IUTAM Opening Address xvii
Tatsumi, T.

TURBULENCE AND MIXING
Session 1

Mixing as an Aggregation Process 1
Villermaux, E. (Invited Lecture)

Downstream Evolution of the Most Energetic Pod Modes in the Mixing Layer of a High Reynolds Number Axisymmetric Jet 23
Jung, E., Gamard, S., George, W.K., Woodward, S.

Micro Structure and Lagrangian Statistics of the Scalar Field with a Mean Gradient in Isotropic Turbulence 33
Brethouwer, G., Hunt, J.C.R., Nieuwstadt, F.T.M.

Dynamics & Control of Jets in Crossflow 45
Blossey, P.N., Narayanan, S., Bewley, T.R.

Characteristics of Supercritical Transitional Temporal Mixing Layers 59
Okong'o, N., Bellan, B.

On Accurate Experimental Measurements of the Dynamics of Large Scale Sructures in Turbulent Flows 73
Ewing, D., Woodward, S.

Mixing in a Cross-Jet Enhanced by a Coaxial Annular Synthetic Jet 85
Sigurdson, L., Diep, J.

TURBULENT MIXING WITH COMBUSTION
Session 2

Challenges in Turbulent Mixing with Combustion 95
Dimotakis, P.E. (Invited Lecture)

One-Dimensional Stochastic Simulation of Advection-Diffusion-Reaction Couplings in Turbulent Combustion 113
Hewson, J.C., Kerstein, A.R., Echekki, T.

Compressibility Effects on the Scalar Mixing in Reacting Homogeneous Turbulence 125
Livescu, D., Madnia, C.K.

The Interaction of Scalar Mixing and Heat Release in a Reacting Shear Layer 137
Pantano, C., Sarkar, S., Williams, F.

Influence of Differential Diffusion on Local Equilibrium and Super-Equilibrium Combustion in Turbulent Non-Premixed Flames 149
Gicquel, O., Hilbert, R., Thévenin, D., Darabiha, N.

Diffusion Edge-Flame Quenching 161
Boulanger, J., Vervisch, L.

A Model Description of the Effects of Variable Fuel-Air Mixture Composition on Turbulent Flame Propagation 169
Jérôme, H., Trouvé, A.

MODELING SCALARS
Session 3

Challenges in Modeling Scalars in Turbulence and LES 181
Meneveau, C., Kang, H.S., Charlette, F., Averill, J., Knio, O., Veynante, D. (Invited Lecture)

PDF of Temperature Fluctuations in Uniformly Sheared Turbulence 201
Ferchichi, M., Tavoularis, S.

A Mathematical Prototype to Validate LES Strategies for Turbulent Flames 211
Bourlioux, A., Khouider, B., Majda, A.J.

Study of Mixing in Swirling Turbulent Jets 221
Szasz, R.Z., Fuchs, L., Caraeni, D.A.

DNS/LES OF FLAMES
Session 4

LES of Partially Premixed Combustion 235
Vervisch, L., Domingo, P. (Invited Lecture)

Nonpremixed Turbulent Flames Investigated with Direct Numerical Simulation using Detailed Chemistry 251
Hilbert, R., Thévenin, D.

Counter-Gradient Scalar Transport in Large Eddy Simulation of Turbulent Premixed Flames 259
Tullis, S., Cant, R.S.

Local Flame Structure in Hydrogen-Air Turbulent Premixed Flames 269
Tanahashi, M., Ito, Y., Fujimura, M., Miyauchi, T.

Three-Dimensional Direct Numerical Simulations of Turbulent Flames Using Realistic Chemistry Modeling 279
Thévenin, D., Hilbert, R., Gicquel, O.

Simulation of a Buoyancy-Driven Jet Diffusion Flame 287
Charentenay, J., Thévenin, D., Zamuner, B.

Visualization of the Fuel Stripping Mechanism For Wake-Stabilized Diffusion Flames in a Crossflow 295
Johnson, M.R., Kostiuk, L.W.

CONTROL
Session 5

Progress in Control of Mixing and Large Eddy Simulation of Turbulent Combustion 305
Moin, P. (Invited Lecture)

Large Eddy Simulation of a Non-Premixed Turbulent Burner Using a Dynamically Thickened Flame Model 315
Légier, J.P., Poinsot, T., Varoquié, B., Lacas, F., Veynante, D.

Characterisation of an Air-Blast Injection Device With Forced Periodic Entries 327
Giuliani, F., Gajan, P., Diers, O., Ledoux, M.

Numerical Analysis of Hydrogen/Air Jet Diffusion Flame 337

Mizobuchi, Y., Tachibana, S., Shinjo, J., Ogawa, S., Takaki, R.

Turbulent Mixing With Sprays 347
Ben-Dakhlia, R., Charentenay, J., Darabiha, N., Giovangigli, V.

Large Eddy Simulations of Combustion Instabilities in a Swirled Combustor 357
Ducruix, S., Poinsot, T., Candel, S.

An Experimental Investigation of Mixing and Combustion Characteristics on the Can-Type Micro Combustor With a Multi-Jet Baffle Plate 367
Choi, H., Nakabe, K., Suzuki, K.

Effects of One, Two and Quadruple Tabs on the Mixing and Entrainment in the Near Field of Round Jets 377
McIlwain, S., Holme, T., Waterman, S., Pollard, A.

PERSPECTIVES FOR TURBULENT MIXING AND COMBUSTION
Session 6

Mixing and Combustion, Perspectives 387
Janicka, J., Kempf, A., Mengler, C., Sadiki, A. (Invited Lecture)

Using Large Eddy Simulations to Understand Combustion Instabilities in Gas Turbines 405
Poinsot, T., Schlüter, J., Lartigue, G., Selle, L., Krebs, W., Hoffmann, S.

Large Eddy Simulation of Turbulent Sprays 415
Menon, S., Sankaran, V.

Numerical Simulations of Partially Premixed Turbulent Combustion 427
Haworth, D.C., Jiménez, C., Cuenot, B., Poinsot, T., Blint, R.J.

Turbulent Combustion of Sprays Under Supercritical Conditions 439
Tripathi, A., Scouflaire, P., Candel, S.

Author Index 447

List of Participants 449

Opening Address

Turbulence and Combustion

Tomomasa TATSUMI
International Institute for Advanced Studies
9-3 Kizugawadai, Kizu-cho, Kyoto 619-0225, Japan

It is my great honour and pleasure to give an Opening Address in this IUTAM Symposium on "Turbulent Mixing and Combustion" on behalf of the IUTAM.

1. IUTAM

The present symposium is one of the IUTAM Symposia sponsored by the IUTAM, or the International Union of Theoretical and Applied Mechanics. Every two years, the IUTAM collects proposals for such symposia from the scientists in the various fields of Theoretical and Applied Mechanics all over the world and selects about eight to ten of them per year as the IUTAM Symposia. The IUTAM Symposia are reserved to invited participants and its refereed Proceedings are published in series mostly by the Kluwer Academic Publishers. To our great pleasure, now the IUTAM Symposia are generally considered as the top-level international symposia in the respective research fields and their Proceedings as the first-class literature on the subjects.

The IUTAM was founded in 1947 shortly after the end of the Second World War. It has, however, an excellent prehistory as the ICAM, or the International Congress of Applied Mechanics, whose beginning dates back to 1922, when an International Conference on Hydro- and Aerodynamics was held at Innsbruck, Austria, under the joint chairmanship of two young scientists, Theodore von Kármán in Aachen and Tullio Levi-Civita in Rome [1]. It was just after the First World War and concurrent with the great revolutions in Physics such as Relativity Theory and Quantum Mechanics. Although less eminent than these revolutions, there also existed novel developments in Hydro- and Aerodynamics such as Turbulence by Osborn Reynods, Shock Wave by G.I. Taylor and Boundary Layer by Ludwig Prandtl. Thanks to the introduction of these novel concepts, all unphysical singularities existed in old Hydrodynamics of an inviscid fluid have disappeared and given place to the physical realities including Turbulence, Shock waves and Bound-

A. Pollard and S. Candel (eds.), IUTAM Symposium on Turbulent Mixing and Combustion, xvii-xxii.

ary layers which can be described by the Navier-Stokes equations of the viscous fluid.

Von Kármán's idea was based on the recognition of a large discrepancy between such rapid developments in the new fields of Hydro- and Aerodynamics on one hand, and only a limited space devoted to such problems at scientific meetings on the other. He proposed to break the dependence of these new fields from their mother disciplines, Mathematics, Physics and Engineering. His proposal was strongly supported by Levi-Civita and they succeeded in organizing the above-mentioned International Conference.

This Conference, which was a highly successful one, was the begining of the ever-growing series of the ICAM which have been held almost regularly every four years at various cities in the world. After the interruption due to the Second World War, the ICAM was succeeded by the IUTAM as the ICTAM, or the International Congress of Theoretical and Applied Mechanics. It is well known that the 20th ICTAM was held in the year of 2000 at Chicago, USA.

2. Fluid Mechanics

About eighty years since the time of Innsbruck Conference, no one can deny that enormous developments have been achieved so far in various fields of Theoretical and Applied Mechanics, particularly in its core field, Fluid Mechanics.

As mentioned before, the big progress in Fluid Mechanics during the first half of the 20th century has been achieved by the introduction of new concepts such as turbulence, shock wave and boundary layer. In the later half of this century, the progress has been largely promoted by the employment of another series of novel concepts such as Local Equilibrium of Turbulence by A.N. Kolmogorov, Soliton by Norman Zabusky and M.D. Kruskal, Chaos by E.N. Lorenz and Fractals by Benoit Mandelbrot. The large contribution by the high-speed numerical computation is also to be noted.

Thus it seems to be well established that all fluid motions including highly complex flows at large Reynolds numbers can be described precisely and in detail as the solutions of the Navier-Stokes equations, either analytical or numerical, although it is still a matter to be proved mathematically. In this sense, von Kármán's idea of proposing Hydro- and Aerodynamics as an independent research sector from their mother disciplines seems to be well justified.

3. Turbulence and Combustion

Now let us consider our central subject "Turbulent Mixing and Combustion". Although I am not very familiar with this subject it was fortunate for me to learn from an excellent lecture on "Turbulent Mixing" given by Prof. Paul Dimotakis at the 20th ICTAM [2].

To my understanding, "Turbulent Mixing" provides the initial and boundary conditions for "Combustion" and, on the other hand, it finds its simplified version in the diffusion in homogeneous turbulence. Thus it can be said that in order to understand "Combustion" properly we have to know about "Turbulent Mixing", and in order to understand "Turbulent Mixing" we have to know about "Homogeneous Turbulence". Then, how far do we know about "Homogeneous Turbulence"? Concerning this point, people's views are at variance, but I think that the theory of Homogeneous Turbulence must be made as simple and clear as possible in order that it can be applied to Non-Homogeneous Turbulence of more complex nature. In this sense, let me present here my simple but hopefully correct view of "Homogeneous Turbulence" on this occasion.

4. Velocity Distributions of Turbulence

It is well known that the most complete statistical information of turbulence is provided by the probability distributions of its velocity $\mathbf{u}(\mathbf{x}, t)$ at all values of the coordinate $\mathbf{x}$ and the time t. The simplest of such distributions are the one- and two-point velocity distributions which are written for homogeneous turbulence as,

$$f(\mathbf{u}_1, t) = < \delta(\mathbf{u}(\mathbf{x}_1, t)|\mathbf{u}_1) > \tag{0.1}$$

$$f^{(2)}(\mathbf{u}_1, \mathbf{u}_2; \mathbf{r}, t) = < \delta(\mathbf{u}(\mathbf{x}_1, t)|\mathbf{u}_1)\delta(\mathbf{u}(\mathbf{x}_2, t)|\mathbf{u}_2) >, \mathbf{r} = \mathbf{x}_2|\mathbf{x}_1 \tag{0.2}$$

where $< \cdot >$ denotes the mean value with respect to a certain initial distribution.

The equation governing the one-point distribution f is already given by Lundgren [3] and Monin [4] independently, but the equation is not closed since it includes the two-point distribution $f^{(2)}$ as another unknown. This unclosedness is not solved even if we proceed to the equation for $f^{(2)}$ since the latter includes another higher-order distribution $f^{(3)}$ an so on, and this is indeed nothing but the central difficulty of the turbulence problem. So far, a number of attempts have been made for solving this difficulty mostly in terms of the moment-expansion formalism, but the distributions f and $f^{(2)}$ themselves have not been dealt with except for numerical and phenomenological arguments.

In this context, let me propose a closure assumption for connecting f and $f^{(2)}$ [5]. First, introduce a couple of the cross-velocities, that is, the velocity-sum $\mathbf{u}_+ = (\mathbf{u}_1 + \mathbf{u}_2)/2$ and the velocity-difference $\mathbf{u}_- = (\mathbf{u}_2 - \mathbf{u}_1)/2$ and express $f^{(2)}$ in term of these velocities as,

$$f^{(2)}(\mathbf{u}_1, \mathbf{u}_2; \mathbf{r}, t) = 2^{-3} g^{(2)}(\mathbf{u}_+, \mathbf{u}_-; \mathbf{r}, t) \tag{0.3}$$

and then factorize $g^{(2)}$ into the distributions of the velocity-sum g_+ and the velocity-difference g_- as,

$$g^{(2)}(\mathbf{u}_+, \mathbf{u}_-; \mathbf{r}, t) = g_+(\mathbf{u}_+, \mathbf{r}, t) g_-(\mathbf{u}_-, \mathbf{r}, t) \tag{0.4}$$

assuming the independence of g_+ and g_-.

Obviously, the cross-independence assumption (3) and (4) is equivalent to the notion of Kolmogorov's local equilibrium hypothesis which assumes the statistical independence of the small-scale components of turbulence from those of large-scale [6]. This relationship is also supported experimentally so far as the distance r is below the order of the inertial length [7].

Applying the assumption (3) and (4) to $f^{(2)}$ in the equation for f and doing some order-of-magnitude analysis, we obtain the following closed equation for f:

$$[\partial/\partial t + \alpha^2(t)\triangle] f(\mathbf{u}, t) = 0 \tag{0.5}$$

where $\triangle = |\partial/\partial \mathbf{u}|^2$ and $\alpha^2(t)$ is a function proportional to the energy dissipation rate,

$$\epsilon(t) = (\nu/2) q(\partial u_i/\partial x_j + \partial u_j/\partial x_i)^2 r \tag{0.6}$$

It should be noted that eq.(5) includes only a single parameter $\alpha^2(t) \propto \epsilon(t)$ but not the viscosity ν as an independent parameter. This means that the statistics governed by eq.(5) satisfies the inertial similarity which depends only upon $\epsilon(t)$, while Kolmogorov's local equilibrium similarity depends in general upon two independent parameters $\epsilon(t)$ and ν.

5. Inertial Similarity

Equation (5) for f has the following self-similar solution in time:

$$f(\mathbf{u}, t) = (t/(4/3)\pi\epsilon_0)^{3/2} \exp[-|\mathbf{u}|^2 t/(4/3)\epsilon_0], \quad \epsilon(t) = \epsilon_0/t^2 \tag{0.7}$$

which represents an isotropic normal (Gaussian) distribution of the one-point velocity $\mathbf{u}$. As mentioned before, the distribution (7) depends only upon an inertial parameter ϵ_0, which does not vanish in the inviscid limit $\nu \rightarrow 0$. This makes a clear contrast of eq.(7) to the normal distribution of weak turbulence which includes ν as the reverse diffusion constant.

Reflecting the lack of the external excitation in eq.(5), the distribution (7) changes in time. At the initial instant $t = 0$, it represents a uniform distribution with vanishing probability density. For $t > 0$, it grows up as the normal distribution with increasing height and decreasing variance in time, and eventually for $t \to \infty$ it tends to a delta distribution $\delta(\mathbf{u})$, representing the dead still state.

6. Inertial Normality

The same argument can be applied to the equation for the two-point velocity distribution $f^{(2)}$ which includes the three-point distribution $f^{(3)}$, and thus $f^{(2)}$ is obtained as the solution of this closed equation. Since, however, $f^{(2)}$ is a function of two-vector variables, it can be expressed more conveniently in terms of its factors in (4), the distributions of the velocity-sum $g_+(\mathbf{u}_+, \mathbf{r}, t)$ and the velocity-difference $g_-(\mathbf{u}_-, \mathbf{r}, t)$. Actually, all existing measurements of the two-point velocity distributions, either experimental or numerical, are mostly presented in terms of g_+ and g_-.

It should be noted that there exist isotropic normal distributions for both g_+ and g_- expressed as

$$g_\pm(\mathbf{u}_\pm, \mathbf{r}, t) = (t/(2/3)\pi\epsilon_0)^{3/2} \exp[-|\mathbf{u}_\pm|^2 t/(2/3)\epsilon_0] \qquad (0.8)$$

Obviously, the normal distributions (8) is identical to the normal distribution (7) except that the variance of (8) is 1/2 of that of (7). This implies that the distributions g_+ and g_- are nothing but the con- volutions of two independent velocity distributions f at the points $\mathbf{x}_1$ and $\mathbf{x}_2$. The independence of the two velocities for arbitrary distance $\mathbf{r}$ is guaranteed by the inertial similarity which makes all the correlation lengths of the order of the Kolmogorov length $\eta = (\nu^3/\epsilon)^{1/4}$ vanishingly small.

On the other hand, the inertial similarity causes a difficulty for the distribution (8) since it does not satisfy the coincidence conditions in the limit of $\mathbf{r} \to 0$,

$$g_-(\mathbf{u}_+, 0, t) = f(\mathbf{u}_1, t), \quad g_-(\mathbf{u}_-, 0, t) = \delta(\mathbf{u}_-) \qquad (0.9)$$

This discrepancy is, however, only apparent, and in reality there exists a dissipation region of finite length, and the distribution (8) changes rapidly but continuously to the limits (9) through this region. This process of transition of g_+ and g_- can be described in the framework of the present approximation (4) if we start from the equations for g_+ and g_-, but this problem will be dealt with in a separate work.

The above-mentioned relationship between the inertial normality and the dissipation region in the velocity distributions of turbulence has close resemblance with that for the inviscid outer flow and the boundary layer in the laminar problem. In either case, the inertial or inviscid regions occupy the major physical space, but they essentially owe their physical reality to the dissipative or viscous regions limited to the infinitesimal space. The inertial normality of the velocity distributions prevailing the major physical space gives a good reason for the practical success of the so-called quasi-normal theories of turbulence. The secret is how to take into account the inertial dissipation ϵ into the approximation scheme and actually the success or not of the theories totally depends upon their degrees of incorporating this inertial effect.

The merit of the inertial normality largely depends upon its simplicity and clarity which provide us with strong power for expanding the research field from Homogeneous Turbulence to Non-Homogeneous Turbulence, including shear flows, thermal effects, electro-magnetic interactions, chemical reactions, and so on and, of course, our central subject "Turbulent Mixing and Combustion".

7. References

1 Battimelli, G.(1988) The early international congresses of applied mechanics. In S. Juhaz ed. IUTAM - A Short History, Springer Verlag, pp.9-13.

2 Dimotakis, P.E.(2000) Recent advances in turbulent mixing. To be published in Proc. 20th ICTAM 2000.

3 Lundgren, T.S.(1967) Distribution functions in the statistical theory of turbulence. Phys. Fluids, 10, 969-975.

4 Monin, A.S. (1967) Equations of turbulent motion. PMM J. Appl. Math. Mech. 31, 1057-1068.

5 Tatsumi,T.(2001) Mathematical physics of turbulence. In T. Kambe et al eds. IUTAM Sym. on Geometry and Statistics of Turbulence. Kluwer Acad. Pub. pp.3-12.

6 Kolmogorov, A.N.(1941) The local structure of turbulence in incompressible viscous fluid for very large Reynolds numbers. Dokl. Akad. Nauk SSSR, 30. 301-305.

7 Sreenivasan K.R. and Dhuruva, B.(1998) Is there scaling in high-Reynolds number turbulence? Prog. Theo. Phys. Suppl. No.130, 103-120.

Chapter 1

MIXING AS AN AGGREGATION PROCESS

Emmanuel Villermaux
IRPHE, Université de Provence, Aix-Marseille 1
Technopôle de Château-Gombert
49, rue Frédéric Joliot-Curie
13384 Marseille Cedex 13, France
villerma@irphe.univ-mrs.fr

MÉLANGE. s. m. *Il se dit de l'aggrégation de plusieurs choses diverses.*

Encyclopédie, Diderot et d'Alembert, Paris, 1753.

Abstract On hand of several experiments, it is shown how a stirred scalar mixture relaxes towards uniformity through an aggregation process. The elementary bricks are stretched sheets whose rates of diffusive smoothing and coalescence built up the overall mixture concentration distribution. The cases studied in particular include the ever dispersing mixture, and the confined mixture in two and three dimensions.

Keywords: Mixing, passive scalar, diffusion, dispersion, turbulence.

1. How interaction leads to uniformity

A mixture is a transient state between the initial segregation of the constituents, and their ultimate homogeneity, when homogeneity is ever reached. Indeed, the overall mixing process of say, a drop of A in a medium of B involves two a-priori antagonists phenomena : a process of dispersion of the A drop in B by which the phases interpenetrate, and a process of aggregation from which homogeneity arises.

These two processes occur simultaneously in a stirred medium and are not in general completely independent.

A. Pollard and S. Candel (eds.), IUTAM Symposium on Turbulent Mixing and Combustion, 1-21.

Figure 1.1. Paul Signac (1863–1935), The Pink Cloud.

Depending on the dispersion properties of the underlying stirring flow, a mixture might evolve quite differently. This is very well illustrated by the painting displayed on figure 1.1, "Le Nuage Rose" (The Pink Cloud) made in Antibes, France, by Paul Signac (1863–1935, he was the leader of the *neo-impressionistes* school). The composition of the central big storm cloud is close to uniform since it merely expands as it is stirred by vigorous motions. On the other hand, one can distinguish at its left a tiny diffuse cloud composed of loose, weakly interacting sheets which fade away in a nearly independent way.

The scalar field in figure 1.1 i.e. the concentration in water droplets in air, is very different in these two, very nicely rendered, limit situations. In the storm cloud, the knowledge of the *average concentration* nearly tells everything about the mixture as the fluctuations are scarce around the mean. In the tiny evanescing cloud, the fluctuations of concentration are of the order of the mean; the mixture is far from homogeneous and it is clear that its whole *distribution of concentration* is needed to characterize it.

As we will show below, the structure of a mixture results from the equilibrium between diffusive smoothing and aggregation. The eye of the artist is already suggesting us that a uniform mixture is obtained when aggregation is enforced by confinement.

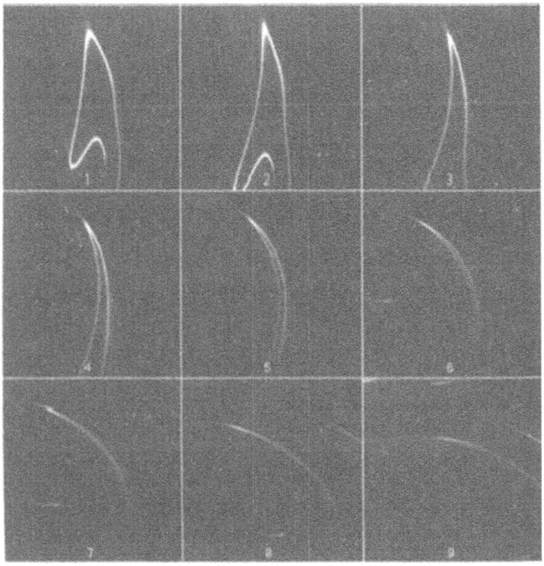

Figure 1.2. A folded sheet seen at consecutive instants of time embedded in a three-dimensional, turbulent flow and undergoing coalescence. The sheet is made visual by a planar two-dimensional cut through the medium. Length and time scales are given in figure 1.4.

2. The Elementary Interaction

Inject a stream of colored dye in a larger agitated stream of transparent fluid. The scalar field rapidly looks like the tiny evanescent cloud of figure 1.1, composed of sheets diluting in the transparent medium as they are stretched by the stirring motions (see for the details of the setup Villermaux and Innocenti 1999, and Girimaji and Pope 1991, Duplat and Villermaux 2000 to understand why a succession of random stretching motions form sheets). Before we analyse the scalar composition of this configuration, it is useful to consider first the elementary interaction between two scalar lamellae, as shown on figure 1.2.

Two pieces of a folded sheet embedded in a saddle point of the underlying displacement field are brought close to each other. The two pieces bear a slightly different concentration of scalar and at one point, the diffusive boundaries of the concentration profiles across the sheets interpenetrate to give rise to a single sheet, whose concentration profile is the *addition* of the concentration profiles of each individual sheet.

This elementary interaction rule is a consequence of the linearity of the Fourier diffusion equation. Let us consider a *single* lamellae first, let C be the scalar concentration in the vicinity of the lamellae and z

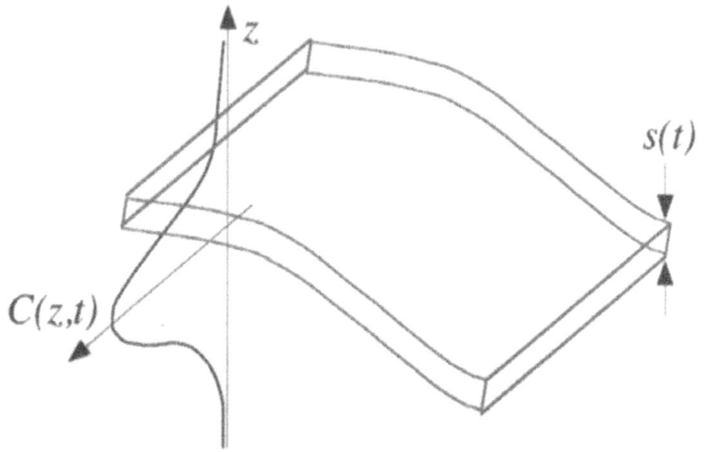

Figure 1.3. Sketch of an isolated stretched scalar lamellae being compressed in its transverse direction, and associated concentration profile.

a coordinate in the direction normal to the isoconcentration surface C. The diffusive uniformisation of the dye is enhanced by the stretching of the underlying motions. The convection-diffusion transport equation for C reduces (see e.g. Ranz 1979, Marble and Broadwell 1977, Marble 1988, Dimotakis and Catrakis 1999) to a one dimensional problem when the radius of curvature of the isoconcentration surface is large compared to the lamellae thickness. In that case the direction z aligns with the direction of maximal compression of the substrate perpendicular to the lamellae. The rate of compression $\frac{\partial \ln s(t)}{\partial t}$ can be measured from the initial thinning rate of a thick lamellae, when diffusive broadening of its concentration profile is still weak. The mass conservation equation then writes

$$\frac{\partial C}{\partial t} + \left(\frac{\partial \ln s(t)}{\partial t}\right) z \frac{\partial C}{\partial z} = D \frac{\partial^2 C}{\partial z^2}. \tag{1}$$

By the change of variables

$$\tau = D \int_0^t \frac{dt'}{s(t')^2} \quad \text{and} \quad \xi = \frac{z}{s(t)}, \tag{2}$$

equation (1) is reduced to a simple diffusion equation

$$\frac{\partial C}{\partial \tau} = \frac{\partial^2 C}{\partial \xi^2}. \tag{3}$$

The lamellae has an initial width s_0, and uniform concentration C_0. Its later concentration profile writes, in the scaled coordinates of equation (2)

$$\frac{C(\xi,\tau)}{C_0} = \frac{1}{2}\left[\operatorname{erf}\left(\frac{\xi+1/2}{2\sqrt{\tau}}\right) - \operatorname{erf}\left(\frac{\xi-1/2}{2\sqrt{\tau}}\right)\right]. \tag{4}$$

The thickness reduction process goes on until the rate of diffusive spreading of the concentration profile across the lamellae balances the rate of compression of the concentration gradient transverse to the lamellae. This instant defines the mixing time t_s after which the maximal concentration in the lamellae $C(0,t)/C_0 = \mathrm{Erf}(1/4\sqrt{\tau})$ starts to decay. The lamellae has reduced to a sheet which vanishes in the diluting medium.

The time t_s is readily estimated by noticing that the maximal concentration $C(0,t)$ decays as soon as $1/4\sqrt{\tau}$ is of order unity. Depending on the compression rate, that is, depending on the flow, the mixing time can take slightly different expressions. Consider a few examples:

1) *The simple shear, the egg beater, the point vortex flow in two dimensions...*

We consider incompressible flows in two dimensions which increase the length of material lines in proportion to time as γt. Many flows in two dimensions present a persistent shear and have this property (see e.g. Ranz 1979, Ottino 1989). The transverse size of a blob is therefore given by $s(t) = \frac{s_0}{\sqrt{1+(\gamma t)^2}}$ and, from equation (2), $\tau = \frac{Dt}{s_0}\left(1+\frac{(\gamma t)^2}{3}\right)$, thus

$$t_s \sim \frac{1}{\gamma}\left(\frac{\gamma s_0^2}{D}\right)^{1/3} \quad \text{and} \quad C(0,t) \sim (\gamma t)^{-3/2} \quad \text{for} \quad t > t_s. \tag{5}$$

2) *The unstable shear layer in three dimensions.*

We consider an incompressible flow in three dimensions which increases isotropically the length of material lines in proportion to time as γt. An example of such a flow is an unstable shear layer which rolls-up in the spanwise direction and produces concomitantly longitudinal vortices in the axial direction (Villermaux and Rehab 2000). The transverse size of a blob is in that case given by $s(t) = \frac{s_0}{1+(\gamma t)^2}$ and, from equation (2), $\tau = \frac{Dt}{s_0}\left(1+\frac{2}{3}(\gamma t)^2+\frac{1}{5}(\gamma t)^4\right)$, thus

$$t_s \sim \frac{1}{\gamma}\left(\frac{\gamma s_0^2}{D}\right)^{1/5} \quad \text{and} \quad C(0,t) \sim (\gamma t)^{-5/2} \quad \text{for} \quad t > t_s. \tag{6}$$

3) *The saddle point flow, the random uncorrelated stretchings....*

We consider an incompressible flow in which the length of a material line increases exponentially in time as $e^{\gamma t}$ or, in other words, whose rate of stretch is independent of its length. It is usually assumed that disordered, turbulent flows have this property, although laboratory measurements rather suggest a weaker (closer to a power law) time dependence (Villermaux and Innocenti 1999). The transverse size of a blob is, in two dimensions, given by $s(t) = s_0 e^{-\gamma t}$ and, from equation (2), $\tau = \frac{Dt}{s_0}\left(e^{2\gamma t} - 1\right)$, thus

$$t_s \sim \frac{1}{2\gamma} \log\left(\frac{\gamma s_0^2}{D}\right) \quad \text{and} \quad C(0,t) \sim e^{-\gamma t} \quad \text{for} \quad t > t_s. \tag{7}$$

The thickness of the sheet remains equal to $s(t_s) = \sqrt{\frac{D}{\gamma}}$ for $t > t_s$.

The times t_s given in equations (5)–(7) are the relevant mixing times of a blob as soon as the inverse of the elongation rate γ^{-1} is smaller than the diffusive time of the blob constructed on its initial size s_0^2/D, that is for $\frac{\gamma s_0^2}{D} > 1$. In this limit, t_s is essentially given by the time needed to deform the blob γ^{-1} and molecular diffusion, although a crucial step in the ultimate uniformization, plays only a weak correction role in the kinetics of the process.

The above concentration profiles and timescales are readily generalized to multiple blobs. In the same way that, owing to the linearity of the diffusion equation (3), the concentration profile (4) was obtained by a summation of the impulse response of the diffusion equation $\frac{C_0 d\xi}{2\sqrt{\pi\tau}} e^{-\xi^2/4\tau}$ over the width of the lamellae $[-\frac{s_0}{2}, \frac{s_0}{2}]$

$$C(\xi, \tau) = \int_{-1/2}^{1/2} \frac{C_0 d\xi'}{2\sqrt{\pi\tau}} e^{-(\xi-\xi')^2/4\tau}, \tag{8}$$

the concentration profile in the presence of two lamellae 1 and 2 as those shown on figure 2, each having a profile $C_1(\xi,\tau)$ and $C_2(\xi,\tau)$ is obtained by

$$C(\xi, \tau) = C_1(\xi, \tau) + C_2(\xi, \tau). \tag{9}$$

The operation is shown on figure 1.4. Also shown on figure 1.4 is the evolution of the distance between the maxima of concentration of the profiles, and the evolution of the maximal concentration of the overall profile. Those seem to be consistent with one of the flows described above (case 2), in particular the maximal concentration decreases fairly like $t^{-5/2}$.

The elementary composition operation (9) is the building block of the evolution of a complex mixture. If the concentration C in a region

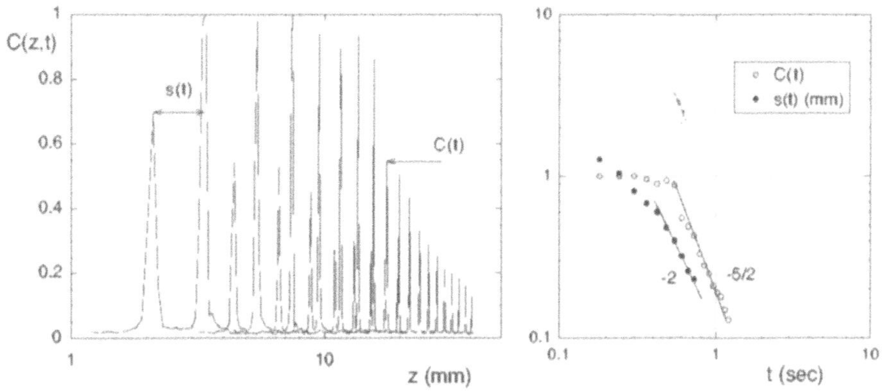

Figure 1.4. Left: Spatial concentration profiles $C(z,t)$ of the folded coalescing sheet shown on figure 1.2. The sheet moves perpendicular to itself as its two pieces gets closer and the figure shows the profiles resulting from the intersection of the sheet with a line fixed in space. Right: The distance $s(t)$ between the maxima of concentration of the profiles, and the evolution of the maximal concentration $C(t)$ of the overall profile as a function of time.

of a flow is the result of, for example, the addition of two different, independent sources 1 and 2, each characterized by a distribution $P_1(C)$ and $P_2(C)$, then the distribution $P(C)$ of the concentration C in the resulting mixture is the sum of the concentration C_1 from source 1 and of concentration C_2 from source 2, C_1 and C_2 being chosen independently in each of the original distributions provided that $C = C_1 + C_2$, that is

$$P(C) = \int_0^C P_1(C_1)P_2(C_2)\mathrm{d}C_1, \quad \text{with} \quad C = C_1 + C_2. \tag{10}$$

The above composition operation defines a convolution between the two original distributions, an operation whose consequences are examined below in some practical situations.

3. Practical mixtures: Limit cases

A mixture may evolve by ever dispersing in an ocean of diluting medium, or may be confined in a closed volume. In the latter case, its average concentration $\langle C \rangle$ is a constant of time, while in the first case, $\langle C(t) \rangle$ diminishes gradually with t (see figures 1.1, 1.5 and 1.7). The comparison between these extreme cases is of interest.

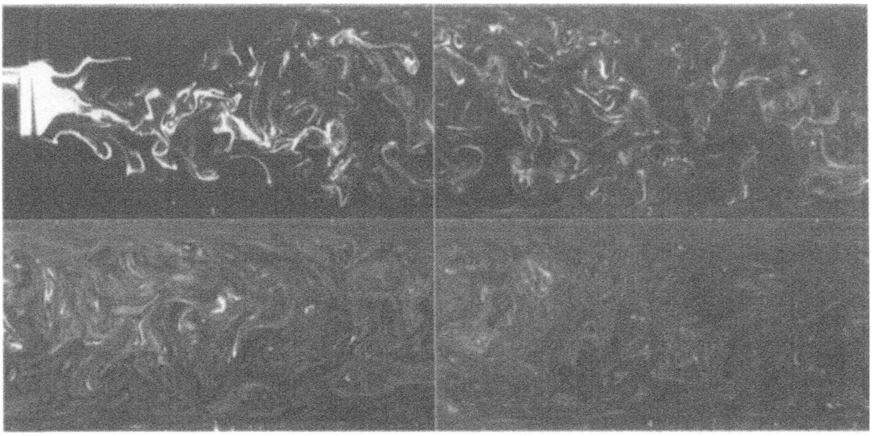

Figure 1.5. Mixing of a dye discharging from a jet of diameter $d = 8$ mm in a square ($L \times L$ with $L = 3$ cm) duct. From 1 to 4, successive instantaneous planar cuts of the scalar field at increasing downstream locations in the duct showing the progressive uniformization of the dye concentration.

3.1. The confined mixture

We let a turbulent jet of water plus dye (disodium fluorecein, $C_0 = 10^{-5}$ mol./l.) discharge in a square, transparent, long duct. The jet and the duct are immersed in a large tank filled with transparent water at rest. The jet exit velocity u is such that $Re = ud/\nu \simeq 10^4$ (see, however, section 4 for the intrinsic role played by Re).

The flow is made visual by means of a plane Argon laser sheet slicing the duct along its axis, and the evolution of the mixture from the entrance of the duct to farther downstream is recorded by a cooled, 12 bits, 1280×1024 pixels camera.

For a given duct cross section, the injection diameter d and the velocity of the co-flow at the entrance of the duct can be varied so that the average concentration of the dye in the channel $\langle C \rangle$ can be set at will. Since the cross-section of the duct and the average velocity of the mixture in the downstream direction are constant, the average concentration is conserved.

As can be seen on figure 1.5, the dye rapidly invades the whole duct cross section and its concentration differences are progressively erased as it travels downstream to relax towards a more or less uniform mixture.

After the dye has filled the channel cross section and evolves around a constant average concentration, the distribution $P(C)$ presents a skewed, bell shape which gets narrower around $\langle C \rangle$ in time (axial distances are converted to time through the average axial velocity with confidence as

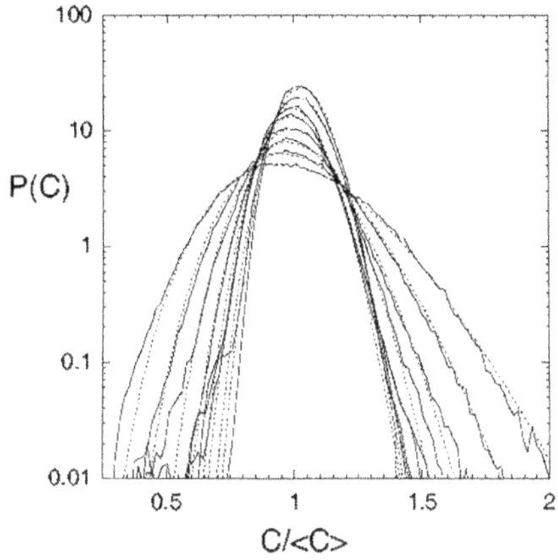

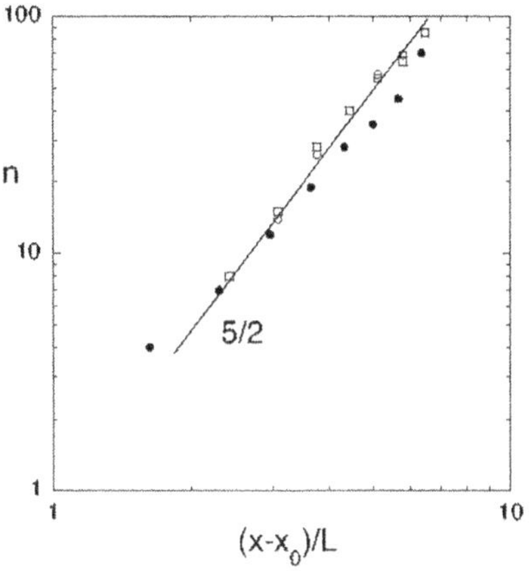

Figure 1.6. Left: Downstream evolution of the concentration distribution $P(C)$ as the dye progresses along the duct as shown on figure 1.5. The concentration distribution of the evolving mixture gets narrower around the average concentration $\langle C \rangle = 0.3$. solid line: experimental distributions, dashed line: distributions given by equation (11). Right: Fitting parameter n of the distributions (11) as a function of the downstream distance $(x-d)/L$. □ , ○: $Re = 10^4$, •: $Re = 5 \times 10^3$.

it is known that radial velocity profiles in a turbulent duct are flat). The shape is very well described by a family of one parameter distributions, namely Gamma distributions

$$P(C) = \frac{n^n C^{n-1} e^{-n\frac{C}{\langle C \rangle}}}{\langle C \rangle^n \Gamma(n)}. \tag{11}$$

The parameter n is adjusted at each downstream location for the Gamma distribution (11) to fit the experimental one. It is seen on figure 1.6 that the fairness of the fit holds for the whole concentration range, down to quite low probability levels, and accounts for the downstream deformation of $P(C)$ through the single parameter n, whose dependence on the downstream location is quite strong: figure 1.6 suggests a power-law dependence with an exponent between 2 and 3. The dependence of n on the jet Reynolds number is, although noticeable, very weak.

The physical content of (11) can be further explored by rewriting the distribution $P(C)$ in a different way. We are interested in the concentration distribution of the fluid particles bearing a *non-zero* concentration level: this avoids the ambiguity of the definition of the integration volume for the normalization of $P(C)$ and allows, as it will be seen below, a generalization to the case where the average concentration is itself a function of time (the huge Dirac Delta at $C = 0$ is certainly irrelevant when discussing the mixing of a small drop of dye in a vast ocean).

The initial concentration distribution is thus a Delta at $C = C_0 = 1$ as we make the reference concentration C_0 equal to 1 for convenience

$$P(C, t = 0) = \delta(C - 1). \tag{12}$$

Now the stream having a concentration equal to 1 is converted into a set of stretched sheets which elongate, fold, and undergo the processes of uniformization and coalescence we have described in section 2.

Consider an isolated sheet having a concentration equal to $1/n$ (we mean *maximal* concentration $C(0, t)$ and disregard the contribution of the intermediate levels of the spatial profile $C(z, t)$ on the global distribution $P(C)$). The number of such sheets it takes to restore the average concentration $\langle C \rangle$ by the addition process we have described in (9) is obviously equal to $n\langle C \rangle$. If it happens that, among an ensemble of $k\langle C \rangle$ sheets that have coalesced, the concentration $1/n$ of the individual sheets is distributed, what is the distribution function $Q(n, k)$ which corresponds to the experimental observation of equation (11)? It is convenient to answer this question to consider the Laplace Transform of $P(C)$

$$P(s) = \int_0^\infty P(C) e^{-sC} \mathrm{d}C. \tag{13}$$

The Laplace Transform of the concentration distribution $\delta(C - 1/n)$ associated to a single sheet of concentration $C = 1/n$ is $e^{-s\frac{1}{n}}$ and the Laplace Transform of the concentration distribution resulting from the addition of $k\langle C \rangle$ of those sheets is $e^{-s\frac{k}{n}\langle C \rangle}$ owing to the convolution property derived in (10). Our question is thus to find the distribution $Q(n, k)$ such that

$$P(s) = \int_0^\infty e^{-s\frac{k}{n}\langle C \rangle} Q(n, k) \mathrm{d}k \tag{14}$$

is the Laplace Transform of the experimental distribution $P(C)$ assumed to be represented by (11), namely

$$P(s) = \left(1 + \frac{s}{n}\langle C \rangle\right)^{-n}. \tag{15}$$

This question receives a very simple answer, since it is found that

$$Q(n, k) = \frac{k^{n-1} e^{-k}}{\Gamma(n)}, \tag{16}$$

meaning that n is distributed according to a Poisson law, of parameter k: the average value $\langle n \rangle$ and the variance $\langle n^2 \rangle - \langle n \rangle^2$ of n are both equal to k so that the dispersion of the distribution $Q(n,k)$ relative to its mean is $1/\sqrt{k}$.

Equations (14)–(16) are nothing but a reformulation of the experimental observation (11), so that the distribution $Q(n,k)$ can be considered as equivalent to the concentration distribution $P(C)$. Its interpretation is the following: in the course of their travel in the flow, fluid particles or lamellae experience a succession of different stretchings. Those, as we saw in section 2, are responsible for the decay of the concentration $C = 1/n$ of the lamellae. The distribution $Q(n,k)$ tells us that when $k\langle C \rangle$ sheets have coalesced, they were *not* all bearing the same concentration $1/k$. If it were the case, the distribution $P(C)$ would amount to a single peak at $C = \langle C \rangle$. Instead, among the aggregate of $k\langle C \rangle$ coalesced sheets, the individual concentrations of the sheets $1/n$ have a mean given by $1/k$ to restore $\langle C \rangle$, but are distributed, reflecting the different histories they have experienced in the flow. The distribution of concentration of the sheets $1/n$ is related to the distribution of n itself, namely $Q(n,k)$.

Note that if a very large number of sheets (measured by k) is needed to restore the average concentration $\langle C \rangle$, then

$$Q(n, k \to \infty) \to \delta(n-k). \tag{17}$$

This implies that the individual concentration of the sheets $1/n$ is very small, meaning also that they have spend a long time in the flow, and experienced a large collection of stretchings. The more a sheet spends time in the flow and accumulate stretchings, the more its individual history is alike the history of its neighbors and the more all the sheets resemble to each other: at the limit of very large times (i.e. very large k and equivalently very small $1/n$), all the sheets bear the same concentration $1/k$, they have coalesced $k\langle C \rangle$ times, the medium is homogeneous at concentration $\langle C \rangle$ i.e. $P(C, n=\infty) = \delta(C - \langle C \rangle)$, as it can be seen from (11).

It remains to give a fluid mechanics interpretation of n. This will be done in the next section, after we have examined the case of an ever dispersing mixture.

3.2. The ever dispersing mixture

The setup realizing an ever dispersing mixture is very similar to the previous one. There is no duct any more confining the mixture, the scalar is instead injected in a sustained, large scale turbulent flow in

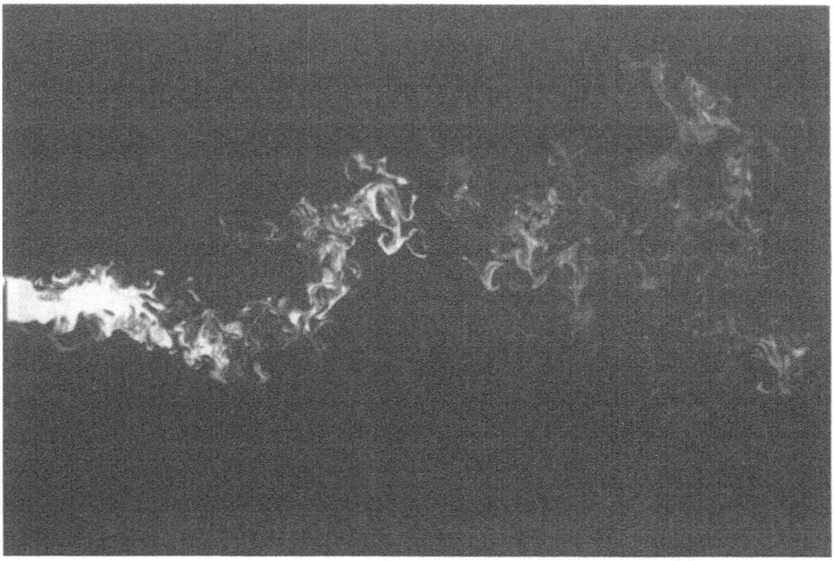

Figure 1.7. A snapshot of a dispersing plume made by the injection of a dye through a small tube of $d = 8$ mm in diameter on the axis of a larger turbulent jet whose integral scale is $L = 8$ cm at the injection location. $Re = u'L/\nu = 10^4$.

which it is free to disperses while it mixes (Villermaux and Innocenti 1999, Villermaux et al. 2001). The scalar is injected continuously in the far field and on the axis of a turbulent jet via a small tube whose diameter d is smaller than the local integral scale L (typically $d/L \approx 1/10$). The injection point behaves neither as a source, nor as a sink of momentum, in the mean, and the properties of the flow (stirring scale L, r.m.s. velocity u') are constant during the uniformization period of the scalar. These experiments have been repeated with three types of scalars namely temperature in air ($Sc = \nu/D = 0.7$), temperature in water ($Sc = 7$) and the concentration of disodium fluorescein in water ($Sc = 2000$), thus allowing to quantify the role of the intrinsic scalar molecular diffusive properties on the process.

As it is seen on figure 1.7, the scalar support takes the form of a set of stretched sheets, possibly coalescing as they spread out and fade away. The distribution of concentration $P(C)$ has a contrasted evolution compared to the confined case: it exhibits a cusped shape at the origin (the concentration levels close to $C = 0$ are by far the most probable) and a tail for larger values of C which can be approximated by an exponential fall-off (figure 1.8). The exponential tail steepens with distance x from the injection point as

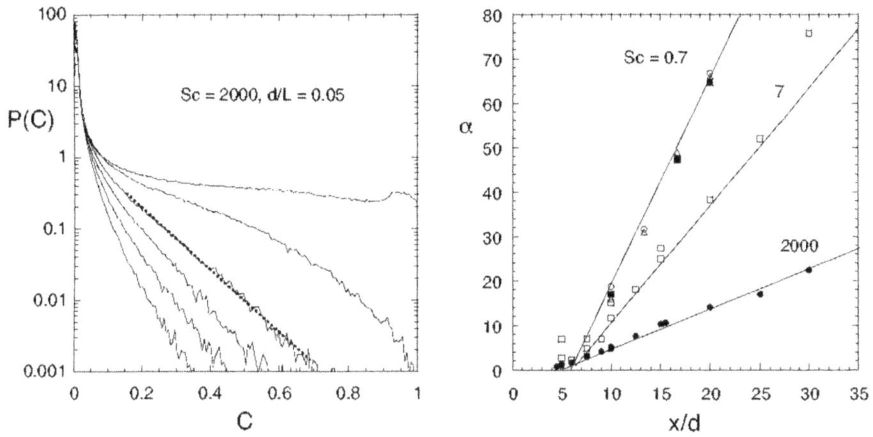

Figure 1.8. Left: Evolution of the concentration distribution $P(C)$ of the ever dispersing mixture of figure 1.7 for increasing distances x from the injection point. The tail of $P(C)$ exhibits an exponential fall-off $\exp(-\alpha C)$. Right: The argument α for different downstream locations $\frac{x}{d}$ and three Schmidt numbers. $\bullet$: $Sc = 2000$, $Re = 6000$ and 12000, $d/L = 0.05, 0.1, 0.6$. $\square$: $Sc = 7$, $Re = 6000$, $d/L = 0.05, 0.1, 0.16$. $\blacksquare$: $Sc = 0.7, d/L = 0.08, Re = 23000$; $\bigcirc, \triangle$: $Sc = 0.7, d/L = 0.08, Re = 45000$.

$$P(C) \sim \exp(-\alpha C) \quad \text{with} \quad \alpha = \frac{x}{d}\mathrm{f}(Sc) \tag{18}$$

where $\mathrm{f}(Sc)$ is a slowly increasing function of the Schmidt number. A fit consistent with the data is $\mathrm{f}(Sc) \sim \ln(Sc)$, although a weak power-law dependence of the form $\mathrm{f}(Sc) \sim Sc^{1/5}$ is not inconsistent as well. Time t and downstream distance x in (18) are linked by $x = ut$ where u is the mean velocity of the jet flow at the injection point.

The use of three different injection diameters indicates that d is actually the relevant lengthscale which sets the argument of the exponential fall-off in (18), when $d/L \ll 1$. Villermaux et al. (2001) have underlined the consequences of this fact on the spectral signature of the scalar field.

The distributed character of the damping factor n (see equation (16)) is probably best illustrated on figure 1.7. At a given location in the flow coexist sheets which still bear the injection concentration and sheets which are almost fading away in the diluting medium. The average concentration is, however, no more constant, and (15) has to adapted to this new situation. When the average concentration is constant, the Laplace Transform $P(s)$ of the concentration distributions is the product of n identical factors $\left(1 + \frac{s}{n}\langle C\rangle\right)^{-1}$, meaning that the distribution $P(C)$ can be represented as the convolution of n identical, exponential distributions $\frac{n}{\langle C\rangle}e^{-nC/\langle C\rangle}$. The damping factor n depends on time, and we call

$\langle C \rangle_i$ the average concentration in the medium for $\langle n \rangle = i$. For instance, if $n \sim t^\alpha$ and $\langle C(t) \rangle \sim t^{-\beta}$, then $\langle C \rangle_i \sim i^{-\beta/\alpha}$. We choose to write $P(s)$ as, again, a product of n factors and we find

$$P(s) = \prod_{i=1}^{n} \left(1 + \frac{s}{n} \frac{\Delta(i\langle C \rangle_i)}{\Delta i} \right)^{-1} \tag{19}$$

where $\Delta i = 1$. One checks that (15) is recovered when $\langle C \rangle_i = \mathrm{Cst}$, and that $\frac{\partial}{\partial s} P(s) \mid_{s=0} = \sum_{i=1}^{n} \frac{\Delta(i\langle C \rangle_i)}{\Delta i} = \langle C \rangle_n = \langle C(t) \rangle$: the average concentration of (19) is the current average concentration. In our example where $\langle C \rangle_i \sim i^{-\beta/\alpha}$, it is clear that the shape of the large concentration wing of $P(C)$ will be dominated by the factor giving rise to the less steep exponential. Since both α and β are positive, this factor corresponds to $i = 1$. Thus, the large concentration wing of $P(C)$ *is* an exponential of the form

$$P(C) \sim \exp\left(-\frac{n}{m(1)} C \right) \tag{20}$$

with $m(1) = 1 - \langle C \rangle_1$.

Equation (19) means that, as time goes on, the rate of aggregation between the sheets diminishes since the average concentration decays. Conversely, the current distribution of concentration is convoluted by distributions which get closer and closer to a Delta at $C = 0$, and is thus progressively left unchanged. The memory of the primary, exponential distribution is thus reinforced in the overall $P(C)$, therefore inducing the behavior (20).

The behaviors of $\langle C(t) \rangle$ and $n(t)$ in the present flow are dictated by the dispersion properties of the scalar plume and require a special discussion whose conclusions are:

When the injection size of the scalar is smaller than the integral, stirring scale i.e. when $d/L \ll 1$, one has

$$\langle C(t) \rangle = \frac{1}{n(t)} \quad \text{and} \quad n(t) = \frac{t}{t_s} \quad \text{with} \quad t_s = \frac{d}{u'} Sc^{1/5}. \tag{21}$$

When the scalar plume has reached the size of the whole stirring domain (as in the channel experiment of section 3.1) and does not disperse anymore,

$$n(t) = \left(\frac{t}{t_s} \right)^{5/2} \qquad \text{with} \qquad t_s = \frac{L}{u'} Sc^{1/5}. \tag{22}$$

Both relations (21) and (22) account for the observed trends in figures 1.6 and 1.8 and their dependence on Sc come from a stirring field of an extentional type in three dimensions (type 2 in section 2).

4. My cup of tea: Changing dimensions

When the average concentration is conserved in the course of the mixing process, we have shown that the concentration distribution $P(C)$ is described by a family of Gamma functions, and that their rate of deformation is prescribed by the rate of decay of the concentration in a single lamellae, measured by the dependence of n on time. In the three dimensional duct experiment of section 3.1, n was found to increase in proportion to the power 5/2 of the downstream distance in the duct (equivalently with time t), and we have interpreted this trend as reflecting the rate of decay of the maximal concentration in a lamellae stretched in the form of a two-dimensional sheet whose dimensions increase linearly with time (section 2).

By changing the dimensionality of the flow from three to two, one might expect that as long as the stirring motions are sufficiently irregular to ensure an addition of the concentration levels at random so that the distribution is, as in the three dimensional case, still generated by self-convolution and described by equation (11), it is by contrast also expected that the dependence of n on time will be different. Indeed, if the scalar is confined in a two-dimensional layer and the stirring protocol is such that material lines grow linearly with time (like in the *egg beater* case discussed in section 2), then the maximal concentration in one filament (the analogue of the sheet in two dimensions) decreases like $t^{-3/2}$ instead of $t^{-5/2}$ in three dimension.

We illustrate this difference on hand of a simple experiment consisting in stirring a blob of dye with a rod in a thin layer of very viscous fluid, by a two-dimensional quasi-periodic protocol (figure 1.9).

The stirring protocol consists in slicing the medium in the plane of the fluid layer with a small rod. A number of parallel cuts is made in one direction, and then the same number at right angle, this operation defining one cycle (figure 1.9). This very much looks like the motion of a spoon, stirring a drop of milk in a cup of tea except that the stirring rod is smaller that the size of the initial drop and that the fluid is very viscous (the appropriate image would rather be a straw in a milk shake).

At low Reynolds number (the typical Reynolds number of the motion of the rod is 10^{-1}), the fluid is deformed by the passage of the rod on a scale which is given by its own size (see e.g. Batchelor 1967). The length of material lines after one cycle is therefore equal to the distance traveled

Figure 1.9. The stirring protocol of a drop of ink deposited at the surface of pure Glycerol by a small rod. The sequence displays $2\frac{1}{2}$ stirring cycles.

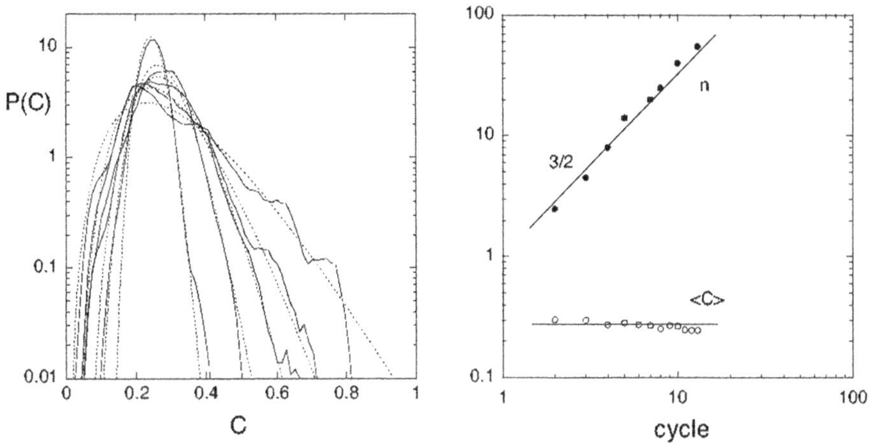

Figure 1.10. Left: Evolution of the dye concentration distribution $P(C)$ after 3, 4, 5, 6 and 13 cycles according to the protocol shown on figure 1.9. The concentration distribution of the evolving mixture gets narrower around the average concentration $\langle C \rangle$. solid line: experimental distributions, dashed line: distributions given by equation (11). Right: Fitting parameter n and average concentration of the distributions (11) as a function of the number of cycles.

by the rod in the medium during a cycle, and it is actually observed that the net contour length of the deformed scalar drop increases in proportion to the number of cycles.

The maximal rate of stretch is obtained for fluid particles close to the rod trajectory, while the protocol leaves nearly unstretched fluid parcels which therefore keep a concentration close to the initial concentration as the number of cycles is increased. Concomitantly, fluid particles are brought together in the wake of the rod, and coalesce. The amplitude of the slicing movements is constant through the cycles, so that the average concentration of the dye must be conserved; these are the ingredients required for the aggregation scenario we have described in section 3.1 to occur.

The dye concentration distributions $P(C)$ displayed on figure 1.10 are actually found to be reasonably well described by the Gamma functions family (11), although, those being obtained on a single image cycle after cycle, their statistical convergence is worst than in the duct experiment (figure 1.6). The average concentration $\langle C \rangle$ is approximately constant.

Since material lines grow linearly with time in this two-dimensional flow, the *egg beater* paradigm tells us that the maximal concentration in one filament must decay like $t^{-3/2}$. It is therefore expected that the number of convolutions in (11) will follow

$$n \sim (\text{number of stirring cycles})^{3/2}\,, \tag{23}$$

to keep $\langle C \rangle$ constant cycle after cycle, a trend which is not inconsistent with the data reported on figure 1.10.

5. Conclusion

The elementary act which alters the concentration distribution in a scalar mixture is the addition of the concentration levels between nearby elements coalescing by diffusion. As soon as the mixture is stirred in a more or less random manner, the addition is made at random among the levels available in the current overall distribution. This operation implies that the concentration distribution evolves by self-convolution and, when it actually does, gives an a-posteriori precise definition of what "random" means.

This evolution process was first proposed in this context by Curl (1963) and the since then celebrated *coalescence–redispersion* model, although a very similar construction was imagined by von Smoluchowski (1917) to represent the kinetic aggregation of colloidal particles. Pope (1985) and Dopazo (1994) give a general exposition of this and others stochastic models of mixing.

The addition of random variables erases the differences from the mean as a general property of the Central Limit Theorem (Feller 1971). The final distribution $P(C)$ reflects more the evolution process by addition itself than the details of the initial concentration distribution. However, this is only true at the limit of a very large number of composition operations. The distribution is particularly trivial in this uniform limit since it reduces to a Dirac Delta around the average concentration. The transient state of $P(C)$ is of more both practical, and conceptual interest since it is itself a mixture between the erasing process of addition of concentration between the scalar sheets, and the process giving birth to these sheets. The present experiments demonstrate that the signature of both of these ingredients appears in the global statistics $P(C)$, although it has been tempting in the past to retain one of them only (Pumir et al. 1991, Shraiman and Siggia 1994). The distribution of the damping factor $Q(n,k)$ which selects the Gamma shape of $P(C)$, the difference between the confined and the ever dispersing mixture, the fact that the diffusional properties of the scalar persist through the construction of $P(C)$ at any time are strong evidences supporting this *mixed* point a view.

It is instructive to see that an experiment performed at a very low Reynolds number produces composition fields very similar to those obtained at much larger Reynolds number (figures 1.6 and 1.10). The reason is that the mechanisms (an additive, random addition of the concentration levels) are the same. In that sense, the motion of the rod in the two-dimensional viscous fluid plays the role of the small-scale stirring motions present in high Reynolds number flows. Their role, if one accepts the interpretation given in section 3.1, is to ensure the independence –in a statistical sense– of the addition of the concentration levels, therefore inducing the self-convolution construction of the overall distribution $P(C)$. It is pleasant to think that the impact of turbulence on mixing amounts to such a simple paradigm.

References

Batchelor, G. K. *An Introduction to Fluid Dynamics.* Cambridge University Press, 1967.

Curl, R. L. 1963 Dispersed phase mixing: I. Theory and effect in simple reactors. *AIChE J.* **9**, (2), 175–181.

Dimotakis, P. E. and Catrakis, H. J. Turbulence, fractals and mixing. in *Mixing: Chaos and Turbulence*, CHAT, H., VILLERMAUX, E. AND CHOMAZ, J. M. (Eds.), Kluwer Academic/Plenum Publishers, New York, 1999.

Dopazo, C. Recent developments in pdf methods. In *Turbulent Reacting Flows*, Chapter 7. Libby, P. A. and Williams, F. A. Eds., Academic Press, 1994.

Duplat, J. and Villermaux, E. 2000 Persistency of material element deformation in isotropic flows and growth rate of lines and surfaces. *Eur. Phys. J. B* **18**, 353–361.

Feller, W. *An Introduction to Probability Theory and Its Applications.* John Wiley & sons, 1970.

Girimaji, S. S. and Pope, S. B. 1990 Material-element deformation in isotropic turbulence. *J. Fluid Mech.* **220**, 427–458.

Marble, F. E. 1988 Mixing, diffusion and chemical reaction of liquids in a vortex field *Chemical reactivity in liquids: fundamental aspects*, Moreau, M. and Turq, P. Editors, Plenum Press.

Marble, F. E. and Broadwell, J. E. 1977 The coherent flame model for turbulent chemical reactions *Project SQUID, Tech. Rep. TRW-9-PU.*

Ottino, J. M. 1989 *The kinematics of mixing: stretching, chaos, and transport.* Cambridge University Press.

Pope, S. B. 1985 Pdf methods for turbulent reacting flows. *Prog. Energy Combust. Sci.* **11**, 119–192.

Pumir, A., Shraiman, B. I. and Siggia, E. D. 1991 Exponential tails and random advection. *Phys. Rev. Letters* **66**, (23), 2984–2987.

Ranz, W. E. 1979 Application of a stretch model to mixing, diffusion and reaction in laminar and turbulent flows. *AIChE Journal* **25**, (1), 41–47.

Shraiman, B. I. and Siggia, E. D. 1994 Lagrangian path integrals and fluctuations in random flows. *Phys. Rev. E* **49**, 2912–2927.

von Smoluchowski, M. 1917 Versuch einer mathematischen Theorie der Koagulationskinetik kolloider Lsungen. *Z. Phys. Chem.* **92**, 129–168.

Villermaux, E. and Innocenti, C. 1999 On the geometry of turbulent mixing. *J. Fluid Mech.* **393**, 123–145.

Villermaux, E. and Rehab, H. 2000 Mixing in coaxial jets. *J. Fluid Mech.* **425**, 161–185.

Villermaux, E., Innocenti, C. and Duplat, J. 2001 Short circuits in the Corrsin-Oboukhov cascade. *Phys. Fluids* **13**, (1), 284–289.

Discussion

Norbert Peters, Stanford University "I wonder if there is a limit of infinite large Reynolds number you would not have more than 2 sheets that coalesce that many sheets that coalesce and then you would go the limit of the sub limit theorem and obtain a Gaussian?"

Reply: This is an excellent question. I have only considered two particle possibilities. Your question is what occurs if you consider 3 particles 4 5 and so on. In fact one can show that in that the case the probability distribution in the confined case for instance this the situation the limited situation where the mean gradients you mentioned is more likely to occur when the sheets are very close to each other in fact in that case instead of a pure gamma distribution you get the sum of gamma distribution and the sum of gamma distribution leads also to a central limit theorem so you will end up with a delta function around the mean. So this dose not change qualitatively the situation but quantitatively of course you have to consider separately 3 particle, 4 particle 5 particles processes and so on absolutely. But the physics is described as this process.

Stephen Pope, Cornell University "I think it is very nice to see a very simple deterministic model able to explain these statistics. There are a couple of assumptions regarding this model that I think require a bit more discussion. Firstly you have this formula this convolution giving the PDF of C as the sum of C1 and C2 that implies that C1 and C2 is statistically independent. OK? Secondly, you make the assumption that the strain of these sheets goes linearly with time which seems to be successful. That's not an obvious assumption in that if you had a uniform constant strain field then you would go exponentially with time. Where as the linear is a long time ahead of the shear flow. I wonder if you have any comment on why that's the assumption that works?"

Reply: OK. Again two very interesting questions. First the convolution operation is suggested from the experiments and I do agree that this amounts to consider that concentration levels are chosen independently. No correlation. This is a suggestion from the experiments. If you have two sources as I show, if you have two sources and if you recall the confrontation field in the presence of the two sources, this is strictly a convolution of the two initial periods, strictly. I mean that if there were a correlation it would be sensitive at the very very very low probability levels and this very probability within the experimental error. So if you, this is again a suggestion from the experiments, this is not an assumption from me this is what suggests the experiments. I built up my model from these observations. Now the question of the kind of solicitation applied to the sheets. Yes, of course if we are to consider a saddle point flow we have purely extension of flow. So what would be the difference of the case I consider, we would have in that case instead of a power 1/5th we would have log in the mixing time for instance. So a power 1/5th and the log you can not see the difference for Schmidt numbers let say 1000 you can not distinguish between the weak power low and the logarithm. So the difference between an extensional flow or a linear shear flow is at this level but again it does not change the physics at all.

Mark Glauser, Syracuse University "I have a question about bringing in the integral lag scales. It seemed to me you talked about the distance between the sheets being important but that would seem to be in a particular triponal flow relevant to the integral spatial scales. How do I bring the spatial lag scales into this analysis? At the end you were starting to bring in the temporal length scales. I was getting a sense of where you were going there but the spatial scales would seem to be important."

Reply: I have been discussing spatial scales. They are always smaller than the integral scales.

Mark Glauser, Syracuse University "OK, so you are smaller than the integral scale.?"

Reply: Yes the blob injection in the punctual injection scale.

Mark Glauser, Syracuse University "So you are always injecting things within an integral scale?"

Reply: Absolutely. In the confined and in the diluting case. Absolutely, Always.

Chapter 2

DOWNSTREAM EVOLUTION OF THE MOST ENERGETIC POD MODES IN THE MIXING LAYER OF A HIGH REYNOLDS NUMBER AXISYMMETRIC JET

Daehan Jung

Department of Mechanical Engineering
Korean Air Force Academy
Cheongwon, Chungbuk 363-849, Korea

Stephan Gamard and William K. George

Turbulence Research Laboratory
Department of Thermo and Fluid Dynamics
Chalmers University of Technology
SE-41 296 Gothenburg, Sweden

Scott H. Woodward

Department of Mechanical and Aerospace Engineering
University at Buffalo
Buffalo, NY 14260, USA

Abstract Experimental results were obtained in the potential core region of an axisymmetric turbulent jet from 2 to 6 diameters downstream, at Reynolds numbers of 78,400, 117,600, and 156,800. Data were collected using the 138 hot-wire probe used by Citriniti and George (2000). The Proper Orthogonal Decomposition was then applied to a double Fourier transform in time and azimuthal direction of the double velocity correlation tensor. The lowest azimuthal mode for all POD modes, which dominated the dynamics at $x/D = 3$ in the previous experiments, dies off rapidly downstream. This is consistent with a trend toward homogeneity in the downstream evolution, and suggests that some residual value may control the growth rate of the far jet. On the other hand, for the

A. Pollard and S. Candel (eds.), IUTAM Symposium on Turbulent Mixing and Combustion, 23-32.

higher azimuthal modes, the peak shifts to lower mode numbers and actually increases with downstream distance. These mixing layer data, normalized by similarity variables for the mixing layer, collapse at all downstream positions and are nearly independent of Reynolds numbers.

Keywords: Proper Orthogonal Decomposition, turbulence, jet, energy.

1. Introduction

The existence of large-scale structures in turbulent flows has been known for a number of years. It is also known that these large-scale structures play important roles in many applications: chemical mixing, noise control, momentum transfer, drag reduction, combustion enhancement, etc. Since the large-scale structures are embedded inside the turbulent motions, it is very difficult to identify and extract them from the chaotic turbulent flows.

The Proper Orthogonal Decomposition (POD) was introduced to the turbulence community by Lumley (1967) to extract the most energetic structures of the flow. Glauser (1987) first applied the POD to the axisymmetric mixing layer without any controlled excitation of the layer, and showed that a low-dimensional description of the time dependent velocity field was possible.

In an earlier version of this experiment, Citriniti and George (2000) obtained the dynamics of the flow from instantaneous realizations of the streamwise velocity field at $x/D = 3$ using 138 simultaneously-sampled hot-wire anemometer probes. They showed that only a few azimuthal Fourier modes are necessary to represent the evolution of the eigenspectra from the turbulent field. Furthermore, the velocity reconstructions using the POD provided evidence for both azimuthally coherent structures that exist near the potential core, and for counter-rotating, streamwise vortex pairs (or ribs) in the region between successive azimuthally coherent structures, as well as coexisting for short periods with them.

Both the Glauser and Citriniti experiments were performed at only a single downstream position and single Reynolds number. Our goal was therefore to investigate whether the modal character of the flow changed with downstream position and Reynolds number. In this experiment, data are presented at Reynolds numbers of $78,400$, $117,600$, and $156,800$ (based on jet exit velocity and diameter) at downstream positions ranging from $x/D = 2.0$ to 6.0.

2. Experimental Apparatus

The experiments were carried out in the turbulent mixing layer generated of an isothermal, non-reacting, incompressible, and axisymmetric jet. This jet facility was used previously by Citriniti and George (2000) and has been described thoroughly in Jung (2001).

The 138 long hot-wire probe array (figure 2.1), used by Citriniti and George (2000), was used to measure the flow velocities. Each single hot wire probe ($l/d \approx 800$), 1 cm long, was made in-house of unplated 12.7μm tungsten wire (Sigmund-Cohn, Mt. Vernon, NY). This length was necessary to reduce the spatial aliasing. The wires were placed in 6 concentric circles with increasing azimuthal resolution, in order to optimize the energy distribution in the eigenspectra.

Figure 2.1. The 138 Hot-wire probe array

The anemometers were designed in the laboratory and optimized for this specific experiment. Their circuit performance and noise properties compared well to industry standards.

3. POD Decomposition

Mathematically, the Proper Orthogonal Decomposition projects the random velocity $u_i(\underline{x}, t)$ into an orthonormal coordinate system $\phi_i(\underline{x}, t)$.

The projection is optimal in the sense that the first projection captures most of the energy or, mathematically speaking, the projection of the velocity field into the function $\phi_i(\underline{x}, t)$ has to have maximal amplitude. If the projection is being done over a finite field, then the solution can be found by means of the Hilbert-Schmidt theory since the kernel of the equation, $R_{ij} = \overline{u_i(\underline{x}, t)u_j^*(\underline{x}', t')}$, is Hermitian symmetric. When dealing with a periodic or stationary dimensions, as is the case for the axisymmetric jet considered, it is better to first Fourier transform the velocity field in those particular directions, then apply the POD to the transformed field. The results then produce an eigenspectrum, $\lambda^{(n)}(m, f)$, and eigenfunctions, $\phi_i^{(n)}(m, f)$, which are functions of azimuthal mode number, m, and frequency, f, for each POD mode, n. The details can be found in George (1988) or Lumley (1967).

- There exists not one, but a denumerable set of solutions[1]:

$$\int R_{11}(x, r, r', m, f)\phi_1^{(n)*}(x, r', m, f)r'dr'$$
$$= \lambda^{(n)}(x, m, f)\phi_1^{(n)}(x, r, m, f), (n = 1, 2, 3...) \quad (2.1)$$

- The eigenfunctions are orthonormal (and therefore form a basis for this space); the eigenvalues are ordered, i.e. $\lambda^{(1)} > \lambda^{(2)} > \lambda^{(3)}... > 0$ and their sum is equal to the kinetic energy:

$$\begin{aligned} E &= \int_f \int_r \sum_m \overline{u_1(x, r, m, f)u_1^*(x, r, m, f)}\, rdr\, df \\ &= \sum_n \sum_m \int_f \lambda^{(n)}(x, m, f)df \end{aligned} \quad (2.2)$$

The power of the POD can be seen in the last equation, since the eigenfunctions order the flow in an energetic sense and allow the most energetic features of the flow to be seen directly. The method is of little use, however, if the number of eigenvalues and eigenfunctions necessary to reproduce the flow are not in very limited number. This is indeed the case for the jet mixing layer as proven in Glauser and George (1992).

4. Experimental Results

Statistics were computed from 388 blocks of 4096 samples from each probe, corresponding to 400 *sec* record length each, giving a variability

[1] The r in the integral is the Jacobian for the cylindrical coordinate system.

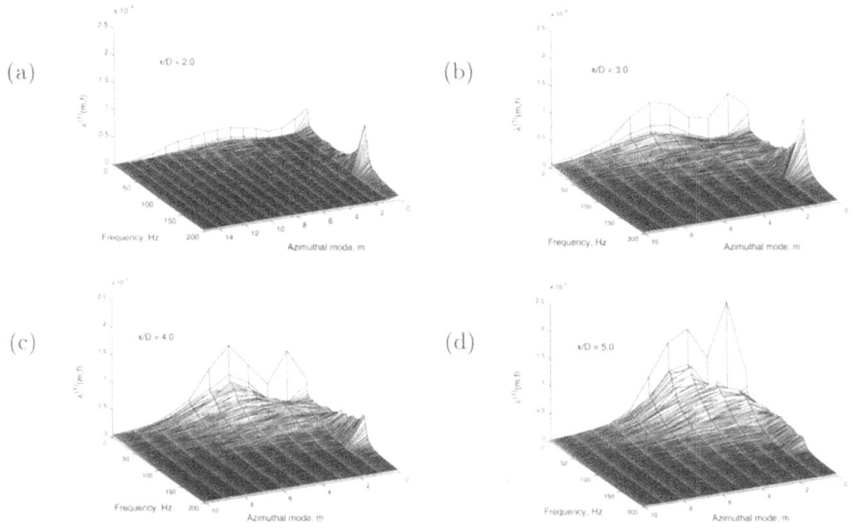

Figure 2.2. The first POD-mode $\lambda^{(1)}(m, f)$, distribution with azimuthal mode number, m, and frequency, f, for $Re_D = 156,800$, at different downstream positions: (a) x/D=2 , (b) 3, (c) 4, (d) 5

of approximately 5%. Normalized mean velocity and intensity profiles were similar to the results of Khwaja (1981).

Due to the distribution of hot-wires on the probe array, the POD can resolve up to 6 eigenspectra in the radial direction, and up to 16 modes in the azimuthal direction. The first eigenspectrum integrated over frequency and summed over azimuthal modes accounts for 66% of the kinetic energy, with the second adding around 18%.

Eigenspectra for $\lambda^{(1)}(\mathbf{m}, \mathbf{f})$

Figure 2.2 presents the eigenspectra for the first POD mode as a function of frequency, f, and azimuthal mode number, m. At $x/D = 2.0$, mode-0 (which contains more than 60 % of the total energy at that position) is spread over a band of frequencies which ranges from about 0 to 150 Hz. The energy in mode-0 moves to lower frequencies as x/D increases, and the total energy in mode-0 decreases.

At $x/D = 2.0$, the energy in the higher azimuthal modes peaks at about $m = 6$, and this energy is concentrated at very low frequencies. The energy in higher modes ($m > 0$) moves to lower modes as x/D increases. Corresponding to the diminution of mode-0 is the emergence of mode-1. This energy in mode-1 is also concentrated at very low frequencies.

Azimuthal dependence only

To emphasize the behavior of the eigenspectra as a function of mode number, we compute the normalized eigenvalue, $\xi^{(n)}(m)$, defined as:

$$\xi^{(n)}(m) = \frac{\int_f \lambda^{(n)}(m, f)\, df}{\sum_n \sum_m \int_f \lambda^{(n)}(m, f)\, df} \tag{2.3}$$

The denominator is the total turbulent kinetic energy in the flow (see equation 2.2). The numerator, on the other hand, is only integrated over frequency.

The azimuthal dependence of the normalized energy distribution of the first POD mode is presented in figure 2.3. The 0th azimuthal-mode, $\xi^{(1)}(0)$, behaves in a manner entirely different than the higher azimuthal modes. The 0-azimuthal mode, which dominated the dynamics at $x/D = 3.0$ in the Glauser (1987) and Citriniti and George (2000) experiments, dies off rapidly downstream for all Reynolds numbers. This is consistent with an approach toward homogeneity in the downstream evolution, and suggests that perhaps some residual value may control the asymptotic growth rate of the far jet. For the higher azimuthal modes ($m > 0$), the peaks shift to lower mode numbers, and their amplitudes actually increase with downstream distance.

This behavior of the azimuthal modes for $m > 0$ is consistent with the observations of George *et al.* (1984), and Khwaja (1981). They noted that the statistical properties of the axisymmetric jet shear layer appeared to scale in shear layer similar variables x and U_E, even though the equations did not admit to such solutions.

Figure 2.4 shows the collapse of the eigenvalues for the first POD mode for $m > 0$ at all Reynolds numbers and positions, when normalized in shear layer variables $\xi^{(n)}/(x/D)$ vs. $m \cdot x/D$. The second POD mode (not shown) also presents reasonable collapse. This is in contrast to the suggestion by Holmes *et al.* (1996) who expected that more complicated modal structures might evolve with increasing Reynolds number. This collapse confirms the Reynolds number independence of the large scale structures for these high Reynolds numbers experiments as presumed by Glauser (1987) and Citriniti and George (2000).

Interestingly enough, when compared with measurements made farther downstream [Gamard et al., 2001], the eigenspectra at the end of the potential core appears to have already reached their asymptote. This is well before the single-point statistics of the jet collapse in far jet similarity variables (typically $x/D > 30$). This could provide an important

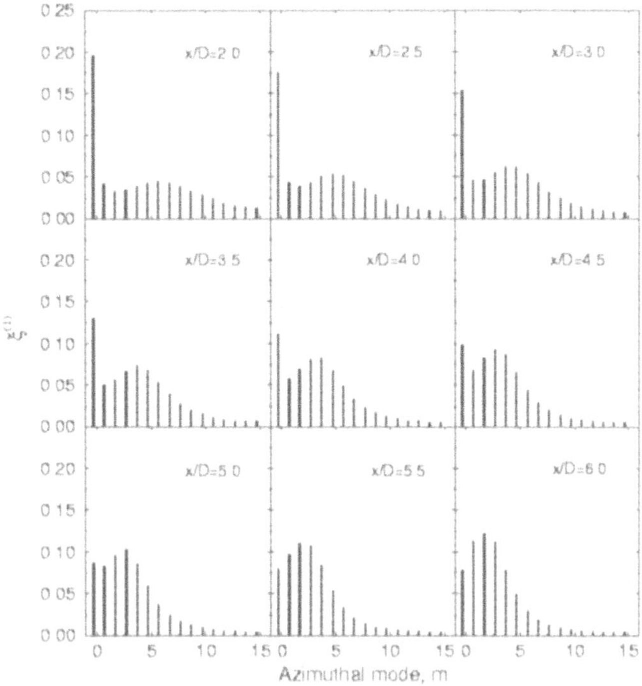

Figure 2.3. Normalized first POD-mode energy distribution, $\xi^{(1)}(m)$, with azimuthal mode number, m, at $Re_D = 156,800$.

clue as to why and how equilibrium similarity governs the far jet, and also why the jet growth rate may reflect the upstream conditions.

5. Summary and Conclusions

The large-scale structure of an axisymmetric mixing layer was investigated using the POD for Reynolds numbers of 78,400, 117,600, and 156,800 at $x/D = 2.0$ to 6. From application of the POD in the radial direction, it was observed that the first POD mode contains more than 60 % of the turbulent energy at all downstream positions and Reynolds numbers. The first 2 POD modes together contain more than 80 %. The energy in mode-0 moves to lower frequencies as x/D increases, and the total energy in mode 0 decreases. Coincident with the diminution of mode-0 is the emergence of mode-1.

The behavior of mode-0 is consistent with an approach toward homogeneity in the downstream direction, and suggests that perhaps some residual value may control (or reflect) the growth rate of the far jet.

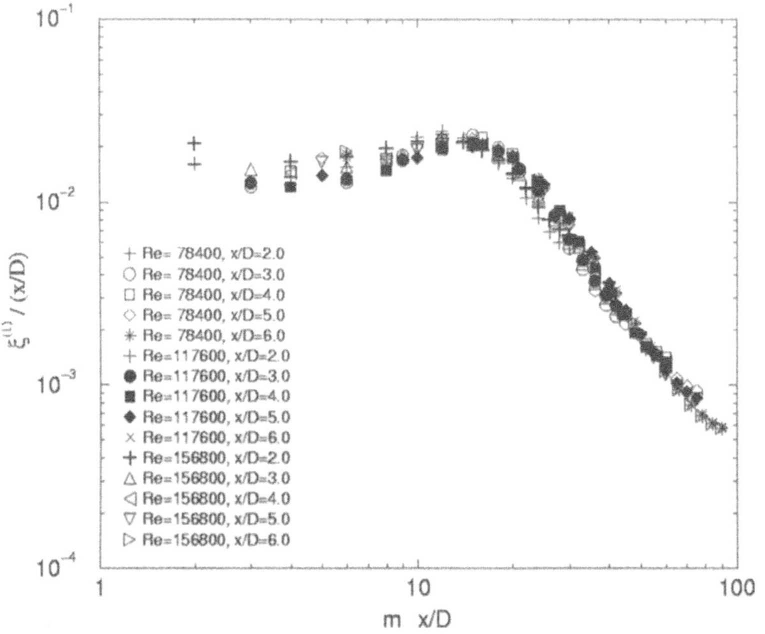

Figure 2.4. The first POD-mode energy distribution normalized in shear layer variables, $(m \cdot x/D)$, at $Re_D = 78,400$, $117,600$, and $156,800$.

The behavior of mode-0 and mode-1 as x/D increases is similar to that predicted from the inviscid instability theory for laminar parallel axisymmetric flows. In particular, Batchelor and Gill (1962) show that for a top-hat profile all modes are unstable, but mode-0 grows the fastest. By contrast, once the profile is fully-developed, mode-0 is stable, and mode-1 grows the fastest, at least for a parallel flow. Similar conclusions were reached by Michalke (1984) as well, but for spatially growing disturbances. These instability results are strikingly similar to the behavior of the POD modes observed here and even the eigenfunctions appear to be similar.

For azimuthal mode greater than $m = 0$, the energy shifts from higher modes to lower modes as x/D increases. In fact, the eigenspectra collapse when scaled in shear layer similarity variables; i.e. λ/xU_0^2 vs. mx/D. Thus, the POD modes in the shear layer are behaving exactly like the energetic single-point statistics of the jet mixing layer.

Acknowledgments

This work was initiated at the Turbulence Research Laboratory of the State University of New York at Buffalo, and continued with the move of TRL to Chalmers University of Technology in Gothenburg, Sweden. The support of the US Air Force Office of Scientific Research under grant number F49620-98-1-0143 is gratefully acknowledged.

References

Batchelor, G. and Gill, E. (1962). Analysis of the instability of axisymmetic jets.

Citriniti, J. and George, W. (2000). Reconstruction of the global velocity field in the axisymmetric mixing layer utilizing the proper orthogonal decomposition. *Journal of Fluid Mechanics*, 418:137–166.

Gamard, S., Jung, D., Woodward, S., and George, W. K. (2001). Application of a 'slice' pod to the far field of an axisymmetric turbulent jet. *submitted to Physics of Fluids.*

George, W., Beuther, P. D., and Arndt, R. (1984). Pressure spectra in turbulent free shear flows. *Journal of Fluid Mechanics*, 148:155–191.

George, W. K. (1988). Insight into the dynamics of coherent structures from a proper orthogonal decomposition. In *Symposium on Near Wall Turbulence*, Dubrovnik, Yugoslavia.

Glauser, M. N. (1987). *Coherent Structures in the Axisymmetric Turbulent Jet Mixing Layer.* PhD thesis, State University of New York at Buffalo.

Glauser, M. N. and George, W. K. (1992). Application of multipoint measurements for flow characterization. *Experimental Thermal and Fluid Science*, 5:617–632.

Holmes, P., Lumley, J. L., and Berkooz, G. (1996). *Turbulence, Coherent Structures, Symmetry and Dynamical Systems.* Cambridge.

Jung, D. (2001). *An Investigation of the Reynolds Number Dependence of the Axisymmetric Jet Mixing Layer using a 138 Hot Wire-Probe and the POD.* PhD thesis, State University of New York at Buffalo.

Khwaja, M. S. (1981). Investigation of the turbulent axisymmetric jet mixing layer. Master's thesis, State University of New York at Buffalo.

Lumley, J. L. (1967). The structure of inhomogeneous turbulent flows. In Yaglom, A. M. and Tatarsky, V. I., editors, *Atmospheric Turbulence and Radio Wave Propagation*, Moscow, USSR. Nauka.

Michalke, A. (1984). Survey on jet instability theory. *Prog. Aerospace Sci.*, 21:159–199.

Discussion

Mark Glauser, Syracuse (Chair) "I would like to make a comment on this. This is really an extension of the work I have been doing over the last several years and it is quite interesting that we are finding that we can collapse these data in the downstream direction and also very little change with Reynolds number in the overall structure of jets. In fact, with some of my students, we have gone up to Mach=.6, .85 and substantially higher Reynolds numbers than presented here and we are seeing similar kind of structure. So from the point of view of using these POD modes for control etc is quite exciting."

Sebastien Candel, EMC2, Paris "I was wondering of the 138 hot wires placed in the flow modify the flow and perhaps favour the zero-th order mode. Is there any possibility to know if this is the case?"

Reply: Blockage effect is less than 0.5% of the area ratio. There is nothing.

Emmanuel Villermaux, Marseille "This is a follow on from the previous question. As an experimentalist, you are using 138 hot wires, where you able to make them operate all at the same time?"

Reply: Yes, we collect the data simultaneously at all points that means. We built all of them in the lab.

Emmanuel Villermaux, Marseille "Congratulations!"

Chapter 3

MICRO STRUCTURE AND LAGRANGIAN STATISTICS OF THE SCALAR FIELD WITH A MEAN GRADIENT IN ISOTROPIC TURBULENCE

G. Brethouwer, J.C.R. Hunt* and F.T.M. Nieuwstadt

Laboratory for Aero and Hydrodynamics, Delft University of Technology

Leeghwaterstraat 21, 2628 CA Delft, The Netherlands

g.brethouwer@wbmt.tudelft.nl

**Also Department of Geological Sciences and of Space & Climate Physics*

University College London

17 Gordon Street

London WC1H 0AH, UK

Abstract This paper presents a numerical study of the relation between the small scale velocity and scalar field in fully developed isotropic turbulence, with a forced random large scale motion and an imposed steady and linear mean scalar gradient. Simulations have been performed with a range of Reynolds numbers from $Re_\lambda = 22$ to 130 and Schmidt numbers from $Sc = 1/25$ to 144.

Keywords: DNS, turbulent mixing, passive scalar

1. Introduction

Conventionally one distinguishes between the large scale advective transport of the scalar, the so-called macro mixing, and the small-scale distortion of the scalar field by the straining and shearing turbulent motions. Scalar gradients become larger and iso-scalar surfaces grow as a consequence of the latter process resulting in an intensive molecular transport of the scalar. This is the so-called micro mixing and is important for chemical reactions.

A. Pollard and S. Candel (eds.), IUTAM Symposium on Turbulent Mixing and Combustion, 33-43.

There have been several approaches to research on the macro and micro mixing process. Here we consider two of them. The first approach is to consider the motion of fluid particles which proved to be a powerful tool to study some important aspects of turbulent mixing. Fung *et al.* (1992) investigated the separation of particle pairs in a turbulent-like flow and concluded that turbulent structures play an important role in the separation of particle pairs. In straining regions pairs can separate very rapidly leading to an intermittent behaviour of the fluid particle dispersion.

The second approach is to focus on the evolution of the local velocity and scalar fields using precise numerical simulations or experiments. Visualisations of the micro structure show that regions of steep scalar gradients have a sheet-like structure (Buch & Dahm 1998, Vedula *et al.* 2001). The visualisations of Holzer & Siggia (1994) suggest that the steep scalar gradients are generated by straining motions. The micro structure of the scalar field is anisotropic if there is a mean scalar gradient (Warhaft 2000) in contradiction with the traditional view.

We study the micro mixing of a passive scalar with a mean gradient and with a wide range of Schmidt numbers in isotropic turbulence by means of direct numerical simulations. The goal of our investigation is finding an answer to the following questions: What is the micro structure of the scalar field and how does it depend on the Schmidt number? How is the micro structure affected by the turbulent structures? Can we find links between the Lagrangian view of the mixing process and the Eulerian statistical information?

The outline of this paper is as follows: We first explain the numerical simulations and then present the results. We end this paper with the conclusions.

2. The direct numerical simulation

For the full details of the numerical simulations we refer to Brethouwer (2000). Here we address only the main points.

We have performed direct numerical simulations (DNS) of three dimensional isotropic turbulence. The incompressible Navier-Stokes equations are solved by a pseudo-spectral method. To obtain statistically stationary turbulence we have forced the largest scales. Simultaneously with the Navier-Stokes equations we also solve the advection diffusion equation for a passive and conserved scalar with Schmidt numbers from 1/25 up to 144 and with a linear mean scalar gradient. Apart from Eulerian statistics, we have computed Lagrangian statistics of the scalar mixing process by tracking 4096 fluid particles.

3. Results

First, we study the micro structure of the scalar field. In figure 3.1 the instantaneous scalar field is shown. At $Sc = 144$ we observe very thin and elongated structures which indicate cross-sections through sheet-like structures in the scalar field. Many of the elongated structures are more or less straight but at some places they are rolled up. At smaller Schmidt numbers the appearance of elongated structures becomes less obvious. At $Sc = 0.7$ the existence of scalar sheets is hardly noticeable anymore. Nevertheless, we can observe lumps of almost uniform scalar concentration bounded by sharp edges with large jumps of the scalar concentration. This suggests that regions with a large scalar gradient have a sheet-like structure, even if Sc is of the order unity. Figure 3.2, which shows the norm of the scalar gradient on a two-dimensional plane at $Sc = 25$ and 0.1, supports this suggestion. Large scalar gradients are concentrated in elongated regions which can be considered as the footprints of sheet-like structures. These sheet-like structures are representatives of the steep cliffs in the scalar field found by Holzer & Siggia (1994). Holzer & Siggia found also that the cliffs are positioned roughly normal to the mean gradient. Figure 3.3 confirms this observation. If the scalar gradient is small there is no preferred direction. Steep scalar gradients are approximately aligned with the mean gradient. This implies that the steep cliffs have a preferred direction normal to the mean gradient.

We now study the spatial dimensions of the cliffs. Figure 3.3 shows the correlation function $\langle\nabla\theta(\boldsymbol{x})\cdot\nabla\theta(\boldsymbol{x}+\boldsymbol{r})\rangle/\langle\nabla\theta\cdot\nabla\theta\rangle$ with $\boldsymbol{r}$ parallel to $\nabla\theta(\boldsymbol{x})$. The correlation function is only determined for points in the computational domain where the scalar gradient is large, i.e. $g > 5\langle g\rangle$ so that these points are representative for the cliffs. The spatial distance $\boldsymbol{r}$ is normalised with the Batchelor scale η_B. The scaled spatial correlation function is according to the figure almost universal for the whole range of Reynolds and Schmidt numbers. The correlation drops off very fast and is close to zero beyond $r = 3\eta_B$ which we can interpret as the thickness of the sheets. This proves that the cliffs found in our simulations are very sharp which is in agreement with other studies (Warhaft 2000).

Next the question how these steep cliffs are related to the turbulent flow is addressed. Figure 3.4 shows the smallest eigenvalue of the strain rate tensor, here called the compressive strain rate, conditioned on the scalar gradient at $Sc = 144$ and $1/25$. We see a positive trend of the conditional compressive strain rate when the scalar gradient increases at both Schmidt numbers. At large $\langle g\rangle$ both curves even overlap. Thus large scalar gradients are related to high strain rates, irrespective of Sc.

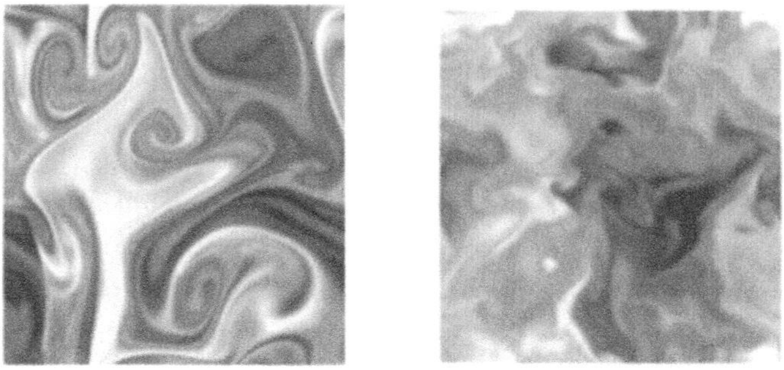

Figure 3.1. The scalar concentration distribution on a two-dimensional plane at $Sc = 144$, $Re_\lambda = 22$ (left) and $Sc = 0.7$, $Re_\lambda = 85$ (right).

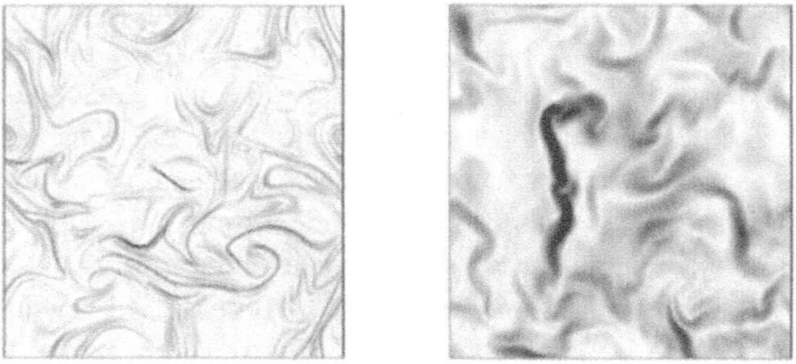

Figure 3.2. The norm of the scalar gradient on a two-dimensional plane at $Sc = 25$, $Re_\lambda = 46$ (left) and $Sc = 0.1$, $Re_\lambda = 95$ (right). The direction of the mean gradient is in horizontal direction. A dark colour to a large gradient.

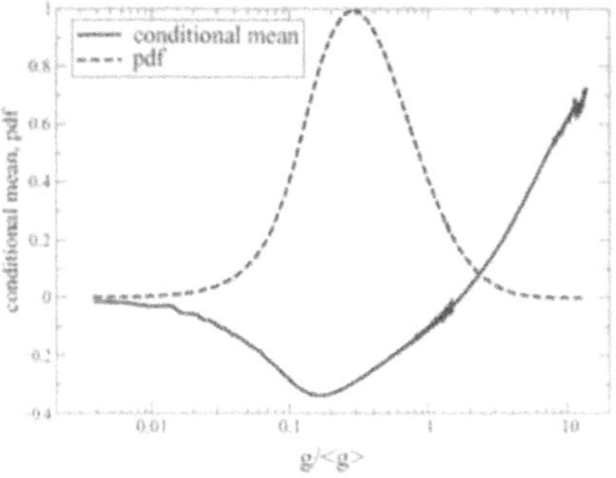

Figure 3.3. $\left\langle \frac{\nabla\theta \cdot \boldsymbol{G}}{|\nabla\theta||\boldsymbol{G}|} \middle| \frac{g}{\langle g \rangle} \right\rangle$ which is the mean value of the cosine of the angle between $\nabla\theta$ and the imposed mean gradient $\boldsymbol{G}$ conditioned on $g/\langle g \rangle$, and the pdf of the norm of the scalar gradient fluctuations $g/\langle g \rangle$.

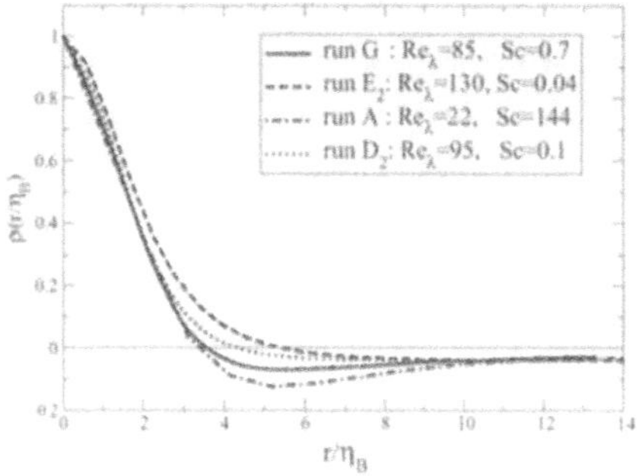

Figure 3.3 (continued)
The spatial correlation $\langle \nabla\theta(\boldsymbol{x}) \cdot \nabla\theta(\boldsymbol{x} + \boldsymbol{r}) \rangle / \langle \nabla\theta \cdot \nabla\theta \rangle$. The vector $\boldsymbol{r}$ is parallel to the scalar gradient at $\boldsymbol{r} = 0$ and is scaled with the Batchelor scale.

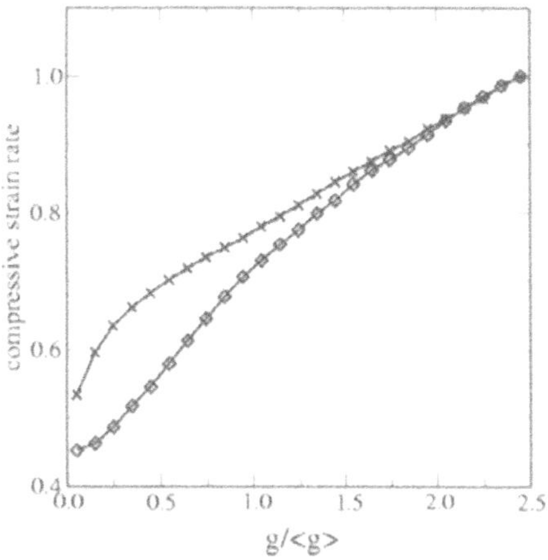

Figure 3.4. $\langle -s_\gamma | g/\langle g\rangle\rangle$ which is the mean compressive strain rate for $Sc = 144$ (×) and $Sc = 0.04$ (□) as a function of $g/\langle g\rangle$.

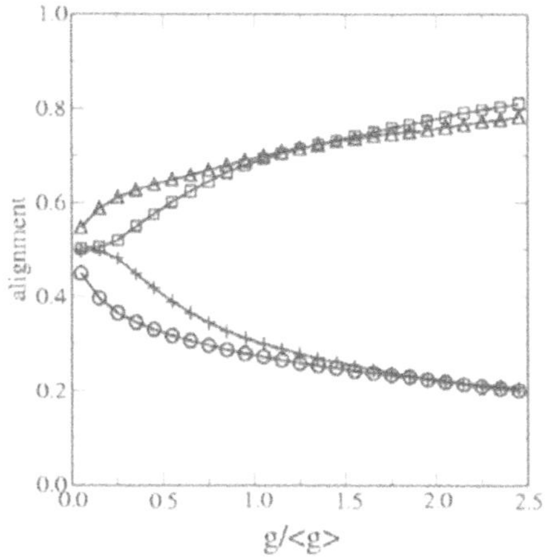

Figure 3.4 (continued)
$\langle (|\boldsymbol{e}_\gamma \cdot \boldsymbol{n}|) \mid g/\langle g\rangle\rangle$: the cosine of the angle between the scalar gradient and the direction of compressive strain conditioned on the g for $Sc = 144$ (△) and $Sc = 0.04$ (□).

Figure 3.4 presents the alignment between the scalar gradient and the direction of the compressive strain conditioned on the scalar gradient. If these directions are completely uncorrelated then the mean alignment is equal to 0.5 while for perfect alignment the value is 1. The figure demonstrates that for both Schmidt numbers the average alignment goes to 0.5 if the scalar gradient goes to zero, which implies that for small values of the scalar gradient the gradient and the direction of compressive strain are uncorrelated. If the scalar gradient is large it is preferentially aligned with the direction of compressive strain for $Sc = 1/25$ and $Sc = 144$. This agrees with the observations of Pumir (1994). The response of the scalar field to the turbulent velocity field is remarkably similar for low and high Schmidt numbers and is now studied in more detail.

In figure 3.5 the norm of the scalar gradient g when it is conditioned on the strain rate $s = (\epsilon/\nu)^{1/2}$ or the norm of the vorticity ω is presented. The conditional value of g increases when the strain rate increases. This result is not surprising if we look at the equation for g given by

$$\frac{1}{g}\frac{dg}{dt} = -\sum_i s_i \lambda_i^2 + \frac{1}{2g^2}\kappa\nabla^2 g^2 - \frac{1}{g^2}\kappa\nabla(\nabla\theta) : \nabla(\nabla\theta) \qquad (3.1)$$

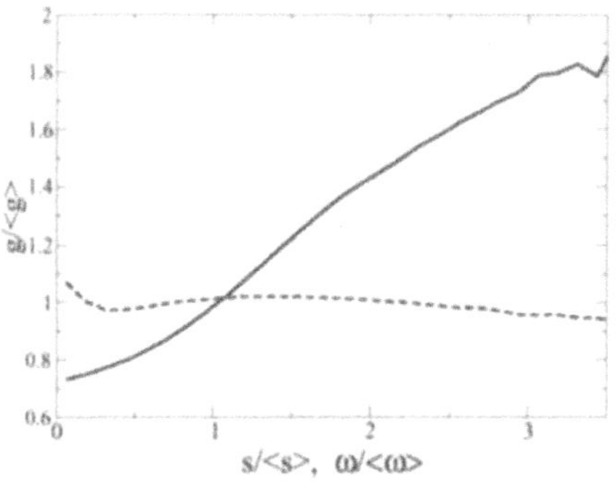

Figure 3.5. $\langle g|s\rangle$ (solid line) and $\langle g|\omega\rangle$ (dashed line) which is the average value of the scalar gradient conditioned on the strain rate s and the vorticity ω. All quantities are scaled with their mean value. $Re_\lambda = 59$ and $Sc = 0.7$.

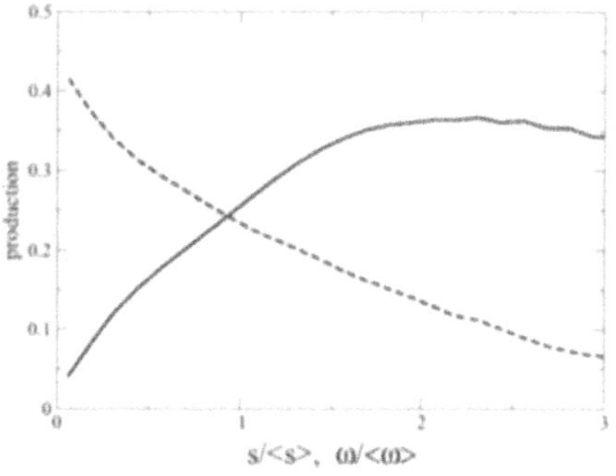

Figure 3.5 (continued) $\langle \frac{\sum_i s_i\lambda_i^2}{s_\gamma} | \frac{s}{\langle s\rangle}\rangle$ (solid line) and $\langle \frac{\sum_i s_i\lambda_i^2}{s_\gamma} | \frac{\omega}{\langle \omega\rangle}\rangle$ (dashed line). $Re_\lambda = 59$ and $Sc = 0.7$.

where $s_i = s_\alpha, s_\beta, s_\gamma$ denote the eigenvalues of the strain rate tensor and κ the diffusivity of the scalar. Furthermore, we define the eigen vectors $\boldsymbol{e}_\gamma, \boldsymbol{e}_\beta, \boldsymbol{e}_\alpha$ of the strain rate tensor, the unit vector $\boldsymbol{n}$ in the direction of the scalar gradient $\nabla\theta$ and the alignment vector $\lambda_i = \boldsymbol{n}\cdot\boldsymbol{e}_i$ which is the cosine of the angle between the scalar gradient and $\boldsymbol{e}_i$. The production term of g thus depends on the strain rate and on the alignment between $\nabla\theta$ and the principal axes of strain.

Equation (3.1) shows that the vorticity has not a direct influence on g. This agrees with the results of figure 3.5 where it is demonstrated that there is no correlation between the norm of the vorticity and the scalar gradient in agreement with the results of Pumir (1994).

Steep scalar gradients are approximately aligned with the direction of compressive strain (here defined as $\boldsymbol{e}_\gamma$) as we have seen. This correlation can be most easily understood if equation (3.1) is considered. The first term on the right hand side, the production term, is maximal when the scalar gradient fully aligns with $\boldsymbol{e}_\gamma$ which implies $\lambda_\gamma = 1$ so that the production term equals $-s_\gamma$. In this case the ratio $\sum_i s_i\lambda_i^2/s_\gamma$ has its maximal value which is equal to 1. We can therefore interpret the normalised production rate $\sum_i s_i\lambda_i^2/s_\gamma$ as the alignment between the scalar gradient and the direction of compressive strain.

In figure 3.5 we show the normalised production rate of the scalar gradient conditioned on the strain and conditioned on the vorticity. We observe that the normalised production term increases when s increases, at least for $s < 2\langle s \rangle$. The vorticity has the opposite effect. The normalised production term increases when the vorticity decreases, with the maximal value when the vorticity is zero. In other words, if the strain rate is high there is relative strong alignment between $\nabla\theta$ and e_γ. The vorticity weakens this alignment.

4. Lagrangian statistics of mixing

More insight may be obtained if we consider the mixing process in a frame moving with the turbulent flow.

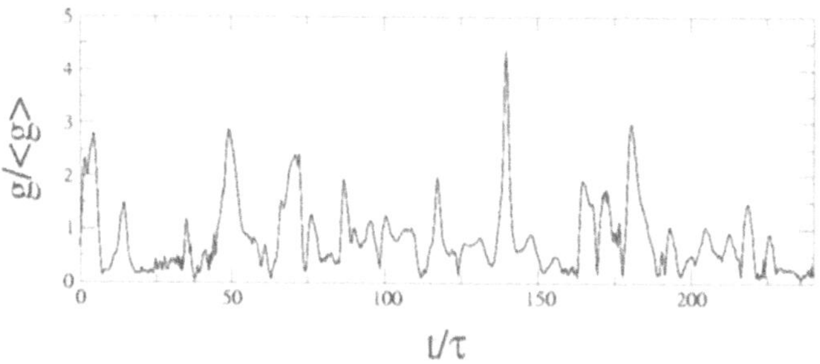

Figure 3.6. Time series of the scalar gradient.

To get a first impression of the mixing process in a Lagrangian frame we have plotted in figure 3.6 a typical Lagrangian time series of the scalar gradient for one fluid particle. The time is normalised with the Kolmogorov time scale to emphasize the rapid change of the scalar gradient. The scalar gradient g show short moments of high values implying rapid molecular mixing. For instance, at $t/\tau = 140$ we see a strong increase of the scalar gradient with a peak value several times the mean value of the scalar gradient, after which it decreases again very rapidly. To obtain more insight into scalar mixing it is of interest to study these events of rapid molecular mixing in more detail.

It is not practical to study all individual events. Therefore, we have defined an ensemble average. The procedure to obtain this ensemble average is as follows. We have recorded 4096 Lagrangian time series. In the time series of the fluid particles we detect the events when the instantaneous scalar gradient exceeds a certain threshold value where for this threshold we have chosen a value 3 times the mean gradient $\langle g \rangle$. Each event is labeled by the time instant at which the scalar gradient g reaches its maximum value. This moment is considered to be the point of reference where we set $t = 0$. We then compute the scalar and velocity

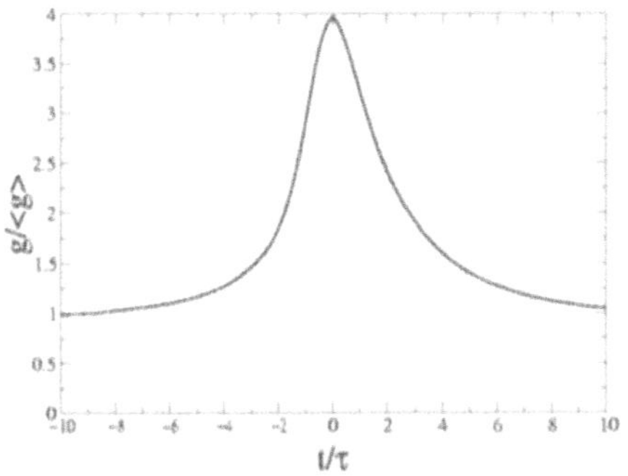

Figure 3.7. The Lagrangian evolution of g. The time is normalised with the Kolmogorov time scale τ and g with its mean value.

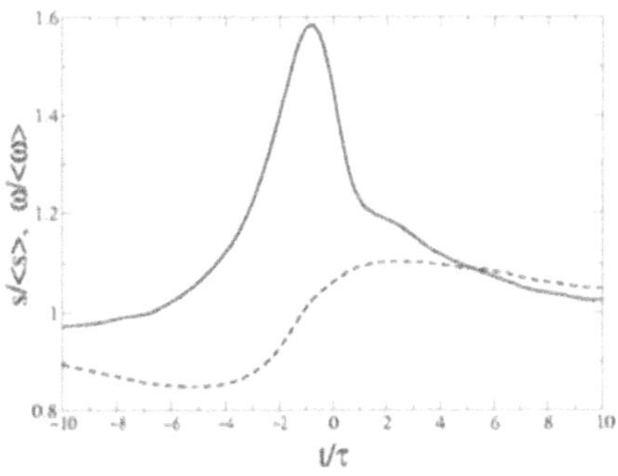

Figure 3.7 (continued)
The Lagrangian evolution of s (solid line) and ω (dashed line). The time is normalised with the Kolmogorov time scale and s and ω with their mean value.

parameters in the short period of time before and after $t = 0$. Finally, we calculate the ensemble average of all detected events with all events centered at $t = 0$.

Figure 3.7 presents the Lagrangian evolution of the the scalar gradient, obtained by applying the procedure described above, showing the intermittent behaviour of mixing. The average duration of the events of rapid molecular mixing is a few Kolmogorov time scales. In about three Kolmogorov time scales the scalar gradient increases by a factor of three to its maximum value. Thereafter the scalar gradient rapidly decreases to a moderate value.

The evolution of the strain rate and vorticity during the detected events is shown in figure 3.7. Similar to the scalar gradient the strain rate shows also a short moment of relatively high intensity. The Lagrangian evolution of the vorticity is different from the evolution of the strain rate and scalar gradient. The vorticity increases around $t = 0$ but the relative increase of the vorticity is much smaller than the increase of the strain rate and the scalar gradient. We see thus that around $t = 0$ the strain rate is much larger than the vorticity implying that the fluid particles are in regions dominated by straining motions.

Figure 3.7 demonstrates also that the maximum value of the strain in terms of its mean is about 1.6 which is considerable smaller than the value we have found for g. The relative increase of the strain rate is thus much smaller than the relative increase of the scalar gradient.

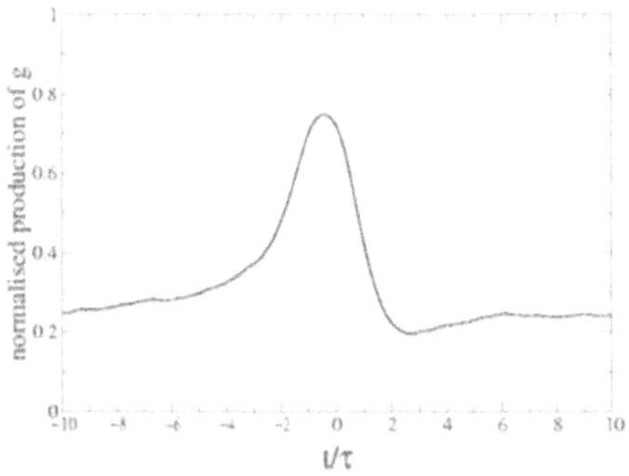

Figure 3.8. The Lagrangian evolution of the normalised production rate of g given by $\sum_i s_i \lambda_i^2 / s_\gamma$.

One of the causes of the strong reaction of the scalar gradient to the increased strain can be the change of the orientation of the scalar gradient with respect to the principal axes of strain during the detected events. We can investigate this by considering the Lagrangian evolution of the production rate of the scalar gradient normalised with its maximal value. This is shown in figure 3.8. Recalling figure 3.5 we have observed that the mean normalised production rate increases when the strain increases and decreases when the vorticity increases. Because before $t = 0$ the strain rate increases considerably and the vorticity hardly we can expect an increase of the normalised production rate. This can indeed be observed. Shortly before $t = 0$ the normalised production rapidly increases, which implies that the scalar gradient rapidly rotates to the direction of the compressive strain. The maximal value reached by the normalised production rate is about 0.75 which is a relative high value. Note that this parameter can be at most equal to one. This suggests that the scalar gradient is almost completely aligned with the direction of compressive strain at $t = 0$. The rotation of the scalar gradient to the direction of the compressive strain results in an extremely large increase of the production term of the scalar gradient, because both the strain rate and λ_γ increase.

5. Conclusions

Visualisations demonstrate the existence of sharp cliffs in the scalar field at $Sc \gg 1$ and $Sc \ll 1$. The cliffs are roughly aligned normal to the mean scalar gradient and the thickness of the cliffs scales with the Batchelor length scale over the whole range of Schmidt numbers.

Turbulent structures are important for the generation of steep cliffs. We have observed that the cliffs are mostly found in regions dominated by strain. The scalar gradient is approximately aligned with the direction of compressive strain in these regions characterised by a high strain rate and low vorticity. Because the scalar gradients are also aligned with the mean scalar gradient we are lead to the conclusion that the direction of compressive strain forming these cliffs, is aligned with the mean gradient as well.

A Lagrangian study demonstrates the very rapid generation of the steep cliffs followed by a fast destruction. The straining motions responsible for the generation of these cliff have also a short life time.

Acknowledgments

We gratefully acknowledge T. Dubois for providing his numerical code. We thank A.R. Kerstein, A. Tsinober and P.K. Yeung for their help.

References

Brethouwer, G. (2000) Mixing of passive and reactive scalars in turbulent flows. PhD thesis, Delft, the Netherlands.

Buch, K.A., Dahm, W.J.A. (1998) Experimental study of the fine-scale structure of conserved scalar mixing in turbulent shear flows. Part 2. $Sc \approx 1$. *J. Fluid Mech.* **364**, 1–29.

Fung, J.C.H., Hunt, J.C.R., Malik, N.A., Perkins, R.J. (1992) Kinematic simulation of homogeneous turbulence by unsteady random Fourier modes. *J. Fluid Mech.* **236**, 281–318.

Holzer, M., Siggia, E.D. (1994) Turbulent mixing of a passive scalar. *Phys. Fluids* **6**, 1820.

Pumir, A. (1994) A numerical study of the mixing of a passive scalar in three dimensions in the presence of a mean gradient. *Phys. Fluids* **6**, 2118–2132.

Vedula, P., Yeung, P.K., Fox, R.O. (2001) Dynamics of scalar dissipation in isotropic turbulence. *J. Fluid Mech.* **433**, 29–66.

Warhaft, Z. (2000) Passive scalars in turbulent flows. *Annu. Rev. Fluid Mech.* **32**, 203–240.

Discussion

Emmanuel Villermaux, Marseille "You observe sheets, and their steepness of their gradients are a few units of the Bachelor scale. You have varied the Reynolds number, the Schmidt number, have you tried different scalings?"

Reply: No, only the Bachelor length scales have been used to scale the thickness of the sheets.

Emmanuel Villermaux, Marseille "You are sure this thickness is dependent on the Reynolds number? The Bachelor scaling suggests dependence on $Re^{3/4}$."

Reply: I have shown four different cases of Schmidt and Reynolds number and in all cases they scale perfectly with the Bachelor length scale.

Emmanuel Villermaux, Marseille "There is another way to look at this and this is to coarse grain the field. That is you use the resolution of your picture and then you look at the composition of the medium, once you have coarse grained your field, to investigate the elementary break of the field, have you tried this?"

Reply: No.

Emmanuel Villermaux, Marseille "If you do, you usually end up with different conclusions"

Reply: OK.

Chapter 4

DYNAMICS & CONTROL OF JETS IN CROSSFLOW

Direct numerical simulations & experiments

Peter N. Blossey
Dept. of Applied Mathematics,
University of Washington, Seattle, WA 98195-2420.
bloss@amath.washington.edu

Satish Narayanan
United Technologies Research Center,
411 Silver Lane, East Hartford, CT 06108
narayas@utrc.utc.com

Thomas R. Bewley
University of California, San Diego,
La Jolla, CA 92093-0411
bewley@turbulence.ucsd.edu

Abstract The structure of a jet in crossflow and its mixing have been studied through detailed direct simulations, and the findings were verified by experiments. Using a newly developed pseudo-spectral code, simulations have been performed both with and without sinusoidal forcing at Strouhal numbers in the range $St_d = fU_j/d = 0.1$–0.64. Based on the simulation results, which show substantial entrainment and mixing enhancements with lower frequency forcing, experiments have been performed for unforced and forced jets with jet Strouhal numbers up to $St_d = 0.2$. The simulations and experiments indicate that forcing at the lower end of the frequency range considered in the numerical results, namely $St_d = 0.1$–0.26, is most effective for increased mixing and entrainment. Finally, we assess the benefits of the controlled flow states using a variety of time-averaged and unsteady mixing metrics.

Keywords: Jet in crossflow, transverse jet, forced jet.

A. Pollard and S. Candel (eds.), IUTAM Symposium on Turbulent Mixing and Combustion, 45-57.

Introduction

The jet in crossflow has wide application in industrial and environmental flows such as dilution jets in combustors and smokestacks in the atmosphere. Mixing is central to these flows for cooling or reaction in combustors and for pollutant dispersion in the atmosphere.

In a jet in crossflow (also called a transverse jet), fluid emerges from an orifice at a large angle into a stream. The shear layer along the front (or upstream) edge of the jet develops oscillations which roll up into large pockets of jet fluid and travel along the upper edge of the jet, contributing to a high degree of intermittency there. As the trajectory of the jet bends to follow the crossflow direction, a counter-rotating vortex pair (CVP) forms and results in further entrainment of crossflow fluid.[1]

The forcing of jets exhausting into quiescent fluid can have a substantial impact on the structure and mixing of the jet. Optimal forcing frequencies for increased mixing (explored experimentally) are claimed to be in the range $St_d = 0.2$–0.4 [Crow and Champagne, 1971, Vermeulen et al., 1992]. Unsteadily forced jets in crossflow have also received attention in recent experimental and computational studies [Vermeulen et al., 1990, M'Closkey et al., 2001, Eroglu and Breidenthal, 2001]. Our study combines direct numerical simulations of forced jets in crossflow with a preliminary experimental investigation of these flows in an attempt to determine both the most excitable and the most "beneficial" forcing frequencies. (Note that these two frequencies may not coincide.) To this end, the effect of sinusoidal forcing on mixing by the jet in crossflow is evaluated over a wide control parameter range.

1. Computational & Experimental Setup

Computational Setting. The jet in crossflow is simulated in a periodic channel flow geometry. The Navier–Stokes equations are solved without modeling (direct numerical simulation) and are discretized with Fourier modes in the wall-parallel directions and finite differences in the wall-normal direction. An advection-diffusion equation is solved to compute the evolution of a conserved, passive scalar field. Due to the periodic boundary conditions in the streamwise/crossflow direction, a fringe region is added near the exit of the channel (the last 30% of the domain) to remove fluctuations from the outgoing velocity and scalar

[1] In addition to the jet shear layer and the CVP, two other structures — upright wake vortices and a horseshoe/necklace vortex — can sometimes be seen in the jet in crossflow for particular values of the Reynolds number and the ratio of jet exit velocity to that of the crossflow. Figure 29 in [Kelso et al., 1996] maps out the parameter values where the different structures may be expected to appear.

fields and to provide the desired inflow boundary conditions. These techniques are commonly used in simulations of spatially-evolving disturbances in transitional boundary layers [Nordström et al., 1999]. A hybrid implicit–explicit Crank-Nicolson, Runge-Kutta scheme for time integration allows for the implicit treatment of the wall-normal convective and viscous terms, relieving the severe viscous and CFL time step restrictions imposed by the highly refined mesh near the wall and the flow of jet fluid across it. The scalar discretization employs upwinding [Leonard, 1988] in the wall-normal direction and filtering in the wall-parallel directions to limit the appearance of non-physical oscillations in the scalar field due to the sharp gradients near the jet exit. The jet inflow into the simulation domain is imposed as a velocity boundary condition on the bottom wall of the channel with a inflow velocity profile defined as $v_j(r) = U_j \exp(-(2r/d)^8)$. This profile is then filtered to prevent the excitation of high frequencies which would appear as ringing in the velocity field, while still maintaining a near top hat profile typical for dilution jets in combustion applications. The imposition of the jet inflow as a boundary condition neglects the boundary layer details in the pipe flow leading to the jet exit. While separation in the pipe flow might significantly influence the dynamics of the emerging jet, [Kelso et al., 1996] observed that the flow in the pipe does not separate for velocity ratios $R > 6$ ($R = 6$ in these simulations).

The relevant parameters in the simulation are the domain size $25d \times 12d \times 10d$ and resolution $256 \times 120 \times 96$ (de-aliased) in the streamwise, wall-normal and spanwise directions, respectively; the jet Reynolds number ($Re_j = U_j d/\nu = 3000$); the jet velocity ratio ($R = U_j/U_\infty = 6$) and the forcing amplitude ($F = 0.2$, see equation 4.1). The range of forcing frequencies is $St_d = fd/U_j = 0.1$–0.64. The jet velocity profile:

$$\mathbf{u}_{\text{wall}}(x, z) = [1 + F \ \sin(2\pi \ St_d \ t)] \ v_j(x, z) \ \mathbf{e}_2 \tag{4.1}$$

is modulated so that only the amplitude, not the shape, of the velocity profile changes with forcing.

Experimental Setting. Cold, non-reacting sub-scale experiments of unforced and forced jets in crossflow with a velocity ratio of 6 (as in the simulations) have also been conducted. The Reynolds number based on channel height was 27000, higher than in the simulations, but the jet Reynolds number of 5000 is similar to that used in the simulations. A mechanical actuator (a spinning valve) is used to modulate the jet at several frequencies in the range predicted as effective for mixing by the simulations ($St_d = 0.1$–0.26). The mechanical design and performance of this actuator are described in [Anderson et al., 2001]. Mixing

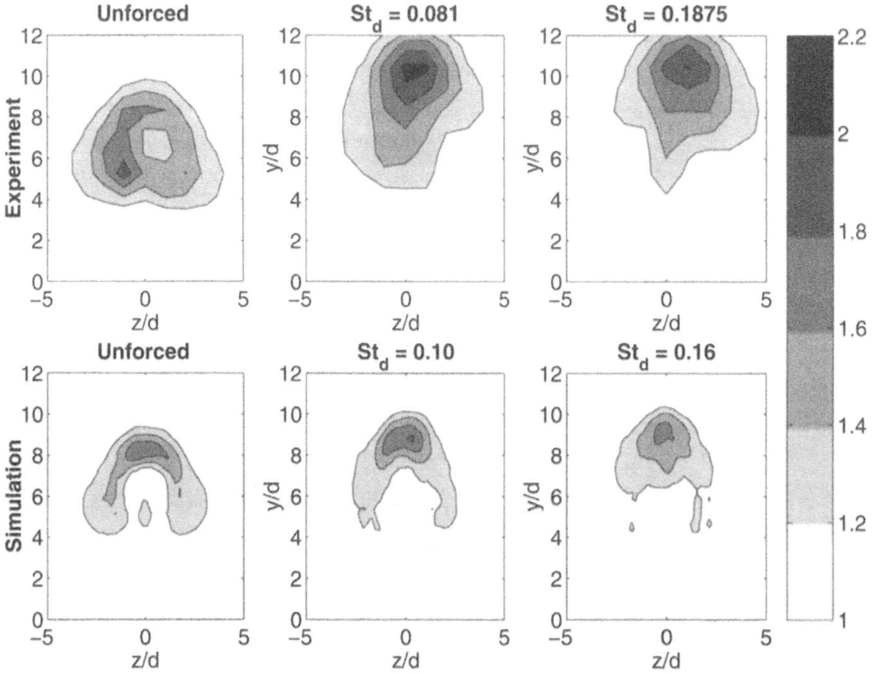

Figure 4.1. Comparison of the mean profiles of the resultant velocity $u_{res} = (u^2 + v^2)^{1/2}$ at $x/d = 4$ in experiments (top) and simulations (bottom).

and entrainment are evaluated by flow visualization and quantitative measurements of the velocity field using TSI single hot-film probes.

The experiments have been conducted in two phases. The first phase explored the "excitable" dynamics in the flow field, via unsteady forcing of the jet emanating into the crossflow over a wide range of frequencies (for a fixed amplitude). The response of the flow has been evaluated using velocity spectra from the hot film probe positioned at various locations both within and outside the jet as it emerges and bends into the crossflow. Forcing of the jet at high frequencies (namely, $St_d > 0.2$) results in an amplified flow response close to the jet orifice with diminishing response downstream, while low frequency forcing ($0.08 < St_d < 0.2$) is observed to produce a response farther downstream that has higher amplification. Note that the jet in crossflow region beyond the first 3–5 jet diameters is of particular interest for mixing enhancement. This first phase of experiments (aimed at the identification of "excitable dynamics") have been followed by the second phase of proof-of-concept, open-loop control experiments to verify and validate the ideas proposed using the direct numerical simulations. The experimental results, in-

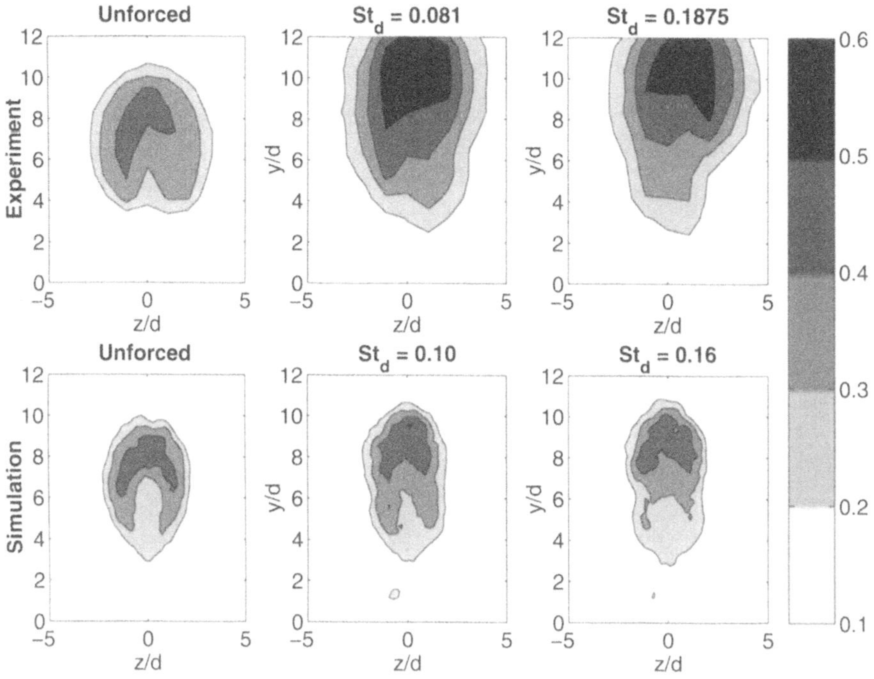

Figure 4.2. Comparison of the rms profiles of the resultant velocity $u_{res} = (u^2 + v^2)^{1/2}$ at $x/d = 4$ in experiments (top) and simulations (bottom).

cluding time-averaged surveys of the scalar and velocity fields and flow visualizations, will be described elsewhere [Narayanan et al., 2002].

2. Results

The ability of the simulations to capture the features of the jet in crossflow is limited to some degree by their resolution and Reynolds number, but they should be expected to capture some of the large-scale features of jets in crossflow measured experimentally at higher Reynolds numbers. The experiments measured the mean and fluctuating velocity using hot film probes parallel to the crossflow and wall-normal directions. These probes therefore measured the velocity $u_{res} = (u^2 + v^2)^{1/2}$, the resultant of the velocities in the streamwise and wall-normal directions (u and v, respectively). These quantities were computed in the simulations and are displayed on matched non-dimensional scales with the experimental measurements in figures 4.1 and 4.2. The resultant velocities and fluctuations from the simulations compare reasonably well to those from the experiments considering the differences in the jet velocity profile (thicker shear layers in the simulations), the level of turbulence in the emerging

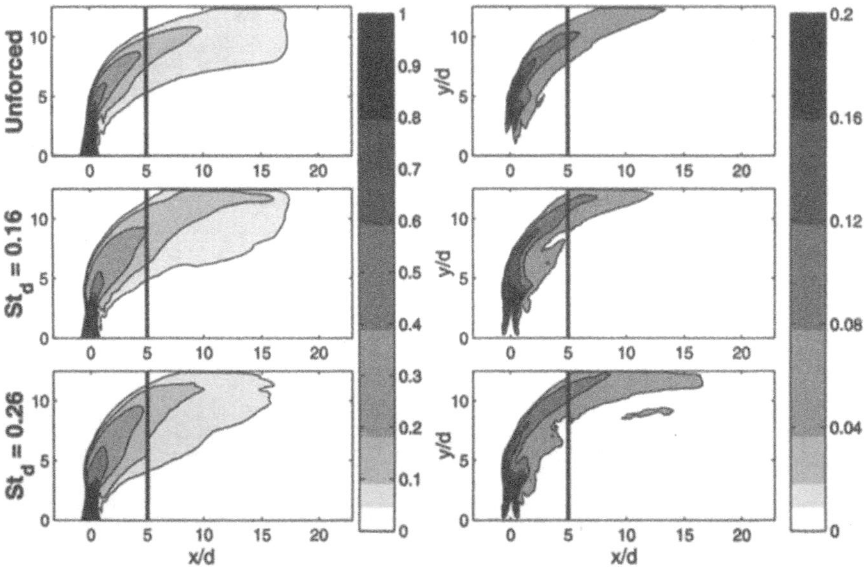

Figure 4.3. Mean (left) and rms (right) scalar profiles from direct simulation of an unforced jet in crossflow (top) and two forced cases at $St_d = 0.16$ (middle) and $St_d = 0.26$ (bottom). Each profile is an $x - y$ slice (parallel to the crossflow direction) through the jet centerline.

jet (zero in the simulations/non-zero in the experiments), the Reynolds number (lower in the simulations), and the artificially-imposed periodic spanwise boundary conditions at $z = \pm 5d$. The forcing levels employed in the experiments are also much higher than those in the simulations. A nearly fully modulated jet is used in the experiments whereas only a partially modulated jet is used in the simulations. The results from the experiments and simulations match closely for the unforced jet, with similar penetration, spreading and magnitude. The forced jets show less penetration, spreading and velocity magnitude in the simulations than in the experiments. The stronger forcing and thinner shear layers in the experiments may lead to larger growth rates for modes in the shear layer and consequently to a larger response to the forcing. Further validation for the simulations comes from the mean concentration profile at $x/d = 5$ in the unforced jet, which corresponds well to that measured by [Smith and Mungal, 1998] for a velocity ratio of $R = 5$.

The structures in the jet shear layer and the counter-rotating vortex pair (CVP) can be seen clearly in the simulations of unforced jet in crossflow. The jet emerges into the crossflow, bends to follow the crossflow and impinges on the upper wall of the channel just downstream of $x/d =$

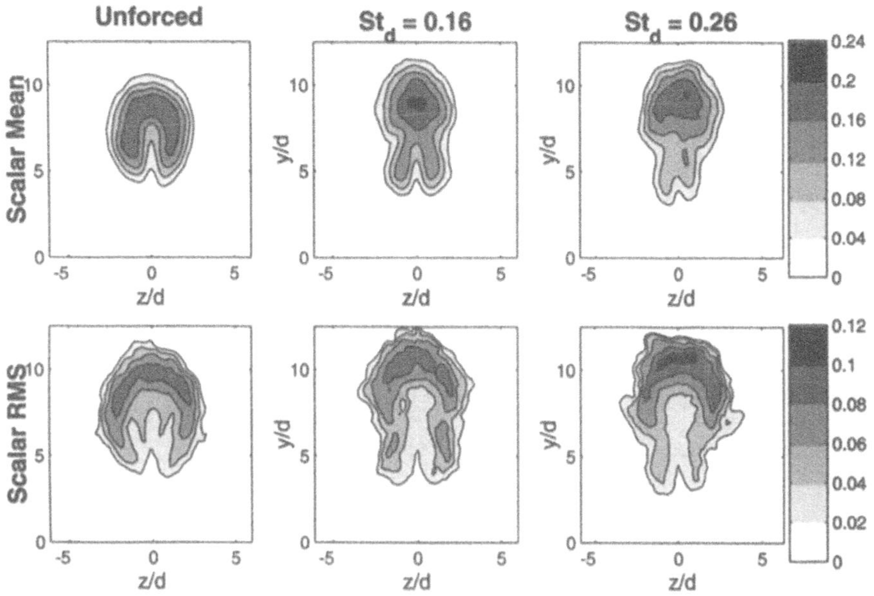

Figure 4.4. Mean (top) and rms (bottom) scalar profiles from direct simulation of an unforced jet in crossflow (left) and two forced cases at $St_d = 0.16$ (middle) and $St_d = 0.26$ (right). Each profile is a $y - z$ slice at $x/d = 5$ (perpendicular to the crossflow direction) whose locations is marked by a vertical line in the $x - y$ slice.

10. Judging by the trajectory of the jet, the near field of the simulations (up to approximately $x/d = 6$) is relatively unaffected by the presence of the channel's upper wall. The mean and fluctuating scalar fields for the unforced jet and the jet forced at $St_d = 0.16$ and 0.26 are shown in figures 4.3 and 4.4. These non-dimensional forcing frequencies were chosen after a survey of the intrinsic dynamics of the jet in crossflow (similar to that in experiments). Surveys of the velocity spectra along the jet trajectory as it bends into the crossflow reveals broadband spectral peaks surrounding $0.3 < St_d < 0.4$ near the jet exit associated with coherent, azimuthal-vorticity-dominated shear layer structures. However, farther downstream (where most of the vorticity dynamics are confined to a counter-rotating vortex pair with predominantly streamwise vorticity) lower frequency activity around $0.1 < St_d < 0.2$ is noted. Although the counter-rotating vortex pair appears symmetric in time-averaged images of the velocity and vorticity fields, the downstream structures are instantaneously far from symmetric and are highly unsteady (influenced strongly by the unsteady vortex dynamics occurring on the upstream edge of the jet). The complex, asymmetric counter-rotating structure

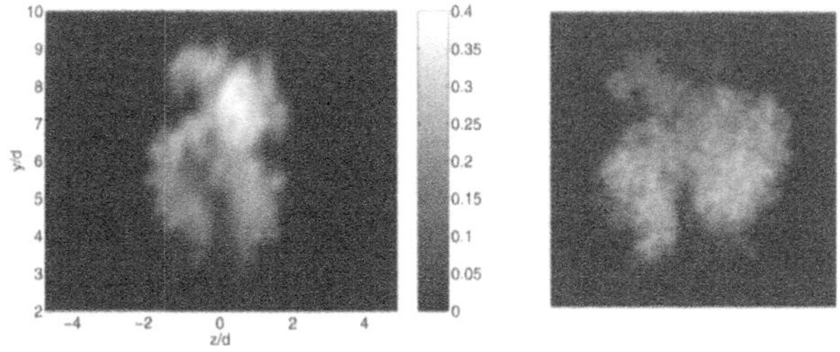

Figure 4.5. Instantaneous profiles of scalar concentration in unforced jet in crossflow at $x/d = 3$ from simulations (left) and experiments (right).

Figure 4.6. Flow visualization from experiments of instantaneous scalar concentration in an $x - y$ plane through the jet centerline for the unforced jet in crossflow (left) and the jet in crossflow forced at $St_d = 0.08$ (center) and $St_d = 0.1875$ (right)

is evident in flow visualizations of the instantaneous scalar fields of the unforced jet in a cross-plane at $x/D_j = 3$ (figure 4.5).

The scalar fluctuations are largest near the end of the jet's potential core and along the top edge of the jet (figure 4.4). The fluctuations along the top edge of the jet correspond to the passage of coherent packets of jet fluid which result from the rollup of structures associated with the jet shear layer. For a range of frequencies similar to but lower than those for round jets, forcing has a substantial impact on the structure and mixing of the jet. Forcing in the frequency range $St_d = 0.1$–0.26 leads to increased spreading of the scalar in the wall-normal direction and greater penetration into the crossflow. In these cases, the mean scalar profiles in a $y - z$ slice are qualitatively different from the unforced case, appearing pinched in the center with legs of fluid stretched below. Fluctuations at the upper edge of the jet are enhanced by the forcing which excites the Kelvin-Helmholtz rollup. Forcing at higher frequencies results in a jet which closely resembles the unforced jet in a time-averaged sense, as in [Schuller et al., 1999, Eroglu and Breidenthal, 2001]. However, forcing at the lower frequencies ($St_d = 0.1$–0.26) results in increased energy far away from the jet exit in these lower frequencies, providing an excitable band of frequencies which are beneficial for mixing.

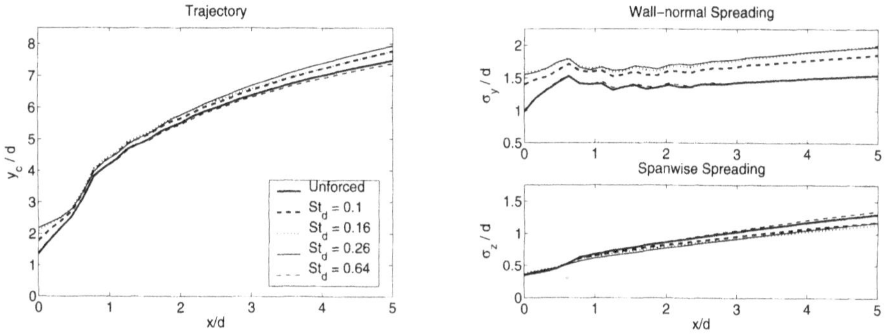

Figure 4.7. Comparison of the trajectory (left) and spreading (right) of the forced and unforced jets in crossflow.

The penetration and spreading of the jet fluid may be characterized by the first and second moments of the mean scalar profiles. The trajectory y_c/d of the jet as a function of the downstream distance x/d is displayed in figure 4.7. With the exception of the highest forcing frequency considered here ($St_d = 0.64$), the forced jets show a slight increase in penetration relative to the unforced jet. The second moment of the scalar profile in the wall-normal and spanwise direction indicates the spreading of the jet in these directions. Forcing also leads to greater spreading of the jet in the wall-normal direction, up to an increase of 30% relative to the unforced jet for $St_d = 0.16$ and 0.26, while the spreading in the spanwise direction is slightly smaller in these forced cases with 10% less spreading for $St_d = 0.16$. The mean scalar profiles shown in figure 4.3 — where low frequency forcing results in scalar profiles elongated in the wall-normal direction — predicted these trends, and these results are confirmed by comparison with the experimental flow visualizations of instantaneous scalar concentration in figure 4.6. In the experiments, the effect of the forcing is more pronounced with greater spreading than in the simulations (as noted before in the discussion of figure 4.1). The excitation of structures in the jet shear layer and the associated increase in intermittency and unsteadiness on the top edge of the jet is shown clearly in the center panel of figure 4.6 (where $St_d = 0.08$).

The jet in crossflow entrains fluid in the same manner as a jet in quiescent fluid — through the structures in the jet shear layer — but also entrains crossflow fluid through the action of the counter-rotating vortex pair (CVP). The entrainment of fluid by the jet may be characterized by the ratio of the volume flux of jet fluid across a given downstream plane $\dot{V}_{jet}(x) = \frac{1}{T}\int_0^T \left(\int_{A_{jet}} u(x,y,z,t)\ dy\ dz\right) dt$ to the volume flux at the jet exit $\dot{V}_0$, where the jet fluid is identified by a scalar concentration

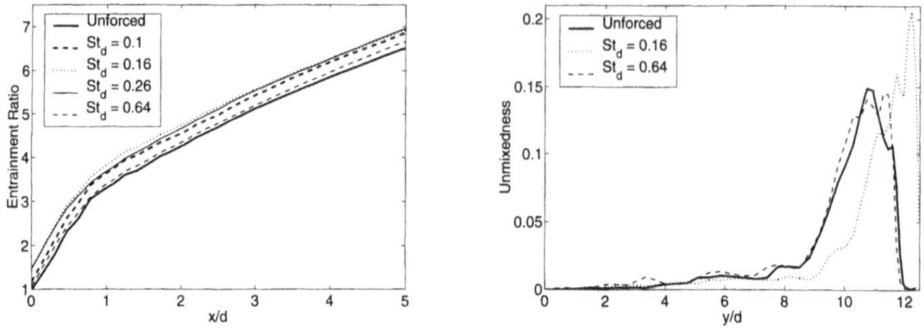

Figure 4.8. Comparison of the entrainment ratio $\dot{V}_{jet}/\dot{V}_0$ and the level of unmixedness (through the centerline of the jet at $x/d = 5$) of the forced and unforced jets.

above 0.01 [Yuan and Street, 1998]. (The scalar concentration is one at the jet exit and zero in the crossflow fluid.) The jets forced at lower frequencies ($St_d = 0.1$, 0.16, 0.26) display enhanced entrainment relative to the unforced jet, entraining more fluid near the jet exit and continuing to entrain fluid downstream at the same rate as the unforced jet.

The level of intermittency and unsteadiness associated with the forcing can be quantified using the 'unmixedness' or 'intensity of segregation' [Danckwerts, 1952, Dimotakis and Miller, 1990]. The unmixedness $\Xi = \overline{c'^2}/\overline{c}(1-\overline{c})$ is maximized (with a value of 1) at a point where equal parcels of purely jet and purely crossflow fluid pass. In figure 4.8, the profile of unmixedness in the wall-normal direction at $x/d = 5$ is shown for the unforced jet and the forced jet with $St_d = 0.16$ and 0.64. The excitation of the jet shear layer structures leads to increased unmixedness near the top edge of the $St_d = 0.16$ jet (also seen experimentally in the center panel of figure 4.6 where $St_d = 0.08$).

3. Discussion and Conclusions

Unsteady forcing in the appropriate frequency range can have a substantial impact on the structure of and mixing by the jet in crossflow. We have simulated the jet in crossflow to examine the response of the jet to sinusoidal forcing at a variety of frequencies in the range $St_d =$ 0.1–0.64. The range of forcing frequencies was chosen to enable excitation of a variety of coherent structures in the flow, including the "near" field unsteady jet shear layer structures (where most prior studies have focused) as well as "far" field unsteady counter rotating vortex pair dynamics. Both the simulations and experiments indicate that the effect of forcing on the downstream jet mixing is most substantial in the range $St_d = 0.1$–0.26, where increased spreading, penetration and entrainment is found. The appropriate metric for mixing in the jets in crossflow is an open question and is by no means sufficiently addressed using traditional time-averaged measures (e.g. jet spread). Increased spreading and entrainment often comes at the expense of increased unsteadiness and intermittency in the jet. The acceptability of this unsteadiness is primarily dependent on the timescale associated with the unsteadiness. In a combustion product cooling application, for example, unsteadiness at timescales much faster than the characteristic convection times for heated products traveling past the combustor exit or the heating/cooling timescales of a material subjected to thermal fluctuations (e.g. a turbine blade) may not result in substantial thermal cycling fatigue. However, slower timescales could lead to substantial cycling and fatigue, even for lower time-averaged thermal stresses. Careful interpretation may lead to greater understanding of the importance of unsteadiness and the potential for increased mixing by forced jets in crossflow.

Acknowledgments

The authors would like to thank Dr. Andrzej Banaszuk for his support and for encouraging us to undertake this effort, Drs. Dave Licsinsky, Igor Mezic and Jeff Cohen for fruitful discussions and comments, and Mr. Bruce True for assistance with the experiments. This work was supported in part by internal funding from the United Technologies Research Center and in part by the Air Force Office of Scientific Research through contract no. F49620-01-C-0021 and F49620-01-1-0048. This work was initiated while the first author was a postdoctoral researcher at the University of California, San Diego.

References

T. J. Anderson, W. Proscia and J. M. Cohen (2001). Modulation of a liquid-fuel jet in an unsteady cross-flow. In *Proc. ASME Turbo Expo '01*. New Orleans, Louisiana, June 2001, paper no. 2001–GT–0048.

S. C. Crow and F. H. Champagne (1971). Orderly structure in jet turbulence. *J. Fluid Mech.*, 48:547–591.

P. V. Danckwerts (1952). The definition and measurement of some characteristics of mixtures. *Appl. Sci. Res., Sec. A*, 3:279–296.

P. E. Dimotakis and P. L. Miller (1990). Some consequences of the boundedness of scalar fluctuations. *Phys. Fluids A*, 2(11):1919–1921.

A. Eroglu and R. E. Breidenthal (2001). Structure, penetration, and mixing of pulsed jets in crossflow. *AIAA J.*, 39(3):417–423.

R. M. Kelso, T. T. Lim and A. E. Perry (1996). An expermental study of round jets in a cross-flow. *J. Fluid Mech.*, 306:111–144.

B. P. Leonard (1988). Simple high accuracy resolution program for convective modeling of discontinuities. *Int. J. Numer. Meth. Fl.*, 8:1291–1318.

R. T. M'Closkey, J. M. King, L. Cortelezzi and A. R. Karagozian (2001). The actively controlled jet in crossflow. *J. Fluid Mech.* submitted.

S. Narayanan, P. Barooah and J. M. Cohen (2002). Experimental study of the coherent structure dynamics & control of an isolated jet in cross flow. In *Proc. 40th AIAA Aerosp. Sci. Mtg.* Reno, Nevada, Jan. 2002.

J. Nordström, N. Nordin and D. Henningson (1999). The fringe region technique and the Fourier method used in the direct numerical simulation of spatially evolving viscous flows. *SIAM J. Sci. Comput.*, 20(4):1365–1393.

T. Schuller, J. King, A. Majamaki and A. R. Karagozian (1999). An experimental study of acoustically controlled gas jets in crossflow. *Bull. Amer. Phys. Soc.*, 44(8):111.

S. H. Smith and M. G. Mungal (1998). Mixing, structure and scaling of the jet in crossflow. *J. Fluid Mech.*, 357:83–122.

P. J. Vermeulen, C. F. Chin and W. K. Yu (1990). Mixing of an acoustically pulsed air jet with a confined cross-flow. *J. Propul. Power*, 6(6):777–783.

P. J. Vermeulen, P. Rainville and V. Ramesh (1992). Measurements of the entrainment coefficient of acoustically pulsed axisymmetric free air jets. *J. Eng. Gas Turb. Power*, 114:409–415.

L. L. Yuan and R. L. Street (1998). Trajectory and entrainment of a round jet in crossflow. *Phys. Fluids*, 10(9):2323–2335.

Discussion

A. Pollard, Queen's University "You mentioned that you used a periodic boundary condition, and that you were filtering out the large scale unsteadiness. To ensure a proper input into the calculation domain, would it not be advisable to use a consistent inlet condition?"

Reply: What we filter out is that you have fluctuations associated with the jet, so what we filter is, at the end of the domain you filter the scalar and turbulence field to drive them to a laminar profile and no scalar at the inflow.

A. Pollard, Queen's University "But surely you are forcing the solution as it evolves downstream, for the large-scale recirculation zone is going to penetrate to the exit plane, normally. I am wondering if you have conditioned the flow, or more precisely, over-conditioned the flow?"

Reply: That's possible and that is something that we can consider; yeah, we are not allowing the mean flow to pulsate that is what we are prohibiting. So we are fixing a mean volume flow through the channel.

A. Pollard, Queen's University "So would your conclusions change if you changed the boundary condition?"

Reply: I would hope that they would not change in the near field; and we are trying to focus on the near field of the jet and if you focus downstream that could change things somewhat. This is an accepted technique for transitional flows and it could have some effect but we hope it is not substantial.

Santanu Sarkar, UCSD "I have another question about the periodic boundary conditions. In the absence of the jet, does it relaminarise in the fringe domain and then is piped back in and becomes turbulent or does it remain laminar. At this pressure gradient, does it go laminar, turbulent, laminar, in the absence of the jet?"

Reply: It is sub-critical and unless you introduce turbulence it will not become turbulent.

Santanu Sarkar, UCSD "So, it is a laminar channel flow."

Reply: Yes, the base flow is laminar in this case although you could run turbulence at these levels if you introduced substantial purturbations.

Sigurdson, University of Alberta "I wanted to ask, that since you are at a Strouhal number of 0.26 you think you are forcing at the preferred mode as Crow and Champagne?"

Reply: I think we are close.

Sigurdson, University of Alberta "So that mode still occurs even though this is a transverse jet?"

Reply: Ah, well, it is close to the preferred mode of the unforced jet, but I can't say it is going to be identical just because the base flow is changed.

Poinsot, CNRS Toulouse "We did similar computations and we found that it was not possible to start the computations at the plane for the jet, we had to calculate inside the jet, basically because the velocity profile is affected by the cross flow. Did you try different velocity profiles?"

Reply: We haven't tried substantially different. One of the things we did was to look at Kelso, Lim and Perry (JFM, 306, 1996, eds) suggested that for velocity ratios as high as 6, there was not substantial separation inside the pipe. But that is a concern at lower velocity ratios where you can have flow separation and flow back into the pipe, but by going to a velocity ratio this high we would be safe in imposing this boundary condition. That is our validation.

Chapter 5

CHARACTERISTICS OF SUPERCRITICAL TRANSITIONAL TEMPORAL MIXING LAYERS

Nora Okong'o and Josette Bellan
Jet Propulsion Laboratory, California Institute of Technology
Pasadena, CA 91109

Abstract Several Direct Numerical Simulation realizations of supercritical, three-dimensional, temporal mixing layers are used to investigate aspects of interest to turbulent combustion. The realizations are based on a model previously developed that accounts for the enlarged transport matrix at supercritical conditions (Soret and Dufour effects), for real gas equations of state, and for variable Schmidt and Prandtl numbers. This model is exercised for two very different sets of binary species, heptane/nitrogen and oxygen/hydrogen, to obtain transitional states for mixing layers excited at different perturbation wavelengths. Visualizations of the transitional states show that all layers develop convoluted regions of high density gradient magnitude (HDGM), which result both from the distortion of the initial density stratification boundary and from mixing. The species mass fraction in these HDGM regions is very weakly dependent of the perturbation wavelength. The existence of the HDGM regions, independent of the perturbation wavelength, indicates that they may be a feature of spatial mixing layers, and furthermore have a similar mass fraction composition. Evaluations of the applicability of the assumed PDF method for describing supercritical transitional flows shows that neither the β density nor the Gaussian are appropriate representations of the conserved scalar, and that the PDFs of the partial densities and the temperature are correlated, invalidating the typical model of the reaction rate as the product of the marginal PDFs.

Keywords: Supercritical transitional mixing layer

A. Pollard and S. Candel (eds.), IUTAM Symposium on Turbulent Mixing and Combustion, 59-71.

1. Introduction

The turbulent mixing of fluids at high pressure is a topic of much interest because it is relevant to many combustion applications. Such applications include general automotive (particularly, Diesel), gas turbine and liquid rockets engines. Liquid rocket combustion conditions present a particular experimental challenge as the operating regime is supercritical with respect to both fuel and oxidizer and the mixture is potentially explosive. In this situation, numerical simulations with validated models can contribute information that would be otherwise impossible to obtain experimentally.

The modeling of supercritical fluid behavior has been so far primarily addressed in the context of drop modeling, bypassing the turbulent aspect of the flows of practical interest. Among such numerous studies, the investigation of Harstad and Bellan [Harstad and Bellan, 2000] is noteworthy because the model was validated using the data of Nomura et al. [Nomura et al., 1996]. The model in [Harstad and Bellan, 2000] is based on Keizer's fluctuation-dissipation (FD) theory [Keizer, 1987] which is totally consistent with non-equilibrium thermodynamics. This set of equations was mathematically closed with real gas equations of state, and high-pressure transport properties.

Fundamental simulations devoted to supercritical shear flows have not been attempted until very recently (see [Miller et al., 2001], [Okong'o and Bellan, 2000] and [Okong'o and Bellan, 2001]). Such fundamental simulations are clearly necessary to unravel the specific aspects of supercritical turbulent shear flows. A specific aspect recently observed by Chehroudi et al. [Chehroudi et al., 1999], [Chehroudi et al., 1999b], and by Mayer and Tamura [Mayer and Tamura, 1996], is the lack of jet atomization under supercritical conditions; instead of drop formation, finger-like structures protrude from the jet.

The present investigation is conducted through Direct Numerical Simulation (DNS) using the validated set of conservation equations of [Harstad and Bellan, 2000], described in Section 2, used here in the context of a three-dimensional temporal mixing layer, as in [Miller et al., 2001], [Okong'o and Bellan, 2000], that is perturbed for the purpose of achieving transition. Aspects of the transitional states pertinent to supercritical turbulent mixing and combustion are emphasized in Section 3. Finally, in Section 4, we summarize our findings and state conclusions.

2. Governing equations

2.1. Conservation equations

The differential conservation equations have been discussed in detail by Harstad and Bellan [Harstad and Bellan, 2000] and by Miller et al. [Miller et al., 2001] for a binary species system, and consist of the continuity, momentum, total energy and species equations. Their important characteristic is the expression of the energy flux and species mass fluxes according to the FD theory, whereby additional to the typical Fourier and Fick's terms encountered at low pressure, the transport matrix is enlarged to contain the Dufour and Soret terms, respectively. The Dufour term is the sum of two contributions each proportional either to the mass fraction gradient, or the pressure gradient, whereas the Soret term contains two contributions each proportional either to the temperature gradient, or the pressure gradient. The addition of the Dufour and Soret terms introduces a new transport process, the thermal diffusion with its corresponding transport property, the thermal diffusion factor, that must be known in order to solve the equations.

To close the system of equations, we select a modified form of the Peng-Robinson equation of state (EOS) [Harstad et al., 1997], [Miller et al., 2001], as this EOS portrays in a simple manner the behavior of real gases. All thermodynamic properties are consistently calculated from the EOS. One such important property is the mass diffusion factor, α_D, which in this formulation multiplies the mass diffusion coefficient, D, in the Fick's law. The significance of α_D is to indicate departures from mixture ideality ($\alpha_D = 1$), meaning that the molecular mixing is impeded.

2.2. Transport properties

In real gases, transport properties are strong functions of the thermodynamic variables and it is thus important to accurately portray the values of $Sc = \mu/\rho\alpha_D D$, the Schmidt number, and of $\mathrm{Pr} = \mu C_p/m\lambda$, the Prandtl number, so as to correctly render mixing and heat transfer (here μ is the viscosity, ρ is the density, C_p is the molar heat capacity at constant pressure, λ is the thermal conductivity and m is the mixture molar weight). To this end, Sc and Pr have been accurately calculated using thermodynamic variable dependent properties, and, for computational efficiency [Miller et al., 2001], at fixed pressure, p, they have been correlated as functions of temperature, T, and mass fractions ($Y_1, Y_2 = 1 - Y_1$). All simulations presented herein have been performed with thermodynamic variable dependent Sc and Pr. On the other hand, for typical fluids μ is too small to enable resolution of all scales in a domain large enough to keep the Knudsen number in the continuum regime, and thus to ensure validity of the Navier-Stokes equations. To palliate this problem, the strategy is to enlarge μ but keep realistic values of velocities, densities and the Reynolds number, Re, so as to depict the correct dynamics while solving the problem in a domain large enough to be in the continuum regime. Finally, depending on the species set, the thermal diffusion factor is found either by comparing simulations with data (see [Harstad and Bellan, 2000] for heptane/nitrogen) or by considering similarity in molecular characteristics (see [Okong'o et al., 2001] for oxygen/hydrogen).

2.3. Configuration, and boundary conditions

The configuration is that of a three-dimensional (3D) mixing layer, depicted in Fig. 1 for one of the sets of studied species. The lower stream contains the higher density fluid (subscript 2), and the initial conditions are supercritical (i.e. $p > p_c$ or/and $T > T_c$, where the subscript c denotes the thermodynamic critical point) for the species in both streams. The boundary conditions are periodic in the streamwise, x_1, and spanwise, x_3, directions, and of outflow type for a real gas (boundary conditions derived by Okong'o et al. [Okong'o et al., 2000]) in the cross-stream direction, x_2.

3. Results

The studied species and their properties are listed in Table 5.1. The initial conditions are listed in Table 5.2. All simulations are conducted with an initial convective Mach number, $M_{c,0} = 0.4$. The simulations are started with streamwise and spanwise vorticity perturbations (whose amplitude is proportional to F_{3D} and F_{2D}, respectively) superimposed on the mean initial velocity profile that is smoothed near the density discontinuity boundary using an error function; see [Miller et al., 2001] for details. The initial vorticity thickness, $\delta_{\omega,0}$ ($\delta_\omega(t) = \Delta U_0 / \left(\partial U / \partial x_2\right)_{max}$ where U is the mean (or $x_1 - x_3$ planar average) flow in the streamwise direction, and $\Delta U_0 = U_1 - U_2$ is the velocity difference across the layer), is 6.859×10^{-3} m for all simulations. The only difference between realizations for the same species set is the value of the perturbation wavelength, λ_1, and the size of the domain that is chosen to accommodate four wavelengths in the streamwise and spanwise directions and two vortex pairings. For each set of species, the value of $\text{Re}_0 = (0.5(\rho_1 + \rho_2)\Delta U_0 \delta_{\omega,0})/\mu_R$ (μ_R is a reference viscosity) for the second simulation is chosen by matching $\text{Re}_0\lambda_1/\delta_{\omega,0}$ of the first simulation. This leads to similar values of $\mathrm{Re}_m = \mathrm{Re}_0 \delta_m/\delta_{\omega,0}$ (where δ_m is the momentum thickness) at transition, which are listed in Table 5.2. The transitional state is chosen to correspond to the peak in the timewise variation of the volume averaged positive spanwise vorticity. Since the initial mean velocity profile is such that all initial spanwise vorticity is negative, the appearance of positive spanwise vorticity is indicative of small scale formation. Braid and between-the-braid visualizations of the spanwise vorticity concur with the choice based on this global quantity.

To ensure that all DNS herein are well resolved, the energy spectra, both streamwise and spanwise, were computed at the transitional time in a manner similar to Kim et al. [Kim et al., 1987]. A sample of the results is displayed in Fig. 2 for C_7H_{16}/N_2, and it clear that there is a ratio of more than ten orders of magnitude between the low and high wavenumbers, thereby showing that most of the energy resides in the large scales and that there is no accumulation of energy in the small scales; for O_2/H_2 this ratio is even larger.

The examination of the transitional states focus on the role of molecular diffusion during mixing and on the utilization of the assumed PDF method for the description of supercritical mixing layers.

3.1. The role of molecular diffusion

Molecular diffusion is a very important process even in fully turbulent reacting flows because it is the mechanism that brings molecules of fuel and oxidizer in contact, thereby allowing chemical reactions. A typical measurement of molecular mixing [Cortesi et al., 1999] is the product thickness, $\delta_p = \int\int\int_V \rho Y_p dV$, based on the evolution of a conserved scalar, $Y_p = 2\min(Y_1, Y_2)$. On Fig. 3, δ_p is illustrated versus $t^* = t\Delta U_0/\delta_{\omega,0}$ for all four realizations listed in Table 5.2. Diffusional mixing is initiated as soon as the layer starts to roll up and it is enhanced following each pairing, as shown by the slope of δ_p. For the C_7H_{16}/N_2 layers, the global molecular mixing seems more effective at the same t^* than for O_2/H_2, this being attributed to the smaller initial ρ_2/ρ_1 that allows earlier pairing; however, this global molecular mixing is not necessarily indicative of local molecular mixing, as will be shown below, and it is precisely the local process that determines the reaction efficiency. For a given set of species, the realization featuring the higher molecular mixing is that having the larger λ_1 (the situation is not immediately clear for O_2/H_2), this being attributed to the longer time necessary to pair during which diffusion has increased effectiveness.

In previous studies of C_7H_{16}/N_2 [Okong'o and Bellan, 2000] and O_2/H_2 [Okong'o et al., 2001] layers excited at a single wavelength it has been shown that a salient characteristic of supercritical mixing layers is the formation of regions of high $\mid \nabla\rho \mid$ (HDGM) which resulted both from the convolution of the initial density stratification boundary and from mixing. These regions were shown to contain partially mixed fluids and to act as material interfaces and damp the formation of small scales, thereby impeding the achievement of a transitional state. To determine the state of molecular mixing at transition, braid and between-the-braid visualizations of $\mid \nabla\rho \mid$ were examined as in [Okong'o and Bellan, 2000] and [Okong'o et al., 2001] to elucidate their composition and state of mixture ideality. Results from a similar analysis are here displayed in Fig. 4a for the braid plane, and epitomize the situation for the between-the-braid plane as well (not shown). The plots illustrate conditional averages of the mass fraction of the heavy fluid (Y_2) with respect to a cut-off value of $\mid \nabla\rho \mid$ which is a fraction of the maximum value of $\mid \nabla\rho \mid$ in that plane, $\mid \nabla\rho \mid_{\max}$; as this cut-off value increases, one probes deeper into the HDGM regions. Noteworthy, for a specified set of species $\mid \nabla\rho \mid_{\max}$ increases with decreasing λ_1, this being attributed to the smaller domain of the simulation which produces flows with higher vorticity. Since the stirring frequency is proportional to the vorticity, a higher vorticity induces more stirring while diffusional processes are unaffected by this change. An enhanced stirring at otherwise unchanged molecular diffusion is responsible for the larger $\mid \nabla\rho \mid_{\max}$. The results portrayed in Fig. 4a show that the fluid in the HDGM regions is mostly composed of the higher density species, with small amounts of the lighter fluid dissolved into the heavier one, and that the deeper one probes into the HDGM regions, the higher is Y_2. For a specified set of species, the average Y_2 is quite insensitive to the value of λ_1, indicating that these features (i.e. HDGM regions and their mass fraction composition) might persist for spatial mixing layers. Moreover, the value of the average Y_2 seems also insensitive to the set of species, indicating that it might be a generic aspect of supercritical mixing layers at a specified p_r. Comparisons between this average Y_2 and its stoichiometric value (0.89 for O_2/H_2 and 0.067 for heptane/air) reveals that the composition of the HDGM regions is very close to stoichiometric in the O_2/H_2 layer, but very far from it in the C_7H_{16}/N_2 layer (the addition of O_2 is believed to change only slightly this aspect), indicating that O_2/H_2 is much better suited for combustion than heptane/air. This conclusion is confirmed by the results plotted in Fig. 4b where conditional averages of α_D are plotted, similarly to the plot in Fig. 4a. Clearly, as one probes deeper into the HDGM region of the C_7H_{16}/N_2 layer, the mixture becomes more non-ideal ($\alpha_D < 1$) meaning that it has difficulty mixing at the molecular level, whereas the HDGM regions of the O_2/H_2 layer contain fluid close to ideality (well mixed regions at the molecular level). Thus, despite the global indications from δ_p that the C_7H_{16}/N_2 layer mixes better than the O_2/H_2 one, the local analysis reveals that it is precisely the opposite, in that it is the slowest characteristic time of the local diffusional processes that governs the fate of the molecular mixing.

3.2. Assumed PDF modeling

When using assumed PDF models, there are two issues of concern: (1) the form of the PDF, and (2) the value of the moments necessary to construct the PDF. Here we are concerned only with the first issue, and will assume that exact values of the moments are available ('best' situation) by constructing the PDF with DNS extracted moments. Since the β density has typically been used for describing the evolution of the conserved scalar from the segregated state of the species to complete mixing while the Gaussian (whose first two moments are 0 and 3) has been used to describe the fully mixed state, the present study focuses on these two PDFs.

As a typical example, in Fig. 5 we display for one of the four realizations the DNS extracted PDF, the Gaussian and the β density calculated with the DNS extracted moments; plots from the other realizations exhibit similar features. The PDFs are calculated at several cross-stream locations of the layer so as to illustrate the mixing features of the layer. Clearly, neither the β density nor the Gaussian are appropriate representations of the conserved scalar due, in particular, to the peaks in the DNS extracted PDF. These peaks are a consequence of the HDGM regions characteristic of these supercritical mixing layers.

Because the examined databases were created without including chemical reactions, they can be used on a 'best' case basis to reveal information about the modeling of the reaction rate, which is a joint temperature - (partial densities) PDF, by the product of the marginal PDFs (a common assumption in assumed PDF reacting flows). Indeed, when including reactions, the PDFs will be additionally correlated, further invalidating the replacement of the joint PDF by the product of the marginal PDFs. Among the four realizations, we illustrate in Fig. 6 the comparison between joint and product of the marginal PDFs which appeared the most favorable. Even in this situation, it is apparent that the comparison does not validate this frequent modeling assumption. The discrepancy between the joint PDF and the product of the marginal ones is attributed to two factors. First, the cubic real-gas equation of state provides increased coupling between the temperature and the partial densities compared to the perfect gas law. Second, the Soret and Dufour terms in the conservation equations represent an additional coupling between the energy and species conservation equations. Based on this evidence, the assumed PDF method is not recommended for utilization in supercritical flow simulations.

4. Conclusions

The peculiarities of supercritical mixing layers are here examined by analyzing the transitional states of four realizations differing either by the set of species, or by the perturbation wavelength used to excite the layer and achieve transition. Particularly, in contrast to gaseous mixing layers, each simulation displays concentrated regions of very high density gradient magnitude which are convoluted due to turbulence. In these regions the fluid is primarily the higher density one with small amounts of the lighter fluid dissolved into it. Departures from mixing ideality, indicating the difficulty in mixing at the molecular level (in contrast to ideal mixtures), are identified for the C_7H_{16}/N_2 layer and are found to increase with increasing density gradient. Moreover, it is shown that even in the absence of combustion, the temperature-species PDFs are well correlated (this being attributed to the higher order equation of state, and Soret and Dufour effects), meaning that their joint PDF is not properly approximated by the product of their marginal PDFs. Because under burning conditions the temperature and species will be even more correlated, this indicates that the traditional reactive flow modeling based on replacing the joint PDF representing the reaction rate by the product of the marginal PDFs, is not appropriate. This means that PDF methods applied to supercritical fluid mixing and combustion must rely on the solution of the PDF equation, as proposed by Pope [Pope, 1985].

5. Acknowledgments

This research was conducted at the Jet Propulsion Laboratory (JPL), California Institute of Technology. Joint sponsorship was provided by the Air Force Office of Scientific Research and by the Army Research Office with Drs. Julian Tishkoff and David Mann serving, respectively, as technical contract monitor, through interagency agreements with the National Space and Aeronautics Administration (NASA); and by NASA Marshall Space Flight Center with Dr. John Hutt as technical contract monitor. Computational resources were provided by the supercomputing facility at JPL.

References

Chehroudi, B., Talley, D. and Coy, E., Initial growth rate and visual characteristics of a round jet into a sub- to supercritical environment of relevance to rocket, gas Turbine, and Diesel engines, AIAA 99-0206, 37th Aerospace Sciences Meeting, Reno, NV, 1999

Chehroudi, B., Talley, D. and Coy, E., Fractal geometry and growth rate changes of cryogenic jets near the critical point, AIAA 99-2489, 35th AIAA/ASME/SAE/ASEE Joint Propulsion Conference, Los Angeles, CA, 1999

Harstad, K. and Bellan, J., An all-pressure fluid drop model applied to a binary mixture: heptane in nitrogen, *Int. J. of Multiphase Flow*, 26(10), 1675-1706, 2000

Keizer, J., *Statistical Thermodynamics of Nonequilibrium Processes*, Springler-Verlag, New York, 1987

Mayer, W. and Tamura, H., Propellant injection in a liquid oxygen/gaseous hydrogen rocket engine, *Journal of Propulsion and Power*, 12(6), 1137-1147, 1996

Miller, R. S., Harstad, K. and J. Bellan, J., Direct numerical simulations of supercritical fluid mixing layers applied to heptane nitrogen, *J. Fluid Mech.*, 436, 1-39, 2001

Nomura, H., Ujiie, Y., Rath, H. J., Sato, J. and Kono, M., Experimental study on high pressure droplet evaporation using microgravity conditions, *Proc. of the Comb. Institute*, *26*, 1267-1273, 1996

Okong'o, N. and Bellan, J., Direct numerical simulation of a transitional supercritical mixing layer: heptane and nitrogen, submitted, 2000

Okong'o, N., Bellan, J. and Harstad, K., Consistent boundary conditions for multicomponent real gas mixtures based on characteristic waves, submitted, 2000

Okong'o, N., Bellan, J., Linear stability analysis of real gas two- and three-dimensional mixing layers, submitted, 2001

Harstad, K., Miller, R. S. and Bellan, J., Efficient High-Pressure State Equations, *A.I.Ch.E. J.*, 43(6), 1605-1610, 1997

Okong'o, N., Harstad, K. and Bellan, J.,Direct Numerical Simulations of LOX/H2 Temporal Mixing Layers under Supercritical Conditions, submitted, 2001

Kim., J., Moin, P, and Moser, R. Turbulence statistics in fully developed channel flow at low Reynolds number. *J.Fluid Mech.*, vol 177, 133–166, 1987.

Cortesi, A., Smith, B., Yadigarloglu, G. and Banerjee, S., Numerical Investigation of the Entrainment and Mixing Processes in Neutral and Stably-Stratified Mixing Layers. *Physics of Fluids* 11(1), 162–185, 1999

Pope, S. B., PDF methods for turbulent reactive flows *Prog. Energy Combust. Sci.* 11, 119-192, 1985.

Species	m (g/mol)	T_c (K)	p_c(MPa)
Hydrogen	2.0159	33.0	1.2838
Nitrogen	28.013	126.26	3.399
Oxygen	31.9988	154.58	5.0430
Heptane	100.205	540.2	2.74

Table 5.1. Properties of the species investigated.

system	p_r (p)	$T_2; T_1$	ρ_2/ρ_1	$\lambda_1/\delta_{\omega,0}$	Re$_0$	F_{2D}	F_{3D}	Re
C_7H_{16}/N_2	2.17 (60 atm)	600 K; 1000 K	12.88	7.29	500	0.1	0.05	12
C_7H_{16}/N_2	2.17 (60 atm)	600 K; 1000 K	12.88	4.57	800	0.1	0.05	12
O_2/H_2	1.98 (100 atm)	400 K; 600 K	24.40	10.35	550	0.1	0.025	15
O_2/H_2	1.98 (100 atm)	400 K; 600 K	24.40	7.29	750	0.1	0.025	15

Table 5.2. Initial conditions of the simulations and value of the momentum thickness based Reynolds number, Re$_m$, at transition.

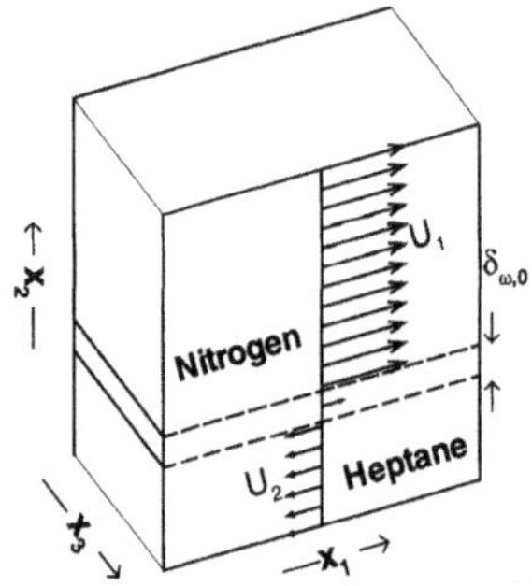

Figure 5.1. Configuration of the three-dimensional mixing layer, depicted for heptane/nitrogen.

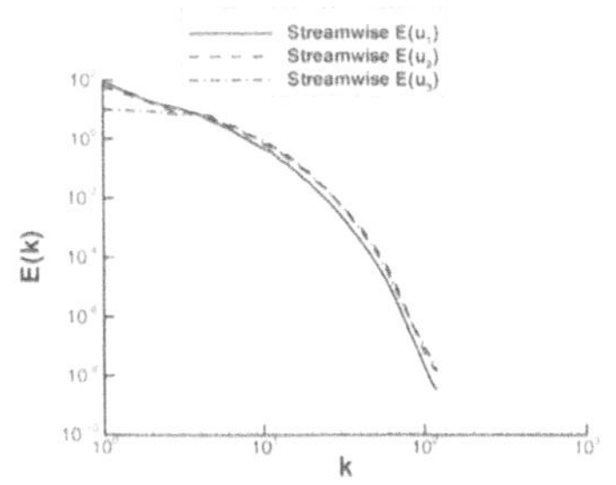

Figure 5.2. One-dimensional energy spectra of the velocity fluctuations for the C_7H_{16}/N_2 simulation with $\lambda_1/\delta_{\omega,0} = 7.29$ at the transitional time ($t^* = 155$).

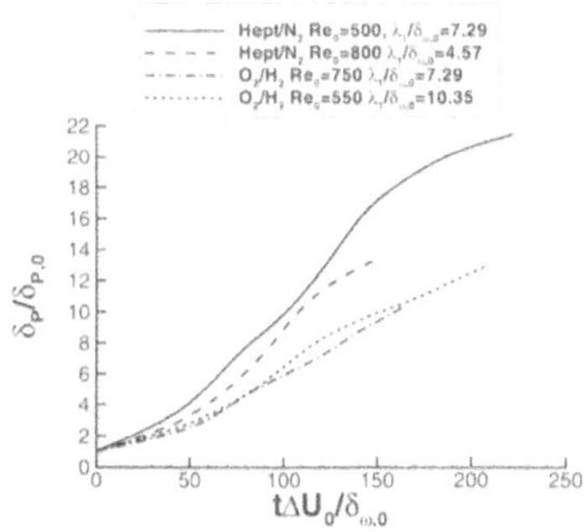

Figure 5.3. Product thickness as a function of time for all four simulations.

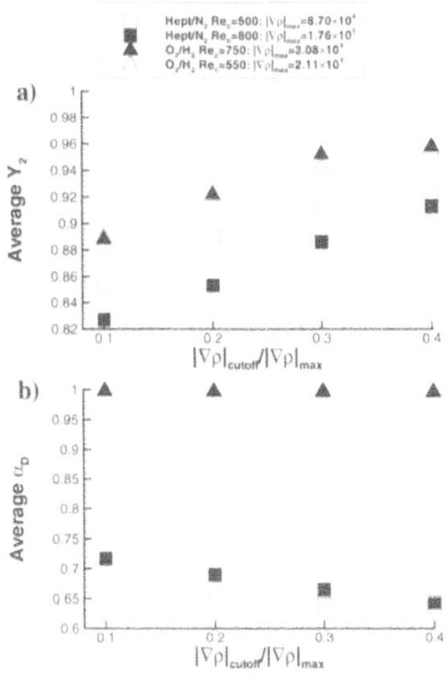

Figure 5.4. Conditional averages in the regions of high density gradient magnitude in the braid plane: (a) mass fractions, and (b) mass diffusion factor.

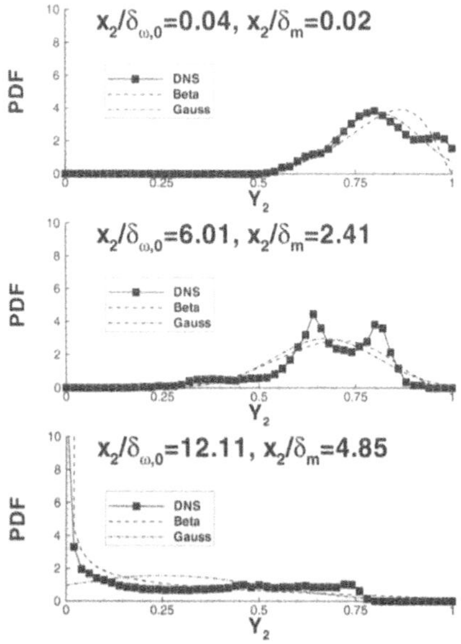

Figure 5.5. Comparison of the DNS extracted PDF with the β density (using the PDF extracted moments) and the Gaussian for the C_7H_{16}/N_2 simulation with $\lambda_1/\delta_{\omega,0} = 7.29$ at the transitional time ($t^* = 155$).

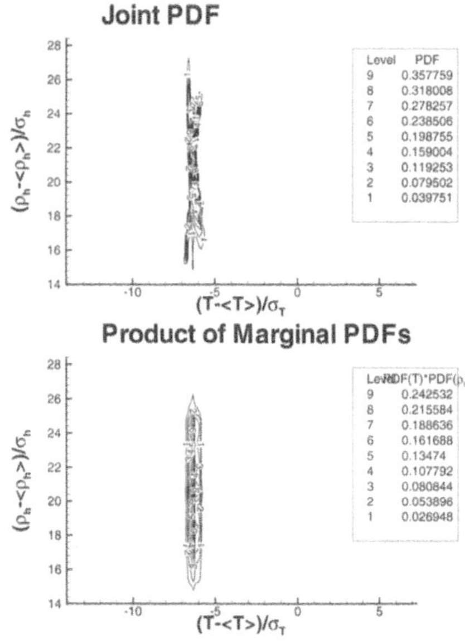

Figure 5.6. Comparison between the joint PDF of the partial density and temperature, and the product of the marginal PDFs for the C_7H_{16}/N_2 simulation with $\lambda_1/\delta_{\omega,0} = 7.29$ at the transitional time ($t^* = 155$).

Discussion

I. Fedioun, CNRS Orleans "What are the initial conditions for your simulations?"

Reply: The initial conditions are basically what we have is only heptane or oxygen in the lower layer, the heptane is a temperature of 600K, the nitrogen in the upper stream is at 1000K and for the oxygen/hydrogen the two simulations that I presented here for the realizations the oxygen is at 400, the hydrogen is at 600K.

I. Fedioun, CNRS Orleans "But, what are the velocity and density profiles?"

Reply: Everything is smoothed by the error function near the boundary of the two streams. We have four realizations and I don't recall all the specific numbers. But I can tell you what the profiles are like if you want.

Chapter 6

ON ACCURATE EXPERIMENTAL MEASUREMENTS OF THE DYNAMICS OF LARGE SCALE STRUCTURES IN TURBULENT FLOWS

Dan Ewing
Department of Mechanical Engineering, McMaster University, Hamilton ON, L8S 4L7, Canada

Scott Woodward
Department of Mechanical and Aerospace Engineering University of Buffalo, Amherst, NY 14260, USA.

Abstract Measuring the dynamics of the large-scale structures in turbulent flows is currently a significant challenge. One technique used to extend current experimental capabilities is to use Linear Stochastic Estimation and Proper Orthogonal Decomposition as complementary techniques to examine the dynamics of the energetic motions. Heretofore, it has not been possible to assess the accuracy of this technique without measuring the full velocity field. It is demonstrated here that the accuracy of the technique can be assessed *a priori* using a transfer function that relates the estimated and actual POD coefficients for the flow.

Keywords: large-scale structures, linear stochastic estimation, Proper Orthogonal Decomposition.

1. Introduction

The large-scale structures in turbulent flows play an important role in determining the mixing in these flow. It is a challenge to study the *dynamics* of these structures experimentally in high-Reynolds-number flows because this requires the measurement of the velocity (or vortic-

A. Pollard and S. Candel (eds.), IUTAM Symposium on Turbulent Mixing and Combustion, 73-83.

ity) field in a two- or three-dimensional region rapidly enough and with sufficient spatial resolution to capture the small-scale motions so they can be filtered. Thus, in many previous investigations the velocity field was only measured across one spatial direction (and time) yielding the dynamics of the roller eddies. Citriniti [1996] recently measured the temporally and spatially resolved velocity field on a plane 3 downstream of a round jet using 138 single hot-wire probes. He then used the Proper Orthogonal Decomposition to extract the dynamics of the large-scale motions and showed that the 'braid' region between the ring vortices in this flow played an important role in the turbulent mixing.

Bonnet *et al.* [1994] proposed that the experiments to examine the dynamics of the energetic structures could be simplified by using Linear Stochastic Estimation (LSE) and the Proper Orthogonal Decomposition (POD) in a complementary manner. In this approach, the instantaneous velocity is measured at a few points in the region of interest and estimated at the remaining points using Linear Stochastic Estimation (LSE). The POD technique is then used to extract the dynamics of the energetic motions from this estimated velocity field. Bonnet et al. showed this technique could be used to accurately reconstruct the dynamics of the roller eddies in a shear layer using 2 cross-wire probes while Ewing and Citriniti [1998] showed the results deduced by Citriniti [1996] could be reproduced using only 56 measurement points. In both cases, however, the accuracy of the technique could only be assessed by comparing the results deduced using the LSE-POD technique with results from measurements of the full velocity field.

Ewing and Woodward [2000] showed recently that the accuracy of the LSE-POD technique could in fact be assessed *a priori.* In particular, they showed that the estimated and actual POD coefficients could be related using a transfer function that only depends on statistical measures of the flow. The derivation of this transfer function is outlined here and then it is used to assess the accuracy of the LSE-POD technique for the experiments examined by Ewing and Citriniti [1998].

2. The POD-LSE Complementary Technique.

The Proper Orthogonal Decomposition (Lumley 1970) represents the turbulent motions using functions that optimally represent the turbulent kinetic energy in the flow. The optimal modes in the homogeneous directions are Fourier modes while the optimal modes in the non-homogeneous directions are solutions to the integral eigenvalue problem

$$\int R_{i,j}(\cdot,\cdot')\phi^n(\cdot')d\cdot' = \lambda^n \phi^n(\cdot), \tag{6.1}$$

where $R_{i,j}(\cdot, \cdot')$ is the two-point velocity correlation tensor and $\cdot$ is the coordinate(s) in the non-homogeneous direction(s). The solutions of this equation are orthogonal and can be normalized to form an orthonormal basis (Lumley 1970). Thus, λ^n represents the contribution of the *nth* mode to the total turbulent kinetic energy in the region of interest.

The dynamics of the energetic modes can be examined by projecting the *instantaneous* velocity field onto the orthogonal functions to determine the *instantaneous* coefficients of the modes given by

$$a_n(t) = \int u(\cdot, t)\Phi^{n*}(\cdot)d\cdot, \tag{6.2}$$

and reconstructing the velocity field produced by these modes; i.e.,

$$u_{rec}(\cdot, t) = \sum_{n=1}^{N_{rec}} a_n(t)\Phi^n(\cdot). \tag{6.3}$$

This low-order reconstruction of the N_{rec} most energetic modes removes (or filters) the small-scale that obscure the large-scale motions in high-Reynolds-number flows.

It is important to note that the small-scale motions must be resolved so their contribution can be filtered without their contribution being aliased into the large-scale motions. This requires the velocity field to be simultaneously sampled at a large number of points in the flow rapidly enough to resolve the dynamics of the motion. Bonnet *et al.* proposed simplifying the experiment by measuring the velocity at a smaller number of points and using Linear Stochastic Estimation (LSE) to estimate the velocity over the remaining points; i.e.,

$$u_i^e(\cdot) = \sum_{l=1}^{L} A_{ij}^l(\cdot)u_j^m(\cdot_l, t), \tag{6.4}$$

where $u_i^e(\cdot)$ is the estimated velocity field, $u_j^m(\cdot_l, t)$ are the L measured velocities, and $A_{ij}^l(\cdot)$ are LSE coefficients determined by minimizing the error in the estimated velocities.

The estimated velocity field is then used to estimate the *instantaneous* coefficients for the POD modes; i.e.,

$$a_n^e(t) = \int u_i^e(\cdot, t)\Phi_i^{n*}(\cdot)d\cdot \tag{6.5}$$

that can be used to estimate the velocity field produced by the energetic modes given by

$$u_i^{rec,e}(\cdot, t) = \sum_{n=1}^{N_{rec}} a_n^e(t)\Phi_i^n(\cdot). \tag{6.6}$$

3. Accuracy of the Estimated Coefficients

The LSE-POD technique reduces the experimental capability required to examine the dynamics of the large-scale structures but heretofore it has not been possible to assess the accuracy of the result without measuring the full velocity field. Ewing and Woodward [2000] showed the accuracy of the technique could be assessed by relating the estimated and actual POD coefficients. Substituting equation 6.4 into equation 6.5 the estimated coefficients can be written as

$$a_n^e(t) = \int \sum_{l=1}^{L} A_{ij}^l(\cdot) u_j^m(\cdot_l, t) \Phi_i^{n*}(\cdot) d\cdot . \tag{6.7}$$

The instantaneous velocity at the measured points can be written as

$$u_j^m(\cdot_l, t) = \sum_{\gamma=1}^{\infty} a_\gamma(t) \Phi_j^\gamma(\cdot_l). \tag{6.8}$$

Substituting this relationship into equation 6.7 and rearranging it can be shown that

$$a_n^e(t) = \sum_{\gamma=1}^{\infty} H_n^\gamma a_\gamma(t), \tag{6.9}$$

where H_n^γ is the transfer function between the coefficients and is given by

$$H_n^\gamma = \int \left[\sum_{l=1}^{L} A_{ij}^l(\cdot) \Phi_j^\gamma(\cdot_l) \right] \Phi_i^{n*}(\cdot) d\cdot . \tag{6.10}$$

In actual applications the POD technique is discretized and the transfer function is given by

$$H_n^\gamma = \sum_{m=1}^{M} \left[\sum_{l=1}^{L} A_{ij}^l(\cdot_m) \Phi_j^\gamma(\cdot_l) \right] \Phi_i^{n*}(\cdot_m) w(\cdot_m), \tag{6.11}$$

where M is the total number of measured points and $w(\cdot_m)$ is the weighting function.

The transfer function only depends on statistical measurements of the flow and thus can be computed without measuring the full velocity field. The transfer function includes both magnitude and phase information. Thus, it can be used to assess the relative contribution of each of the actual POD coefficients to the estimated coefficient and the phase angle of their contribution.

4. Experimental Data Base

Ewing and Citriniti [1998] applied the LSE-POD technique to an experimental data base measured by Citriniti (1996) in the axisymmetric shear layer 3 diameters downstream of a round jet with an exit Reynolds number of 80,000. The exit velocity profile was top hat to within 0.1% with a turbulence intensity of 0.35%. The streamwise velocity component in the axisymmetric shear layer was measured using an array of 138 single hot-wire transducers positioned in 6 radial rings at $r/D = 0.15, 0.28, 0.41,\ 0.54, 0.67,$and 0.80, where D is the outlet diameter of the jet. The wires in the 6 rings were equally spaced in the azimuthal direction. There were 6 wires on the inner-most ring and 12, 24, 32, 32, and 32 wires at the other 5 radial positions, respectively. The signals from the wires were filtered at 800 Hz and simultaneously sampled at 2,048 Hz in 300 blocks of 1024 points.

The flow is homogeneous in the azimuthal direction and stationary in time so these directions were transformed from the problem. The POD modes in the radial direction were solutions to (Citriniti 1996):

$$\int_0^\infty F(r,r',m,f)\Psi^n(r',m,f)r'dr' = \lambda^n(m,f)\Psi^n(r,m,f), \qquad (6.12)$$

where $F(r,r',m,f)$ is the cross-spectrum of fluctuating streamwise velocity, m is the azimuthal mode number, and f is the frequency. The coefficients of the modes are given by

$$a^n(m,f) = \int_0^\infty \hat{u}(r,m,f)\Psi^{n*}(r,m,f)rdr, \qquad (6.13)$$

and the reconstructed the Fourier coefficients are given by

$$\hat{u}_{rec}(r,m,f) = \sum_{n=1}^{N_{rec}} a^n(m,f)\Psi^n(r,m,f). \qquad (6.14)$$

The instantaneous reconstructed velocity field was then computed by inverse Fourier transforming this reconstructed Fourier coefficient in time and the azimuthal direction.

5. Application of the LSE-POD Technique

Citriniti [1996] found that the essential dynamics of the large-scale structures could be recovered using only the first POD mode in the radial direction (*i.e.*, $N_{rec} = 1$) and only the coefficients of azimuthal mode numbers 0 and $3-6$. Ewing and Citriniti [1998] showed that the dynamics of the large-scale structures in the axisymmetric shear layer

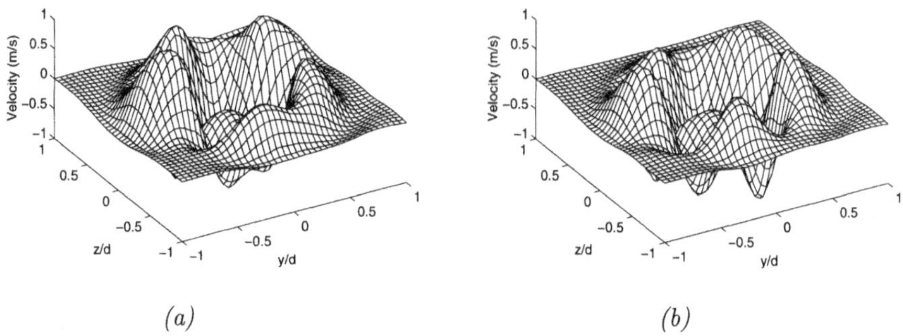

Figure 6.1. Comparison of the velocity field from the large-scale motions recontructed from (a) the original velocity field and (b) the estimated velocity field. The height corresponds to the value of the instantaneous streamwise velocity.

could have been accurately reproduced using the LSE-POD technique by estimating the Fourier coefficients $\hat{u}(r, m, f)$ using only the wires on the third and fifth rings in Citriniti's array. For example, a comparison of the actual and estimated reconstructed fields at one instant is shown in figure 6.1.

The magnitude and phase of the transfer function were computed for this cases using only the statistical measurements of the flow. The results for the first POD mode and aziumthal mode numbers 0 and 6 are shown in figures 6.2 and 6.3. The curves $\gamma = 1$ represent the relative contributions from the correct POD mode in the field while the curves $\gamma = 2$ and 3 represent the fraction of the second and third mode aliased into the estimated coefficient for the first mode. In both cases, over 90% of the first actual coefficient is included in the estimated coefficients in the range of frequencies that make the dominant contribution to the energy in the modes. The relative contribution (or aliasing) from the coefficients of the second and third mode is small in this range of frequencies. These coefficients also contain only a fraction of the energy in the first mode so the error introduced is small. The phase of the contribution from the actual first coefficient is also accurately estimated ensuring the timing of the different modes in the reconstructed velocity field is correct.

The transfer function is an useful design tool that can be used to compare the relative accuracy of the estimated reconstructed field for different measuring points. A comparison of the magnitude of the transfer function for experiments using rings 3 and 5 and rings 3 and 6 in Citriniti's array are shown in figure 6.4. It is clear the coefficients for azimuthal mode 0 would be accurately estimated in both cases but an experiment using rings 3 and 5 would yield more accurate estimates of the coefficients for aziumthal mode 6.

6. Summary and Concluding Remarks

The LSE-POD complementary technique provides a tool that can be used to experimentally examine the dynamics of the large-scale structures in high-Reynolds-number flows. It was shown here that the accuracy of the velocity fields reconstructed using this technique can be assessed using a transfer function that relates the estimated and actual coefficients of the POD modes. This transfer function only depends on statistical measures of the flow and thus can be computed *a priori*.

The transfer function was computed for experiments in the axisymmetric shear layer examined by Ewing and Citriniti [1998] and it was shown that the LSE-POD technique did in fact accurately estimate the POD coefficients in this case. More importantly it was shown that the transfer function can be used as a design tool to assess the relative accuracy of using different measurements points and develop a quantitative estimate of the error in the results.

Acknowledgments

The contribution of J. Citriniti is gratefully acknowledged and the first author would like to acknowledge support from NSERC Canada.

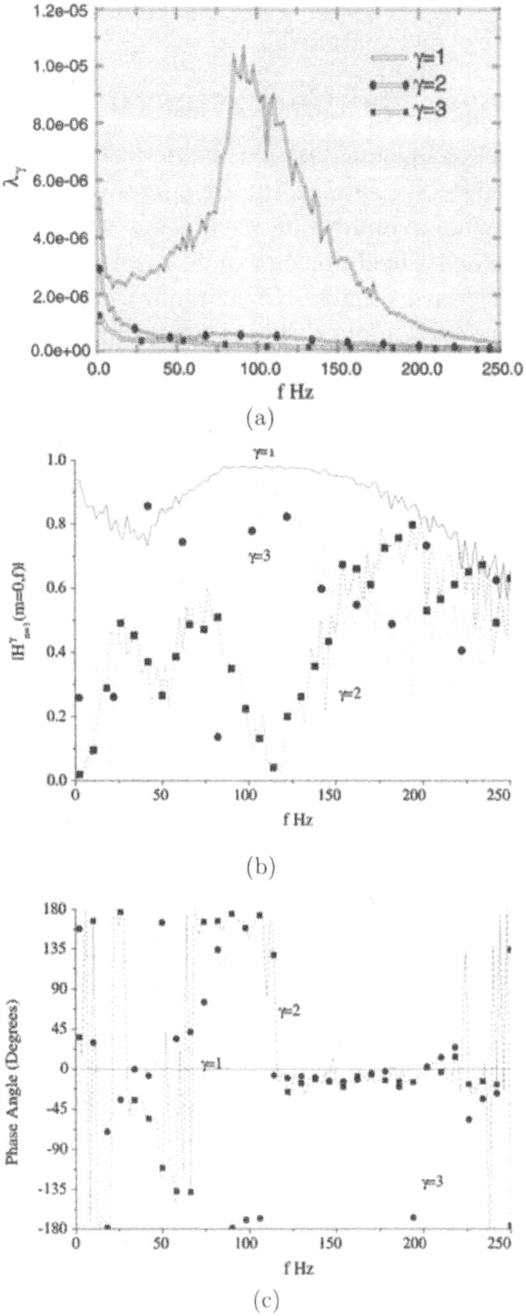

Figure 6.2. Accuracy of the estimated first POD coefficients for azimuthal mode 0 using the rings of hot wires at $r/D = 0.41$ and $r/D = 0.67$. (a) Eigenvalues for modes 1, 2, and 3 in the flow. (b) Magnitude of the transfer function for the first estimated POD coefficient $a_n^e(m = 0, f)$. Contribution from actual POD coefficients —— $\gamma = 1$ ▪ $\gamma = 2$ and • $\gamma = 3$. (c) Phase of the transfer function for the three modes symbols same as in (b)

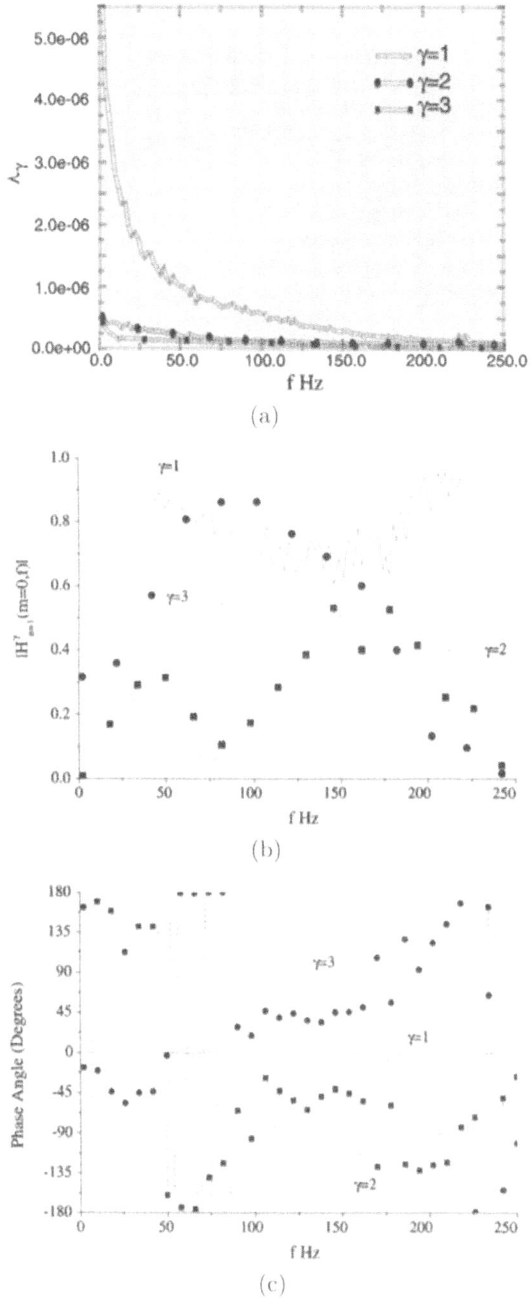

Figure 6.3. Accuracy of the estimated first POD coefficients for azimuthal mode 6 using the rings of hot wires at $r/D = 0.41$ and $r/D = 0.67$. (a) Eigenvalues for modes 1, 2, and 3 in the flow. (b) Magnitude of the transfer function for the first estimated POD coefficient $a_n^e(m = 6, f)$. Contribution from actual POD coefficients —— $\gamma = 1$ ▪ $\gamma = 2$ and • $\gamma = 3$. (c) Phase of the transfer function for the three modes symbols same as in (b)

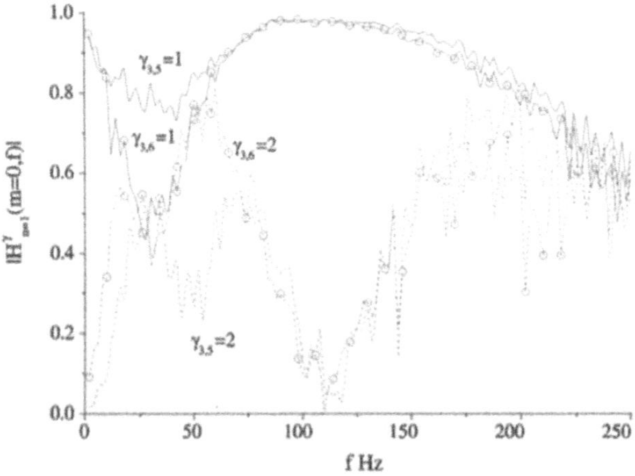

(a)

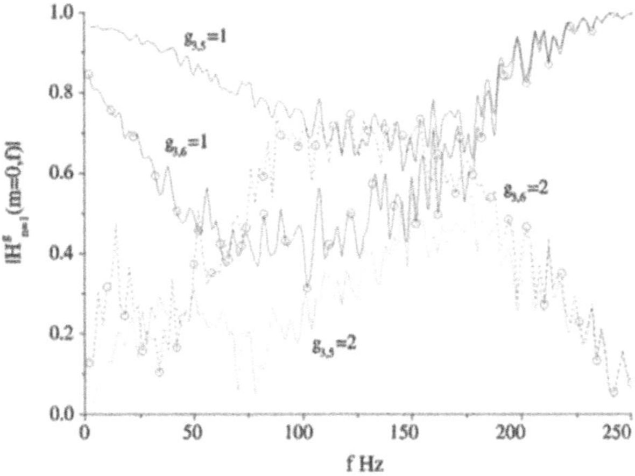

(b)

Figure 6.4. Comparison of the magnitude of the transfer functions for first POD coefficients for (a) azimuthal mode 0 and (b) azimuthal mode 6 computed for two cases. The first case (no symbol) is for rings of wires at $r_3/D = 0.41$ and $r_5/D = 0.67$ and the second cases $\circ$ is for rings of wires at $r_3/D = 0.41$ and $r_3/D = 0.80$. In all cases —— is the contribution from the first POD mode in the field; i.e., $|H_1^1(m = 0, f)|$, and - - is the contribution from the second POD mode in the field; i.e., $|H_2^1(m = 6, f)|$.

References

Bonnet, J. P., Cole, D.R., Deville, J., Glauser, M. N., and Ukeiley, L. S., (1994) Stochastic estimation and proper orthogonal decomposition: Complementary techniques for identifying structure, *Expts. in Fluids*, **17**, 307-314.

Citriniti, J., (1996) Experimental investigation into the large structures of the axisymmetric mixing layer using the Proper Orthogonal Decomposition, *Ph. D. Thesis*, SUNY Buffalo, Amherst, NY.

Ewing, D., and Citriniti, J. (1998) Examination of a LSE/POD complementary technique using Single and Multi-Time Information in the axisymmetric shear layer. *In Simulation and Identification of Organized Structure in Flows.*, J. Soresen, *et al.* Eds., Kluwer.

Ewing, D. and Woodward, S. (2000) Evaluating the accuracy of the LSE-POD technique in an axisymmetric shear layer, *in Advances in Turbulence VIII*, C. Dopazo *et al.* Eds., CIMNE, Barcelona.

Lumley, J. L. (1967) *Stochastic Tools*, Academic Press, New York.

Discussion

Mark Glauser, Syracuse (Chair) "(A comment) I am glad to see Ewing doing something that I should have done, which is an analysis that allows a priori to use the stochastic estimation to cleverly under sample these flows where you want to get a time dependent picture of things."

Chapter 7

MIXING IN A CROSS-JET ENHANCED BY A COAXIAL ANNULAR SYNTHETIC JET[1]

Lorenz Sigurdson
and John Diep
Vortex Fluid Dynamics Lab
University of Alberta
lorenz.sigurdson@ualberta.ca

Abstract

The objective was to discover and evaluate methods of smokestack downwash reduction using unsteady forcing. The flow was essentially a low momentum jet in a crossflow. After preliminary investigations of several concepts, the final wind-tunnel model applied an annular synthetic jet coaxially to a turbulent or laminar pipe flow at the stack exit. The cross-wind Reynolds number based on stack diameter was of the order of 1000. Primary diagnostics were hot-wire measurements and image processing of smoke flow visualization photographs of the plume gas. For a laminar pipe flow and increasing forcing amplitude, the initially turbulent plume can be made to completely relaminarize and then become turbulent again at higher amplitudes. Therefore an increased forcing does not always increase mixing. At higher amplitudes vortices appear at the forcing frequency. An abrupt transition of the vortices' arrangement occurs at a particular amplitude that coincides with an apparent jump in mixing.

Keywords:

Cross-jets, smoke-stacks, flow control, vortex dynamics, synthetic jets, downwash

[1]Part of these results were presented as a poster entitled "Cross-Jet Influenced by a Concentric Synthetic Jet", by Diep, J. and Sigurdson, L., at the Annual Meeting of the Fluid Dynamics

A. Pollard and S. Candel (eds.), IUTAM Symposium on Turbulent Mixing and Combustion, 85-93.

1. Introduction

This paper concerns an attempt at flow control of a jet in a cross-flow. An annular synthetic jet (zero net mass flux) is produced coaxially to the jet plume emerging from a cylinder representing a smoke or flare stack. Results indicate that this type of forcing can reduce the smoke-stack downwash of harmful plume gases to ground level which can occur for high wind velocities. This technique is expected to also be relevant to oil-field flaring where enhanced mixing of the plume with the ambient air could improve combustion; and to non-premixed cross-jets emerging from the walls of combustion chambers. Experiments in other laboratories have looked at the behaviour of two dimensional conventional jets in response to the application of a single synthetic jet on one side [Smith and Glezer, 1998]. To the authors' knowledge the current work is the first time both an annular synthetic jet has been created, as well as used in this manner to force a cross-jet.

2. Experimental Details

The experimental rig consists of a one inch diameter (D) cylinder in a wind tunnel. It is designed with an outer annulus through which a sinusoidal velocity fluctuation is introduced near the jet exit. This annulus extends the length of the cylinder and through the wind tunnel wall to a piston-driven plenum outside. This particular plenum/annulus design concept was proposed by [Coward, 2000] to minimize the length of the high viscous loss annulus. It has a resonance chamber connected to the annulus as is shown in Figure 7.1. The chamber is based on an 8" diameter speaker. The piston is connected to a B&K Vibration Exciter Type 4809, and is excited by an in-house built power oscillator. This shaker table-piston arrangement is used instead of a speaker because of the larger forces it is capable of applying. Sinusoidal movement of the piston creates a similarly varying pressure in the chamber which produces an oscillating flow within the annulus, and resultantly, jetting at the stack tip. The stack tip is constructed of two concentric thin-walled brass tubes. The wall thickness of both tubes is 0.74mm (0.029"), with 25.4mm (1") and 20.6mm (13/16") outer diameters for the inner and outer tube, respectively. This sizing results in a plume area of 57% of the total stack cross section, with a 1.64 mm gap (d_a) between the tubes.

Division of the American Physical Society, Washington, D.C., Nov. 2000. It was selected to be published as a one-page article in the Gallery of Fluid Motion, Physics of Fluids, Sept. 2001.

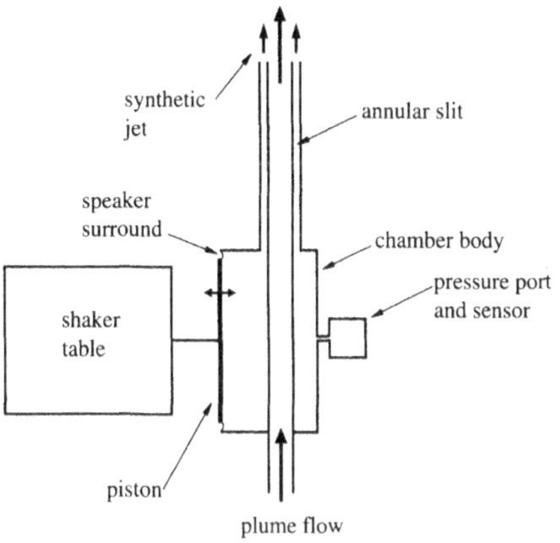

Figure 7.1. Schematic of the Forcing Apparatus (the wind tunnel is not shown)

The apparatus is designed to operate with an axisymmetric laminar or fully developed turbulent plume flow which is seeded with an aerosol for flow visualization. This aerosol is provided by a Radio Shack chemical fog machine. The plume is driven by an small DC centrifugal fan powered by a variable DC power supply. The fog machine fills a plenum with fog that is then allowed to mix with ambient air and cool, as the fog is slightly warm upon leaving the fog machine. This aerosol is then drawn out of the plenum past a valve before entering the fan inlet. To ensure that the plume is fully developed there are approximately 50 pipe diameters of length between this contraction and the exit plane of the plume. Plume flow rate was calculated by using a calibration of the plume centerline velocity, W_{CL}, against the measured pressure drop across the contraction and approximately 30 diameters of straight pipe, P_c.

A suction-type open-circuit wind tunnel was used to create the cross-flow. The stack extends halfway across the test section which has a cross section of 30.5cm by 30.5cm (12" by 12").

The acquisition and processing of all pressure and hot-wire sensor data is done using an 800 MHz Pentium equipped with a 12-bit AT-MIO-16-E National Instruments data acquisition board and running LabVIEW® 4.1.1. Synthetic jet velocity W_{SJ} is characterized by a centerline mean hot-wire reading taken a vertical distance z=3 mm from the orifice exit plane (z/d_a=2). This location is used as a compromise between being too close to the jet orifice where flow reversal takes place, and too far downstream of the jet to measure the jet speed that is affecting the emerging plume close to the exit plane.

Photographs for both this preliminary experiment as well as the final experiment were taken from side-view with the flow going from left to right. The camera used was a Nikon F3 35mm camera using Kodak 3600ASA TMAX black and white film. A General Radio Type 1540 Strobolume was used as a flash.

3. Laminar Pipe Flow Results

This section describes observations made of the special case of a plume from a laminar pipe flow forced with a varying amplitude 60 Hz synthetic jet. During the course of these observations no effort was made to keep the plume flow constant while forcing was applied, but P_c only deviated a maximum of 15% from the average of readings taken over the course of observations. The cross-stream velocity (U_∞) was approximately 0.5 m/s and the unforced centerline velocity of the pipe flow was measured to be 0.9 m/s using the hot-wire. Assuming a parabolic laminar velocity profile within the pipe, the average plume velocity (W_p) can be approximated to be 0.45 m/s, giving a plume flow Reynolds number of 570. The velocity ratio of the plume jet velocity to the cross-stream velocity is $R = 0.9$. The nondimensional forcing frequency based on the stack external diameter and W_p is 3.4. This is 3.4 times the estimated frequency of the naturally occurring structures apparent in the unforced case shown in Figure 3. The vortex structures within the plume are clearly visible. Vorticity in these structures originate from the boundary layer of the plume flow while inside the pipe. This vorticity is shed into the plume jet as vortex rings, which are then tilted and distorted by the velocity gradient between the cross-flow and the slower fluid in the wake of the plume and stack [Moussa et al., 1977]. Some downwashing is evident in this Figure. This would be expected as that is known to be the case for a plume from a turbulent pipe flow with a similar velocity ratio.

The ratio of W_{SJ} to W_p was varied from 0 to 1.3. As low-amplitude synthetic jetting is applied ($W_{SJ} \approx 0.1\ m/s$), the three-dimensional vortex structure on the upstream side of the plume disappears, resulting in the image shown in Figure 3b. Large vortices remaining in this forced plume are reminiscent of the inner vortices appearing on the lower surface of the unforced plume. These large vortices in the forced plume become gradually smaller in size with increasing forcing amplitude. Eventually the plume becomes completely relaminarized when $W_{SJ} \approx 0.3\ m/s$, as shown in Figure 3c. This represents the minimum observed mixing of the plume with the cross-flow. This value of W_{SJ} is nearing the cross-stream velocity.

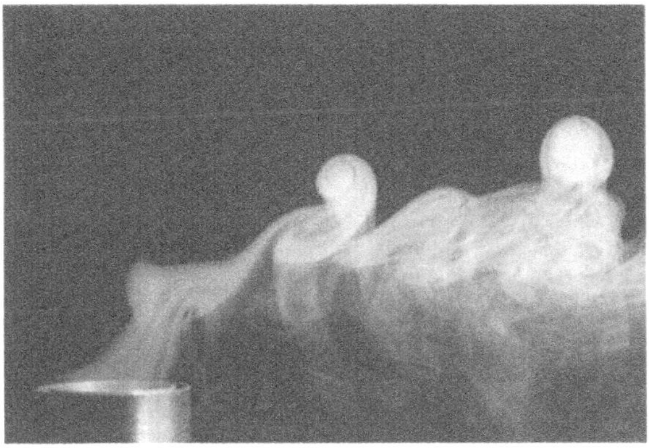

(a) Reference Photograph of Laminar Plume: No forcing, $U_\infty \approx 0.5\ m/s, W_p \approx 0.45\ m/s$.

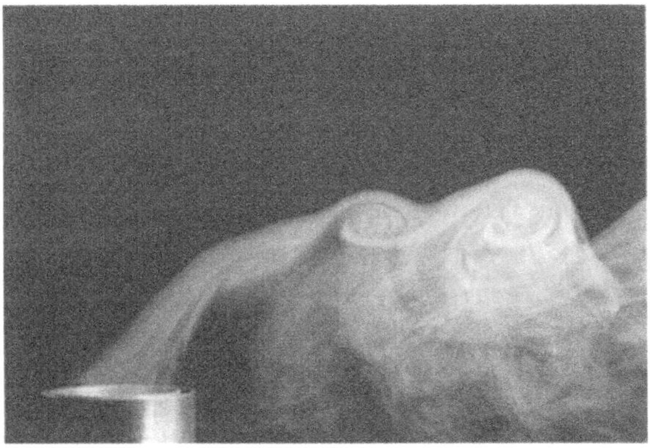

(b) The outer structure disappears, and the inner structure appears to remain, $U_\infty \approx 0.5\ m/s, W_p \approx 0.45\ m/s, W_{SJ} \approx 0.1\ m/s$.

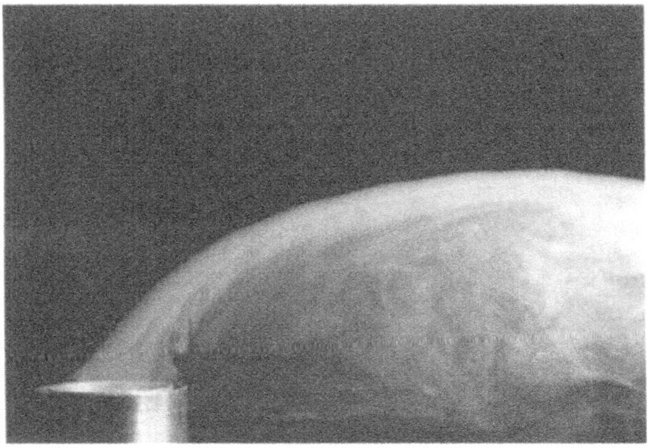

(c) Plume relaminarizes, $U_\infty \approx 0.5\ m/s, W_p \approx 0.45\ m/s, W_{SJ} \approx 0.3\ m/s$.

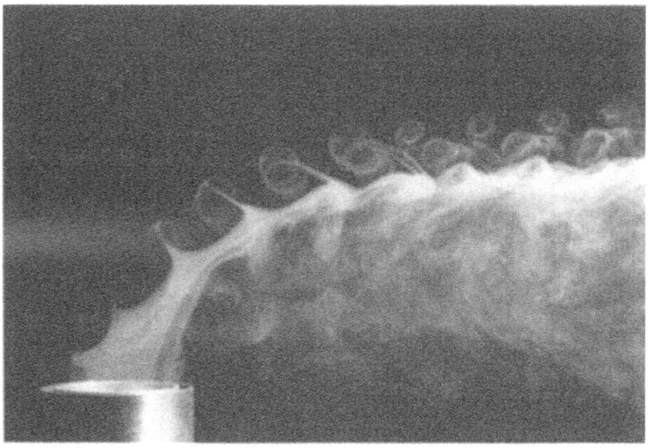

(d) Upstream-pointing mushroom vorticies emerge from upper surface, $U_\infty \approx 0.5\ m/s, W_p \approx 0.45\ m/s, W_{SJ} \approx 0.9\ m/s$.

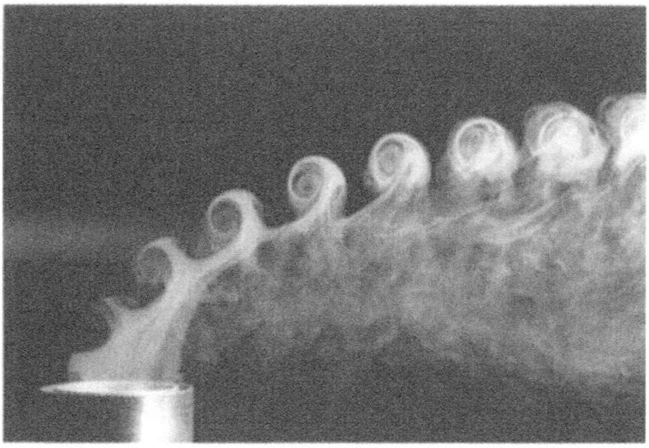

(e) After the abrupt transition, vorticies on the upper surface contain mainly one sign of vorticity and point downstream, $U_\infty \approx 0.5\ m/s, W_p \approx 0.45\ m/s, W_{SJ} \approx 0.9\ m/s$.

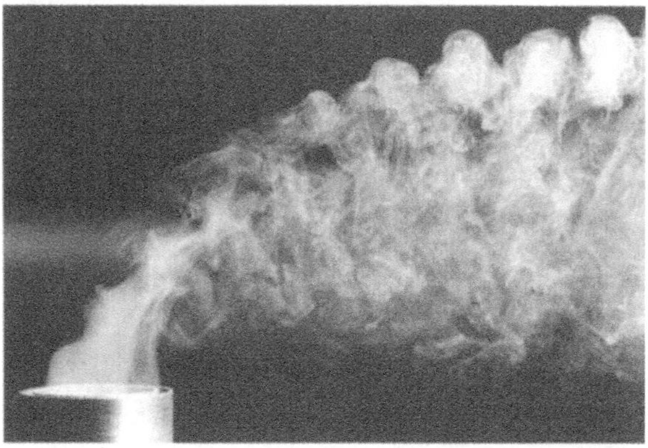

(f) Lifting off and more turbulent, $U_\infty \approx 0.5\ m/s, W_p \approx 0.45\ m/s, W_{SJ} \approx 1.2\ m/s$.

Figure 3: Change in Structure of Laminar Plume With 60 Hz Synthetic Jet Forcing

With increased forcing amplitude, an instability begins to grow from the relaminarized plume. This instability is shown in Figure 3d when $W_{SJ} \approx 0.9$ m/s. This value is close to the centerline value of the pipe flow, measured at 0.9 m/s using the hot-wire. From this side-view there are structures resembling mushroom vortices pointing upstream on the upper surface of the plume. "Pointing" refers to the direction the pairs would self-propagate. The smoke intensity in these structures is rather small, indicating not a lot of plume fluid enters these. Mixing has presumably increased from the laminar case however. With a very small increase in synthetic jetting amplitude, there is an abrupt transition in the vortex structures' arrangement to that seen on the upper surface of the plume in Figure 3e. Between these photographs forcing amplitude was increased by 6% (calculated from chamber pressure measurements) to make sure the transition did not occur while photographs were being taken. This transition point is so sensitive that the apparatus can be set for transition to occur intermittently with no controls being touched. After this transition, the vortices appear to point downstream, and appear to more strongly consist of vorticity out of the page by the right-hand-rule. It also appears that after the transition these vorticies contain more plume fluid than did the upstream-pointing vorticies before transition. This is indicated by the brighter smoke contained in the vorticies. A leap in mixing is implied, of the plume with the cross-stream fluid. As forcing amplitude is further increased, the vortices on the upper surface of the plume continue to lift off and form pinched-off lobes, as seen in Figure 3f where $W_{SJ} \approx 1.2$ m/s. The plume is clearly higher than in the unforced case, as well as showing an increased plume width, which suggests a significantly improved mixing between the plume and surrounding air. With even greater forcing amplitude the plume continues to rise, but becomes too turbulent to discern any structure within it. In any case, downwash is significantly reduced if not eliminated.

4. Conclusions

For the single velocity ratio tested here for a plume resulting from a laminar pipe flow, increasing the synthetic jet velocity causes:

a. relaminarization when it is near the cross-stream velocity,

b. an increasingly complex periodic 3-D vortex structure,

c. an abrupt transition in the vortex structure when the forcing velocity is near the peak in the cross-jet velocity profile,

d. an increased mixing at the highest amplitudes.

The conclusions regarding under what conditions the relaminarization and abrupt transition occur can only be confirmed by testing other velocity ratios of jet to cross-stream.

One important conclusion from the observed relaminarization is that increased forcing does not always increase mixing. There would appear to be an island of calm in the turbulence otherwise observed as the forcing amplitude is increased. It should be noted that the pipe velocity would have to be reduced by an order of magnitude or more to get the jet to be laminar on its own accord.

The abrupt transition noted in c) coincides with a significantly increased mixing of the plume fluid with the ambient fluid. This may not be coincidental but rather be an indication of the reason why the transition occurs. As the forcing becomes more powerful, the vortices created by it are expected to become stronger and entrain plume fluid more quickly. This may be why the sign of the plume fluid now appears to dominate the structure resulting on the upstream side of the cross-jet.

These vortex structures are seen in the photographs as primarily 2-D but are actually quite strongly 3-D. While the two transformations discovered in the plume vortex structure are clear and distinct, no corresponding behaviour in the turbulent plume has been observed, either through visual interpretation or measurements of the photographs. However, the synthetic jet inhibits downwash of the plume into the wake of the stack even at Reynolds numbers much higher than those reported here, when the stack contains turbulent pipe flow.

Acknowledgments

Thanks to Dave Pape for building the rig, Terry Nord for his help with the electronics, Albert Yuen, Don Fuhr, Dallas Chapple and Bernie Faulkner for their advice and assistance, Andrew Coward and Chris Apps for their ideas and help whenever it was needed. Financial support is appreciated from the Natural Sciences and Engineering Research Council of Canada, Luscar Ltd., and the University of Alberta.

References

Smith, B. and Glezer, A. (1998). The formation and evolution of synthetic jets. *Physics of Fluids*, 10(9):2281–2297.

Coward, A. (2000). Personal communications.

Moussa, Z., Trischka, J., and Eskinazi, S. (1977). The near field in the mixing of a round jet with a cross-stream. *Journal of Fluid Mechanics*, 80(1):49–80.

Discussion

Mark Glauser, Syracuse (Chair) "(A comment) The second talk (Sigurdson and Diep) was quite timely, where we have in New York State an Environmentally Quality Systems Centre to look at the interface between the indoor and urban environment and flow control that is having an impact on some of these flows. "

Sébastien Candel, EMC2, Paris "Yes, I have a question for you, Sigurdson, how do you imagine to place this on a stack, is it possible to have large scale synthetic jets and is it possible to have scale up from small scale to large scale?"

Reply: This is work to be continued; in the experiment apparatus, we used a large long annular channel to reach the inside of the wind tunnel so we could reach into the wind tunnel, in an actual smokestack you would build a ring shaped synthetic jet producer that you would place on the top of the smokestack but the details of making that scale of a synthetic jet that is a another student project!

Chapter 8

CHALLENGES IN TURBULENT MIXING WITH COMBUSTION

Paul E. Dimotakis

Graduate Aeronautical Laboratories, California Institute of Technology, Pasadena, CA 91125

Abstract Turbulent combustion combines the complexities of turbulence and mixing, challenges not met in the twentieth century, with the complexity and subtlety of chemical kinetics. This discussion focuses on progress and some turbulent-mixing issues in chemically reacting flows stemming from experimental, modeling, and direct-numerical simulation (DNS) studies. The mixing transition will be discussed. DNS studies of the Rayleigh-Taylor instability in miscible fluids reveal an early-time diffusive growth and a strong sensitivity to initial conditions. Recent experiments address the assumption of isotropy in turbulence and mixing. Experiments in high-speed shear layers elucidate some effects of compressibility on the mixed-fluid field. Issues involving molecular-transport coefficients will also be discussed.

1. Introduction

Turbulence is a three-dimensional, non-linear, unsteady phenomenon of very high dimensionality (Fig.8.1). Accepting Kolmogorov-41 scaling, the ratio of outer to viscous scales is proportional to $Re^{3/4}$, i.e., a minimum of $Re^{9/4}$ grid points for a resolved three-dimensional simulation, and a Re^3 space-time dimensionality and computational effort considering temporal-resolution needs. For the Reynolds numbers of interest, where fully developed turbulence occurs, this renders turbulence beyond direct numerical simulation (DNS) reach, at least for the foreseeable future. Computations must rely on explicit or implicit subgrid-scale (SGS) modeling to represent dynamics and transport below explicitly resolved scales at these Reynolds numbers. This presents a

A. Pollard and S. Candel (eds.), IUTAM Symposium on Turbulent Mixing and Combustion, 95-112.

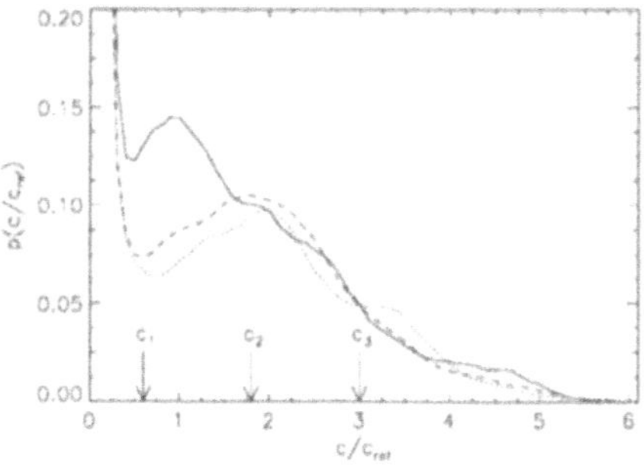

Figure 8.1. Rocket testing in the Los Angeles hills: $Re_{jet} > 10^8$. Photo from *LosAngeles* (1968, Sunset Lane, Menlo Park, CA

particular challenge for mixing, which, for the most part, occurs at the smallest, and therefore unresolvable, scales at high Reynolds numbers.

As others in this Symposium discuss, chemically reacting flows and combustion present us with additional and significant challenges. Even combustion of methane (CH_4), perhaps the simplest hydrocarbon, can require over 200 chemical reactions, and approximately 50 chemical species for a more-or-less complete description of the chemical kinetics. Matters get rapidly worse with increasing carbon number of the parent fuel, or fuel blend. Kinetics models for octane (C_8H_{18}) may entail over 2000 chemical reactions. With each species requiring its own transport equation, the added complexity of multicomponent transport and the coupled chemical kinetics can dwarf the already daunting task of simulating the dynamics of a simple (one-component) fluid. Finally, the large number of kinetic and transport coefficients and their uncertainties renders an assessment of chemical-kinetic models very difficult.

Large density differences in combustion, often in high acceleration fields, as occur in rotating machinery, for example, compund the subgrid-scale modeling difficulty. Most of the models to date rely on assumptions of density uniformity and (near-) isotropy, which are not valid in most combustion environments.

Turbulent mixing with combustion presents us with a truly challenging analysis, simulation, and experimental task. The discussion following addresses some of the issues raised above.

2. The mixing transition in turbulent flows

How high must the Reynolds number be, at a minimum, for turbulence and turbulent mixing to be regarded as fully developed?

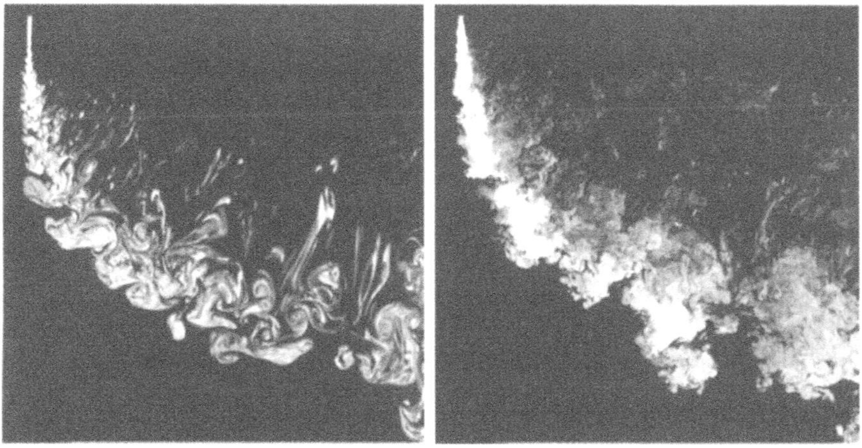

Figure 8.2. Transverse jets in cross-flow: $U_\mathrm{j}/U_\infty \simeq 34$. Jet-fluid concentration in the plane of symmetry. Intensity compensated for downstream concentration decay ($\propto x^{1/2}$). Left: $Re \simeq 1.0 \times 10^3$. Right: $Re \simeq 1.0 \times 10^4$.

Turbulent flows exhibit a transition that can be conspicuous, at outer-scale Reynolds numbers,

$$Re_\delta \equiv \frac{\rho\, U\, \delta}{\mu} \gtrsim 1 - 2 \times 10^4 \,, \qquad \text{or,} \qquad Re_\mathrm{T} \equiv \frac{\rho\, u'\, \lambda_\mathrm{T}}{\mu} \gtrsim 100 \,. \tag{8.1}$$

where δ is an outer-scale length, U is a characteristic velocity (difference) that drives the flow, u' the rms velocity, λ_T the Taylor length scale, and Re_T the Taylor Reynolds number. The latter provides a criterion where an outer-scale Reynolds number is not appropriate. Where both can be defined, the threshold values are consistent with the general approximate relation, $Re_\mathrm{T} \approx Re^{1/2}$. These are *necessary* but *not sufficient* requirements for this transition; *laminar* flows can exist at higher Re's yet (Dimotakis 2000).

This transition was first noted in gas-phase shear layers (Konrad 1976, Bernal *et al.* 1979). For zero-pressure-gradient shear layers, the velocity difference, $U = \Delta U = U_1 - U_2 \simeq$constant and $Re_\delta(x) \propto x$, owing to a linearly increasing thickness of the turbulent region, i.e., $\delta(x) \propto x$. For this flow, a relatively abrupt transition to a more disordered, enhanced-mixing state is observed at a station, x, where the value $Re(x) \simeq 10^4$ is attained (Dimotakis 2000).

The transition is observed in liquid-phase shear layers, as well as gas- and liquid-phase turbulent jets, so it not a Schmidt number effect. Jets are characterized by $Re(x) \simeq Re(0) \neq \mathrm{fn}(x)$, and, as opposed to shear layers, are either globally beyond, or below the mixing transition. A

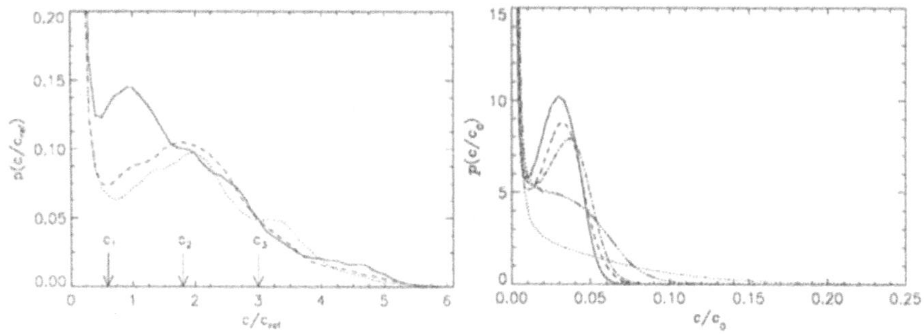

Figure 8.3. Jet-fluid concentration pdfs. Left: Jets in a quiescent reservoir with lines of increasing solidity denoting increasing $Re = 4.5 \times 10^3$, 9×10^3, and 18×10^3 (Catrakis & Dimotakis 1996). Right: Transverse jets with increasing pdf at $c/c_0 \simeq 0.03$ associated with increasing $Re = 10^3$, 2×10^3, 5×10^3, 1×10^4, and 2×10^4 (Shan & Dimotakis 2001).

similar transition was recently documented for liquid-phase transverse jets, in a uniform crossflow (Shan 2001, Shan and Dimotakis 2001). See 8.2. This flow will be discussed further below.

The transition was termed a "mixing transition" because it was initially identified in terms of its consequences on mixing. However, it is a transition of the whole flow dynamics, whose imprint on mixing was most conspicuous. Figure 8.3 depicts the pdf's of far-field ($x/d_{\rm j} = 275$) jet-fluid concentration of jets discharging in a quiescent reservoir [left: $c_{\rm ref} \simeq c_0/200$; c_0 is pure jet-fluid concentration (Catrakis and Dimotakis 1996)] and the far-field ($x/d_{\rm j} = 50$) of transverse jets (right: Shan 2001, Shan and Dimotakis 2001). *Re*'s are based on jet-nozzle parameters, in both cases. Both flows are liquid phase ($Sc \approx 2 \times 10^3$) with all pdf's normalized to differential area fraction, *e.g.*, (Dimotakis and Catrakis 1999),

$$p(c)\,\mathrm{dc} \;=\; \frac{\mathrm{dA(c)}}{\mathrm{A_{tot}}}\;, \tag{8.2}$$

where $A(c)$ is the *area* portion where the concentration is less than c (see also Appendix in Shan & Dimotakis 2001). The differences are interesting and attest to the fact that mixing behavior is, unfortunately, not universal.

Evidence for this transition can be found in a large number of flows, both free and wall-bounded (Dimotakis 2000). This argues for reasons not associated with the large-scale structure, which varies considerably across the various flows. Its apparent universality also suggests a quasi-universal cause. It was proposed that the reason lies in the requirement for an inviscid range of eddies in the flow. This translates into a separation between the Liepmann-Taylor scale, $\lambda_{\rm L}$, and the (inner) viscous scale, λ_ν, estimated as,

$$\lambda_{\rm L} \simeq 5\,\delta\,Re^{-1/2} \qquad \text{and} \qquad \lambda_\nu \simeq 50\,\lambda_{\rm K} \simeq 50\,\delta\,Re^{-3/4}\,, \tag{8.3}$$

where $\lambda_{\rm K} \equiv (\nu^3/\varepsilon)^{1/4}$ is the Kolmogorov scale. The former scale ($\lambda_{\rm L}$) represents the thickness of an internal layer formed as the result of a plunging motion across δ. The prefactor factor of 5 for $\lambda_{\rm L}$ in 8.3 is for a Blasius laminar boundary layer but is also appropriate for free shear layers at near-unity velocity ratios. The latter scale (λ_ν) represents the inner scale where viscous effects begin to act, as diagnosed by where power-law spectra deviate from a near-5/3 exponent (this occurs at $k\,\lambda_{\rm K} \simeq 1/8$, so $\lambda_\nu \simeq 2\pi \times 8 \simeq 50$). As the scale range between δ and $\lambda_{\rm L}$ can be spanned by viscous action and λ_ν is the inner viscous scale, inertial λ-size eddies require, at a minimum,

$$\delta > \lambda_{\rm L} > \lambda > \lambda_\nu > \lambda_{\rm K} \quad \Rightarrow \frac{\lambda_{\rm L}}{\lambda_\nu} > \frac{1}{10}\,Re^{1/4} \gtrsim 1 \quad \Rightarrow Re \gtrsim 10^4\,, \tag{8.4}$$

in accord with observation.

Experimental evidence indicates that Re effects on the dynamics and mixing become substantially weaker as Re increases and, in particular, for $Re > Re_{\rm tr} \approx 10^4$, i.e., beyond the mixing transition. The dependence of flow structure and mixing on Re can be quite strong for $Re < Re_{\rm tr}$, as documented in liquid-phase shear layers (Koochesfahani & Dimotakis 1986, Figs. 7, 9, and 13),and gas-phase jets (*e.g.*, Dimotakis 2000, Fig. 11, and the image sequence in Dimotakis 2001, Fig. 11). and either difficult to discern or logarithmic for $Re > Re_{\rm tr}$ (*e.g.*, Dimotakis 1991, Fig. 24). Consequently, experiment/DNS results at $Re \lesssim Re_{\rm tr} \approx 10^4$ should be viewed with caution, with extrapolations to the higher Re's typically of interest for fully developed turbulence of dubious validity.

Particularly heartening are recent results on mixing from large Eddy Simulation (LES/SGS) modeling work by Pullin (2000). The modeling relies on extensions to the stretched-vortex model (Pullin & Saffman 1994, and Misra & Pullin 1997) that approximates the subgrid dynamics by assuming that unresolved flow in each simulation cell can be represented by a Lundgren (1982) spiral vortex. Mixing is modeled by assuming that such vortices wrap the scalar field to produce the sub-grid scalar fluxes and fluctuations. Figure 8.4 plots the normalized variance computed by Pullin (2000) for a scalar with an imposed mean gradient and compared to the DNS results by Overholt & Pope (1996), for $Sc \equiv \nu/\mathcal{D} = 0.7$, as a function of Taylor Reynolds number. The LES results can be seen to capture the mixing transition, at $Re_{\rm T} \approx 100$,

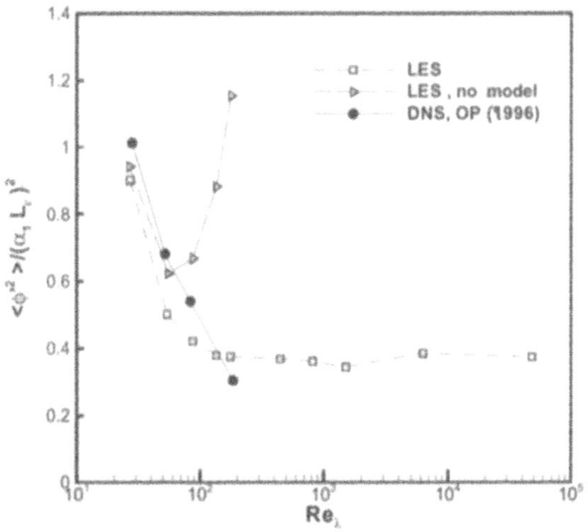

Figure 8.4. LES/SGS results (Pullin 2000, Fig. 3) of the normalized variance of a scalar with an imposed mean gradient, transported by isotropic turbulence. Comparison with DNS results by Overholt & Pope (1996).

even though there is nothing in the SGS modeling to induce this attribute. This model was also used recently to analyze spectra of scalars for non-unity Schmidt numbers (Pullin & Lundgren 2001), successfully capturing the Batchelor (1959), k^{-1} spectrum.

3. Transverse jets: Small-scale scalar-field behavior

Turbulence theories and models often rely on the assumption that small-scale dynamics are (very nearly) isotropic. This permits simplifications in the representation of turbulence statistics, minimizing the parameters and functions that need to be modeled (*e.g.*, Batchelor 1953, Tennekes & Lumley 1972, Hinze 1975, Monin & Yaglom 1975, Pope 2000). More often than not, however, the large-scale flow organization is far from isotropic. Scalar transport, as required for the mixing of entrained species and as occurs in non-premixed combustion, for example, requires anisotropy. A statistically isotropic field can not support a mean flux of any kind. The expectation, or modeling assumption, is often that fluxes will be supported by large-scale anisotropy, while small scales may be approximated as isotropic. These questions were addressed in a recent experimental investigation on transverse jets (Shan & Dimotakis 2001), whence Figs. 8.2 and 8.3, above, derive. Transverse jets are important in a variety of turbulent mixing and combustion contexts. They are relied upon to disperse pollutants from stacks in the atmosphere,

Figure 8.5. Two-dimensional (spatial) scalar power spectra for a transverse jet at $x/d_{\rm j} = 50$. Wavenumber, $\mathbf{k}$, normalized by the local horizontal spatial extent of the turbulent jet, $\delta_{\rm h}$. Contour plot in $\log_{10}$ increments of unity. Right: $Re_{\rm jet} = 1000$. Left: $Re_{\rm jet} = 10{,}000$. Shan & Dimotakis (2001).

sometimes used for effluent discharge in the ocean, and are a candidate fuel-injection configuration for high-speed air-breathing propulsion devices, such as SCRAMJETs, for example.

In the transverse-jet experiments, cuts of the far-field turbulence in planes perpendicular to the streamwise direction permitted spatial scalar spectra and other scalar-field statistics to be compiled, at a fixed $x/d_{\rm j}$ station. Figure 8.5 plots two-dimensional spatial scalar power spectra, computed from spatial cuts of the three-dimensional transverse-jet flow, in a plane perpendicular to the streamwise direction, at $x/d_{\rm j} = 50$, for jet Reynolds numbers, $Re_{\rm jet} = 1000$ and 10,000, for transverse jets with a jet-to-freestream speed ratio of, $U_{\rm jet}/U_\infty \simeq 10$. At least in this flow, contours of the spatial spectra are closer to circular, at low wavenumbers, becoming increasingly elliptical at high wavenumbers. This flow illustrates that the opposite to the usual expectation can occur, i.e., increasing anisotropy with increasing wavenumber.

The reason, in this case, is traceable to the strain-rate field imposed on the small scales by the large-scale, streamwise vortices that dominate the far-field flow of transverse jets. This is illustrated in 8.6, which depicts a space-time image of a scalar field isosurface, i.e., $c(y, z, t;\, x{=}50d_{\rm j})$, compiled from the sequence of transverse cuts, with a framing rate sufficiently high to meet the temporal Nyquist criterion for reconstructing the indicated three-dimensional slice of the four-dimensional, $c(x, y, z, t)$, data set. A scalar isosurface is highlighted by setting a small range of jet-fluid concentrations to be opaque and reflecting (Shan & Dimotakis 2001). At this Reynolds number, $Re_{\rm jet} = 1000$, a small vortex is evident, which is responsible for the asymmetry in the flow. The instability of this flow towards such asymmetrical vortical formation was documented previously (Kuzo 1995).This asymmetry is reflected in the slight (clockwise) tilt of the elliptical contours of the corresponding spectrum (8.5,

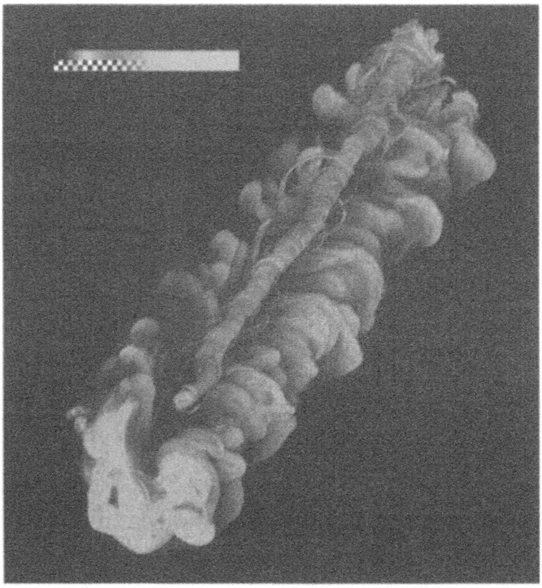

Figure 8.6. Three-dimensional, space-time visualization of a jet-fluid concentration isosurface for a $Re_{\mathrm{jet}} \simeq 1.0 \times 10^3$, $U_{\mathrm{jet}}/U_\infty = 10.1$ transverse jet. Time runs along the axis of the vortices (Shan & Dimotakis 2001; visualization computed in collaboration with S. Lombeyda, Caltech CACR).

left). At higher Reynolds numbers, only two counter-rotating ("kidney") vortices appear, generating a symmetric mean flow about the stream-wise plane that contains the jet nozzle, and correspondingly (untilted) elliptical contours of the spatial spectrum (8.5, right).

4. Rayleigh-Taylor flow

Rayleigh-Taylor instability (RTI) occurs whenever fluids of different density are accelerated in a direction opposite that of the density gradient. The resulting flow eventually becomes turbulent, as secondary instabilities develop and grow. If the fluids are miscible, species diffusion and mixing, that reduce density differences and hence local forcing, can play a dynamic role.

Recent direct numerical simulations (DNS) of RTI between miscible flows allowed investigations on the growth and mixing in Rayleigh-Taylor instability (RTI) flow (Cook & Dimotakis 2001).The Navier-Stokes equations, with matched fluid viscosities for the two fluids, were augmented by a species-conservation equation for binary Fickian diffusion and solved on a $256^2 \times 1024$ grid. Initial error-function density (mole-fraction) profiles were perturbed by displacing the intermediate

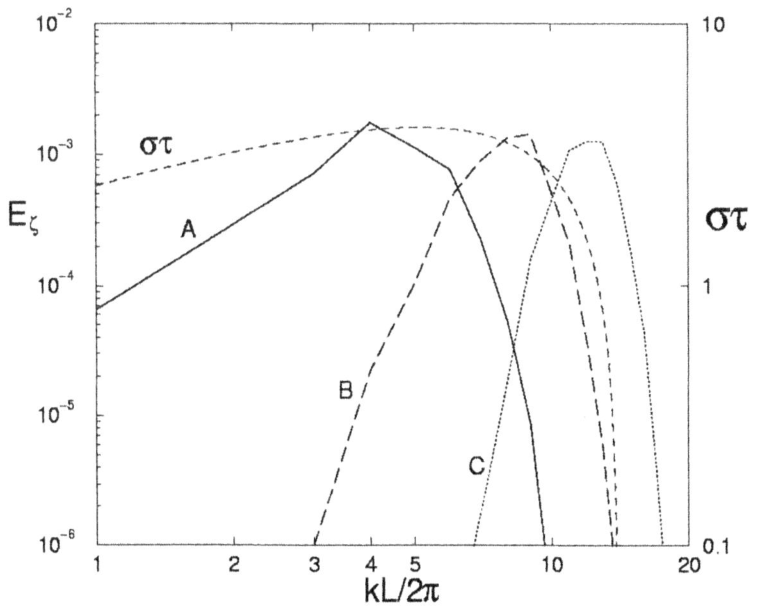

Figure 8.7. Initial perturbation spectra (left scale) for Cases A, B, and C. The linear stability theory (Duff *et al.* 1962) growth coefficient is also plotted (right scale).

isosurface, $X = 1/2$, nominally at $z = 0$, by an amount $\zeta(x, y)$, in units of the initial profile scaling length. Three simulations (Cases A, B, and C) were performed, with matched initial perturbation amplitude, each with a different initial $\zeta(x, y)$ perturbation spectrum, but otherwise identical in every other respect. The initial $\zeta(x, y)$ spectra are plotted in 8.7, along with the non-dimensional linear-stability exponential-growth coefficient, $\sigma\tau$, as calculated with the linear-stability theory of Duff *et al.* (1962)[1], where $\tau = \sqrt{L/Ag}$, with $A = (\rho_2 - \rho_1)/(\rho_2 + \rho_1)$ the Atwood number, and L the transverse extent of the RTI cell.

Models of Rayleigh-Taylor instability mixing zone have it growing quadratically in time, following an initial, linear-instability evolution. For constant acceleration, g, this yields a vertical extent of the RTI mixing zone given by,

$$h(t) \simeq \alpha \mathcal{A} g t^2 . \tag{8.5}$$

α is an empirical constant, with values in the range $0.01 \lesssim \alpha \lesssim 0.07$ (Dimonte & Schneider 2000). Such growth models implicitly assume that the density difference across the advancing fronts remains constant and that the flow structure remains self-similar.

The evolution for Case C is visualized in 8.8. Pure heavy fluid (X=1) is red (top), pure light fluid (X=0) is blue (bottom), and the intermediate (X=0.5)-isosurface (1:1 molar ratio) is light grey. The initial ($t/\tau = 0$) isotropic perturbations are barely discernible on the interface. By $t/\tau = 3.44$, mushroom-like structures have formed on the upper side of the isosurface ($\tau \equiv \sqrt{L/\mathcal{A}g}$; L is the transverse extent of the RTI cell). By $t/\tau = 4.63$, mushrooms have merged into larger structures and a long, thin protuberance ("spike") can be seen penetrating deep into the light fluid.

Figure 8.9 depicts the extent of the RTI mixing zone (8.5), along with indicatrices of $t^{1/2}$ and t^2 growth. An initial, diffusion-dominated regime is the same for all cases, i.e., independent of the initial perturbations, and in good agreement with analytical predictions for such growth. A faster-growth stage occurs at a different time for each of the three cases, reflecting differences in initial modal seeding, with breakout times from the diffusive-growth regime in accord with linear-stability theory (*cf.* 8.7). Late-time Case C growth is well represented as quadratic. Growth for Cases A and B is different, however, indicating that whether a quadratic-growth regime is exhibited by this instability depends on initial modal seeding.

[1]A scaling error for σ led to an incorrect plot of this curve in Cook & Dimotakis (2001, Fig. 2), which has been corrected here. See Cook *et al.* (2001)for a more-complete discussion.

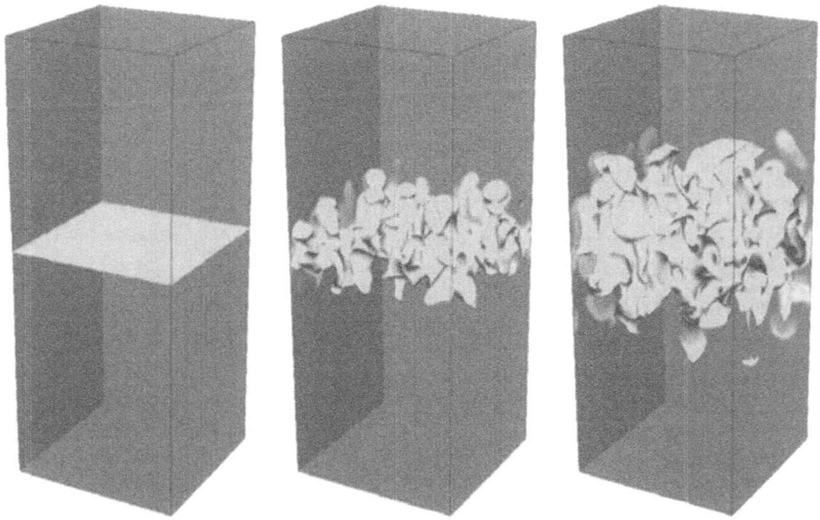

Figure 8.8. Time-evolution of intermediate, (X=0.5)-isosurface (light grey) for Case C. Pure heavy fluid (X=1) is red (top), pure light fluid (X=0) is blue (bottom). Times for the three images displayed: t/τ = 0.0, 3.44, and 4.63 (Cook & Dimotakis 2001, Fig. 4).

Even greater differences between the three cases are found in mixing, as measured by the amount of product that would be formed from a stoichiometric chemical reaction between the two fluids being mixed. These differences persist to the end of these simulations, to $Re = \bar{\rho}\, h\, \dot{h}/\mu \approx$ 3700. While final Re's attained in the simulation are below the anticipated mixing transition discussed above, the flow evolved over a significant multiple of the initial transverse extent, i.e., $h_{\text{final}}/h_{\text{initial}} \approx 10^3$.

5. Mixing in high-speed shear layers

Turbulence and mixing is sustained and promoted by shear stresses and instabilities throughout the turbulent region, responsible, in turn, for the large rates with which species-concentration isosurface area is generated. Shear layers are important canonical flows that capture these dynamics and mixing in shear layers has attracted considerable attention.

High local strain rates predicted by classical turbulence theory suggest that, in high-speed flows, the prerequisites for the applicability of conventional flamelet models (*e.g.*, Peters 1984) are not likely to be met. This was confirmed by detailed kinetic analyses of reactions in shear lay-

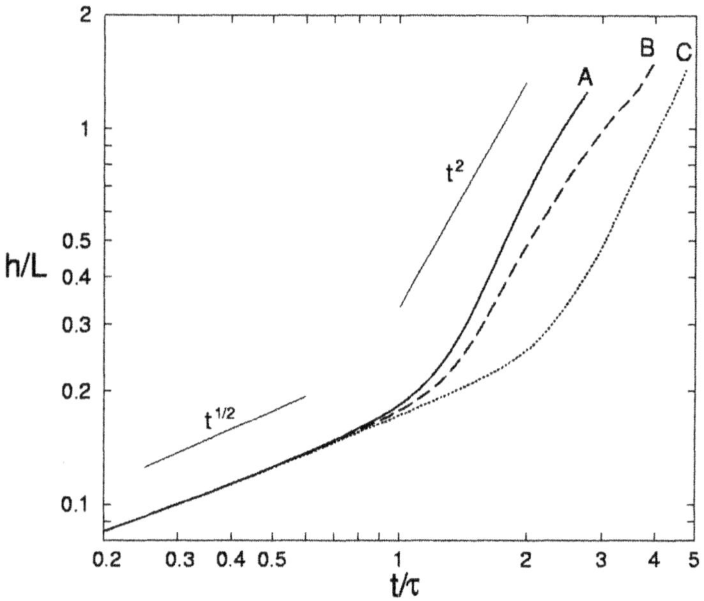

Figure 8.9. Mixing-zone growth for the three cases (Cook & Dimotakis 2001, Fig. 8).

ers, with hydrogen and nitric oxide, in one freestream, and fluorine in another, as discussed in Egolfopoulos *et al.* (1996).

Recent experiments using Rayleigh-scattering techniques corroborate some of the earlier shear-layer mixing models and have shed light on compressibility effects. Figure 8.10 reproduces two Rayleigh-scattering images from Dimotakis *et al.* (2001, Figs. 4b and 6b). The left is from a low Mach number flow with a total convective Mach number, $M_c = (U_1 - U_2)/(a_1 + a_2) \simeq 0.15$, N_2 the high-speed freestream gas, and C_2H_4 the low-speed freestream gas. These are density-matched but have a different Rayleigh-scattering cross section permitting molecular (Rayleigh-scattering) images of mid-span slices to be recorded. The image on the right is from a supersonic high-speed freestream shear layer,

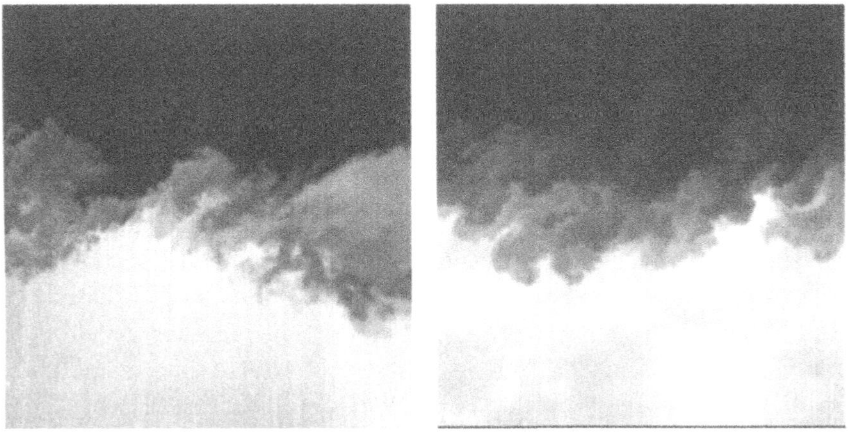

Figure 8.10. Laser-Rayleigh-scattering images of high-speed shear layer flow. Left: Low-compressibility flow ($M_c \simeq 0.15$). $U_1 \simeq 200\,\mathrm{m/s}$ [N_2], $U_2 \simeq 100\,\mathrm{m/s}$ [C_2H_4] Right: High-compressibility flow ($M_c \simeq 0.96$). $U_1 \simeq 1150\,\mathrm{m/s}$ [He], $U_2 \simeq 100\,\mathrm{m/s}$ [C_2H_4].

with $M_c \simeq 0.96$, helium as the high-speed freestream gas, and C_2H_4 the low-speed freestream gas.

At high Reynolds numbers, shear-layer mixing occurs at a range of compositions dictated by the shear-layer entrainment ratio (Dimotakis 1991). For incompressible flow, well-defined interfaces mark the boundary between almost-homogenized mixed fluid and unmixed freestream fluid. For supersonic flow, mixed-fluid compositions are not as homogeneous. Oblique shocks, generated by supersonic convection speeds of the turbulent structures, are evident in the low-speed freestream (Dimotakis *et al.* 2001). Such shocks interact with the turbulence, providing additional (baroclinic) vorticity-generation as well as mixing and kinetic-energy dissipation mechanisms. Not much is known about mixing in this flow regime, with experiment and theory indicating that while shear-layer growth is lower than for incompressible shear layers, the mixed-fluid fraction is higher (Hall *et al.* 1991, Slessor 1998).

6. Mixing and transport in chemically reacting flows and combustion

Mixing and transport in chemically reacting flows and combustion is subject to further complications. Mixing on a molecular/diffusive scale is a manifestation and final stage of the multicomponent transport of participating species/reactants. This is modeled in terms of local gradients of concentration, temperature, and pressure in terms molecular-transport diffusion coefficients typically approximated as unaltered by the presence of chemical reactions (*e.g.*, Hischfelder *et al.* 1964). While this may be adequate in many situations, the details of intermolecular potentials which can be complex for reacting species, are directly linked to transport coefficients (Wang 2000). This is important in the context of flames, for example, where Wang and co-workers have documented a sensitivity in predictions of laminar-flame speeds, for example, to the values of particular binary-diffusion coefficients, *e.g.*, $\mathcal{D}_{H_2-N_2}$, $\mathcal{D}_{O_2-N_2}$, and $\mathcal{D}_{H-N_2}$, that is comparable to, or, in some cases, larger than flame-speed sensitivity to some of the most important reactions, such as the main chain-branching reaction, $H + O_2 \rightarrow O + OH$ (Yang *et al.* 2001), and, as these authors remark, "... chemical diffusion coefficients of free radical species may be highly uncertain because of the formation of a transiently bound complex upon molecular collision."

Transport modeling is also sensitive to other transport-coefficient effects. Volume, or bulk, viscosity, μ_v (Zel'dovich & Raiser 1967, p. 469; Thompson 1972; Giovangigli 1999, p. 107), which multiplies the local dilatation, $\nabla \cdot \mathbf{u}$, to yield a component of the normal stress, is generally neglected, even though, as Giovangigli (1999, p. 107) notes, this is "... a crude approximation that has strictly no justification a priori." In particular, while μ_v is zero for a monatomic gas, it can be appreciable for complex molecules with long relaxation times, *e.g.*, for rotational and vibrational equilibrium. By way of example, the ratio of volume to shear viscosity, μ_v/μ_s, is cited as 32 for H_2 and 1000 for CO_2 (Thompson 1972, p. 30) — important molecules for combustion. The large dilatations in flames suggest that this issue may merit a reexamination in multicomponent-transport modeling.

Both issues render comparison with flame data as a means of validation of chemical kinetic coefficients a particularly tricky endeavor.

7. Conclusions

Of the 19th-century physics problems bequeathed to the 20th, turbulence is the perhaps the only one bequeathed to the 21st. Broadly important and despite significant progress in the 20th century, turbulence must be regarded as the last, largely unsolved, classical physics problem. Crucial issues, such as turbulent drag, transport, and mixing at all scales, are still addressed primarily empirically and satisfactorily only for the simplest scenarios and flow geometries for which experimental data are available.

Turbulent mixing occurs at the smallest spatial scales of the flow rendering this a particularly difficult aspect of turbulent combustion to capture in models and simulations. Despite recent progress in this field, as also discussed by others in this meeting, turbulent mixing with combustion remains a grand-challenge problem.

8. Acknowledgements

I would like to acknowledge discussions with A. Cook, S. J. Chapman, and T. Mattner on the Rayleigh-Taylor instability, with H. Wang on the sensitivity to molecular transport coefficients, and the on-going work on the effects of volume viscosity with H. Lam. This paper and the recent experimental and computational work discussed was supported by AFOSR grants F49620–98–1–0052, F49620–00–1–0036, and F49620–01–1–0006, and DOE contract W–7405–ENG–48, which are gratefully acknowledged. Finally, I would like to thank A. Pollard for his gracious invitation to attend and contribute to this Symposium.

References

Batchelor, G. K. (1953) *The Theory of Homogeneous Turbulence* Cambridge U.P., London.

Batchelor, G. K. (1959) Small-scale variation of convected quantities like temperature in turbulent fluid. Part 1. General discussion and the case of small conductivity. *J. Fluid Mech.* **5**, 113–133.

Bernal, L. P., Breidenthal, R. E., Brown, G. L., Konrad, J. H. and Roshko, A. (1979) On the Development of Three-Dimensional Small Scales in Turbulent Mixing Layers. 2nd *Int. Symposium on Turb. Shear Flows* (Springer-Verlag, NY, 1980), 305–313.

Catrakis, H. J. and Dimotakis, P. E. (1996) Mixing in turbulent jets: scalar measures and isosurface geometry. *J. Fluid Mech.* **317**, 369–406.

Cook, A. W. and Dimotakis, P. E. (2001) Transition stages of Rayleigh-Taylor instability between miscible fluids. *J. Fluid Mech.* **443**, 69–99.

Cook, A. W., Dimotakis, P. E., Chapman, S. J. and Mattner, T. W. (2001) Corrigendum: Transition stages of Rayleigh-Taylor instability between miscible fluids. (in preparation).

Dimonte, G. and Schneider, M. (2000) Density ratio dependence of Rayleigh-Taylor mixing for sustained and impulsive acceleration histories. *Phys. Fluids* **12**, 304–321.

Dimotakis, P. E. (1991) Turbulent Free Shear Layer Mixing and Combustion. *High Speed Flight Propulsion Systems*, in *Progress in Astronautics and Aeronautics* **137** Ch. 5, 265–340.

Dimotakis, P. E. (2000) The mixing transition in turbulence. *J. Fluid Mech.* **409**, 69–98.

Dimotakis, P. E. (2001) Recent advances in turbulent mixing. *Mechanics for a New Millenium* (Kluwer AP, Netherlands), 327–344.

Dimotakis, P. E. and Catrakis, H. J. (1999) Turbulence, fractals, and mixing. *Mixing: Chaos and Turbulence* (Edited by, H. Chaté, E. Villermaux, and J.-M. Chomaz, Kluwer Academic/Plenum Publishers, New York, NY), 59–143.

Dimotakis, P. E., Catrakis, H. J. and Fourguette, D. C. (2001) Flow structure and optical beam propagation in high Reynolds number, gas-phase shear layers and jets. *J. Fluid Mech.* **433**, 105–134.

Duff, R. E., Harlow, F. H. and Hirt, C. W. (1962) Effects of Diffusion on Interface Instability Between Gases. *Phys. Fluids* **5**, 417–425.

Egolfopoulos, F. N., Dimotakis, P. E. and Bond, C. L. (1996) On Strained Flames with Hypergolic Reactants: The $H_2/NO/F_2$ System in High-Speed, Supersonic and Subsonic Mixing-Layer Combustion. *Twenty-Sixth Symposium (International) on Combustion* The Combustion Institute, Pittsburgh, 2885–2893.

Giovangigli, V. (1999) *Multicomponent flow modeling.* Birkhäuser, Boston.

Hall, J. L., Dimotakis, P. E. and Rosemann, H. (1991) Some measurements of molecular mixing in compressible turbulent mixing layers. *AIAA* 22nd *Fluid Dynamics, Plasma Dynamics, and Lasers Conference* Paper 91–1719.

Hinze, J. O. (1975) *Turbulence.* 2nd ed., McGraw-Hill.

Hirschfelder, J. O., Curtiss, C. F. and Bird, R. B. (1964) *Molecular Theory of Gases and Liquids.* John Wiley, NY.

Konrad, J. H. (1976) *An Experimental Investigation of Mixing in Two-Dimensional Turbulent Shear Flows with Applications to Diffusion-Limited Chemical Reactions.* Ph.D. thesis, California Institute of Technology.

Koochesfahani, M. M. and Dimotakis, P. E. (1986) Mixing and chemical reactions in a turbulent liquid mixing layer. *J. Fluid Mech.* **170**, 83–112.

Kuzo, D. M. (1995) *An Experimental Study of the Turbulent Transverse Jet.* Ph.D. thesis, California Institute of Technology.

Lundgren, T. S. (1982) Strained Spiral Vortex Model for Turbulent Fine Structure. *Phys. Fluids* **25**, 2193–2203.

Misra, A. and Pullin, D. I. (1997) A vortex-based subgrid stress model for large-eddy simulation. *Phys. Fluids* **9**, 2443–2454.

Monin, A. S. and Yaglom, A. M. (1975) *Statistical Fluid Mechanics: Mechanics of Turbulence* **II** Ed. J. Lumley, MIT Press, Cambridge, MA.

Overholt, M. R. and Pope, S. B. (1996) Direct numerical simulation of a passive scalar with imposed mean gradient in isotropic turbulence. *Phys. Fluids* **8**, 3128–3148.

Peters, N. (1984) Laminar Diffusion Flamelet Models in Non-Premixed Turbulent Combustion. *Prog. Energy Combust. Sci.* **10**, 319–339.

Pope, S. B. (2000) *Turbulent Flows.* Cambridge U. P., Cambridge, U.K.

Pullin, D. I. (2000) A vortex-based model for the subgrid flux of a passive scalar. *Phys. Fluids* **12**, 2311–2319.

Pullin, D. I. and Lundgren, T. S. (2001) Axial motion and scalar transport in stretched spiral vortices. *Phys. Fluids* **13**, 2553–2563.

Pullin, D. I. and Saffman, P. G. (1994) Reynolds stresses and one-dimensional spectra for a vortex model of homogeneous anisotropic turbulence. *Phys. Fluids* **6**, 1787–1796.

Shan, J. W. (2001) *Mixing and Isosurface Geometry in Turbulent Tranverse Jets.* Ph.D. thesis, California Institute of Technology.

Shan, J. W. and Dimotakis, P. E. (2001) Turbulent mixing in liquid-phase transverse jets. http://resolver.library.caltech.edu/ CaltechGalcitFM:2001.006.

Slessor, M. D. (1998) *Aspects of turbulent-shear-layer dynamics and mixing.* Ph.D. thesis, California Institute of Technology.

Tennekes, H. and Lumley, J. L. (1972) *A First Course in Turbulence.* MIT Press.

Thompson, P. A. (1972) *Compressible-Fluid Dynamics.* McGraw-Hill, New York.

Wang, H. (2000) Effect of transiently bound collision on binary diffusion coefficients of free radical species. *Chem. Phys. Ltrs.* **325**, 661–667.

Yang, Z., Yang, B. and Wang, H. (2001) The influence of H-atom diffusion coefficient on laminar flame simulation. 2nd *Joint Meeting of the U.S. Sections of the Combustion Institute* (Oakland, CA).

Zel'dovich, Ya. B. and Raizer, Yu. P. (1967) *Physics of Shock Waves and High-Temperature Hydrodynamic Phenomena, Vol. II.* Academic Press, New York, NY.

Discussion

Norbert Peters, Stanford University "You forgot about the good news about strained flamelets, that they reduce the temperature thereby reduce substantially pollution formation. It is not the goal of turbulent combustion to have optimal mixing it's a compromise to have enough mixing and low pollution formation."

James Riley, University of Washington "So Paul you gave us an estimate of the band width for a DNS in terms of grid points. Do you have anything comparable for LES?"

Reply: I think that if some of the models like the ones that Dale Pullin has put forth prove to be successful and stand the test of time there is good hope. I am not so sure that we will be able to find universal separate scale models because of what the laboratory is saying. There is a saving grace because of this factor of 50 that I mentioned and the people who do DNS have already discovered this that they don't quite need to have to resolve the Kolmogorov scale to do it. Unfortunately, that doesn't change the scale, it changes the prefactor only and so it is a little bit better than I said but not much better considering the grounds of acute penalty in moving up in Reynolds number. But I think in the long run we have to put our eggs in the LES - SGS basket. This can not be done by brut force.

E. Villermaux, Marseille "The measurements of λ_c the generalized Taylor scale for scalar, what was the units of λ_c in the pictures you showed?"

Reply: I plotted them on the image itself so you could see it relative to the flow, you remember the bars, they were, the length of the bars were 3 times the actual computations so you could see.

Steven Pope, Cornell University "Paul, you quoted the very high levels of strain rate in the turbulent mixing layer. People working in non-premixed turbulent combustion think it's the scalar dissipation rate that is a relevant inverse time scale not the strain rate."

Reply: Yes. We have measured this not at that Reynolds number but we have measured it in other flows and that was used for scaling purposes assuming Kolmogorov scaling and the scalar dissipation rate for Schmidt number 1, the highest values we have measured in those flow where we can resolve those things completely, is actually not that far off from that estimate. Now the most probable mixing rate is somewhat lower than that but nevertheless the distribution is scaled by such numbers, which we find to be approximately log-normal by the way as some of the simulations that was shown this morning seem to also argue.

Sébastien Candel, EMC2, Paris "I think you like very much paradoxes. You try to show this doesn't work, this doesn't work one of yours is the following. You have strain rates, which are very, very high. 100 thousands and then you even multiply that by saying that you in the intermediate range they are even higher. So then the paradox is how can flame live in such a situation. It can not. It is extinguished, but maybe the reason that it is not quite so is that maybe the strain rates are not that high or maybe the time by which they are exposed to the flame are also important so speaking just about the range may not be enough. So what is your comment?"

Reply: That is a very good comment. The first is that the chemically reactive shear layer data that I showed you are for hydrogen and fluorine. They are just about the fastest reactants known to man. If we try to do this with hydrocarbon combustion the hydrocarbons would exit the test section just like they came. You can't light them or keep them lit. In fact high-speed shear layers may not be very likely or very successful environments to try to burn hydro carbons at high speed for precisely the scaling reasons I showed. There maybe other flows for which you will be able to do this. In other flows such as jets I think what keeps the flame lit is the fact that it is possible to find pathways that connect low strain rate regions so that there is a probability of connecting the high reaction rate regions with each other within a slow low strain rate region. But for the shear layer you don't have this luxury. So that is why we don't know how to burn hydrocarbons in high-speed shear layers but the observation is exactly correct. We could not have done this with methane.

Chapter 9

ONE-DIMENSIONAL STOCHASTIC SIMULATION OF ADVECTION-DIFFUSION-REACTION COUPLINGS IN TURBULENT COMBUSTION

J. C. Hewson, A. R. Kerstein and T. Echekki
Combustion Research Facility
Sandia National Laboratories
Livermore, CA 94551-0969, USA

Abstract

The study of turbulent reacting flows invariably involves simplifying assumptions. Here an alternative modeling strategy is adopted that explicitly represents certain nonlinear couplings among the various subprocesses governing turbulent combustion, including unsteadiness and multi-scale interactions. This strategy involves fully resolved simulation at moderately large Reynolds numbers, which is rendered affordable for fully turbulent regimes by formulating a one-dimensional stochastic representation of turbulent flow evolution. The modeling challenges that arise, and the present approach to addressing these challenges, are illustrated by applying the new methodology, denoted one-dimensional turbulence (ODT), to nonpremixed jet flames that exhibit varying degrees of localized extinction and reignition. The role of unsteady strain and molecular transport in ODT in representing extinction and reignition processes in a turbulent environment is noted.

1. Introduction

The problem of turbulent combustion is complicated by the wide range of length scales that are typically important. These length scales include the full range of turbulent length scales, from device scales to molecular mixing scales, as well as the full range of chemical scales. Typically, turbulent combustion models assume some form of scale separation. For example, flamelet models [Peters, 1984, Peters, 2000] assume that the turbulence scales are large compared to the flamelet thickness. PDF-

A. Pollard and S. Candel (eds.), IUTAM Symposium on Turbulent Mixing and Combustion, 113-124.

evolution models [Pope, 1985, Xu and Pope, 2000] implicitly assume that turbulence scales associated with mixing enhancement are embedded within the chemical scales; larger turbulence scales are typically represented by a turbulence model extrinsic to the PDF formulation. Both of the aforementioned modeling techniques have enjoyed considerable success. However, certain phenomena of topical interest are captured in a limited manner within these models. The flamelet models and related models like conditional moment closure [Bilger, 1993, Klimenko and Bilger, 1999] have difficulty accounting for phenomena like extinction that depend on statistical fluctuations about mean mixing rates. The PDF-evolution methods readily account for statistical fluctuations, but must adopt models for the molecular-mixing process. The molecular-mixing process is of preeminent importance in combustion since it is the actual molecular mixing process that allows reactions to take place. This mixing process is also responsible for both extinction and reignition in many typical flames as will be described in this paper.

In the present paper, a recently developed model is presented that has certain features desirable for the study of reacting flows. This model is the one-dimensional turbulence (ODT) model. Results of ODT simulations of Syngas-air and methane-air turbulent jet diffusion flames are discussed from the viewpoint of the model representations of turbulent nonpremixed combustion. The combustion regimes that are simulated involve significant localized extinction and reignition that highlight certain challenges that arise in turbulent combustion modeling. These challenges include the coupling of various subprocesses occurring at multiple scales.

The ODT model has previously been applied to three sets of turbulent jet flames as a part of the ongoing program coordinated through the International Workshop on the Measurement and Computation of Turbulent Nonpremixed Flames [Barlow, 2000]. One set is a series of H_2 flames that was the initial focus of ODT combustion simulation [Echekki et al., 2001]. The second set is a series of Syngas flames, where the fuel is 40% CO/30% H_2/30% N_2 by volume. Direct comparisons of ODT results to the Syngas series are presented elsewhere [Hewson and Kerstein, 2000]. Here, Syngas simulation results are reported for a somewhat different set of conditions that are used to study the effect of mixing rate on local extinction and reignition at a constant Reynolds number. The third set is a series of partially-premixed methane-air flames. Methane-air simulation results are presented here for two experimental cases that also illustrate the coupled mixing-rate and Reynolds-number dependence of local extinction and reignition.

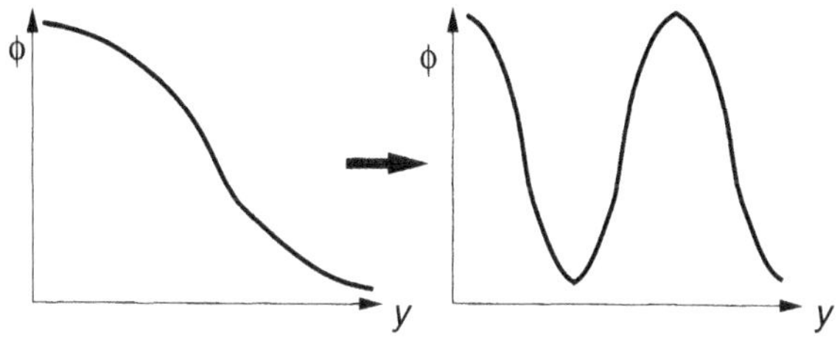

Figure 9.1. Effect of the triplet map on a scalar profile.

2. The one-dimensional turbulence model

The ODT model is described in detail in a number of recent publications [Kerstein, 1999, Echekki et al., 2001, Hewson and Kerstein, 2000], so only an overview is provided here. The ODT model involves two main processes implemented on a one-dimensional domain: an evolution equation for reaction-diffusion and a mechanism for turbulent advection. On the domain, the evolution of the entire thermochemical state and a representative velocity component (or all three components, in a recent formulation [Kerstein et al., 2001]) is resolved across the full range of relevant length scales, from the largest relevant advection scales to the smallest molecular mixing or reaction scales.

To mimic turbulent advection on a one-dimensional domain, a stirring process is introduced that emulates the effect of a turbulent eddy on a segment of the domain. This mixing model comprises a rule for the effect of an eddy on the property profiles and a rule for the rate of eddy events.

The so-called 'triplet map,' shown in Fig. 9.1, is used to model the effect of an eddy on property profiles. It has several desirable properties: (1) it wrinkles and folds property profiles, increasing property gradients and the 'area' (points in one dimension) of scalar interfaces and transferring fluctuations to smaller length scales, (2) it does not introduce gross artifacts such as property-field discontinuities, and (3) it obeys the relevant conservation laws.

The rate and sequence of the stirring operations is determined through the evolution of the velocity field, which is carried on the ODT domain along with any thermochemical state variables. The local instantaneous velocity gradient over a selected interval of the domain provides a characteristic rate for eddies of that size at that location and time. The actual implementation is carried out as a stochastic process by randomly sampling eddies of all sizes at all locations, with sampling frequencies based on the instantaneous characteristic rate of each eddy.

The velocity profiles thus drive the eddies. The eddies, in turn, rearrange the velocity profiles as well as the thermochemical profiles. This two-way coupling of the eddies and the velocity profiles induces complex dynamical behavior within the model, including an eddy cascade that reproduces salient qualitative and quantitative features of the turbulent cascade, in particular the transfer of energy to smaller scales.

The velocity and all state variables evolve by diffusive mixing and chemical reactions (where appropriate) as well as by eddy rearrangements. Flow evolution reduces to a laminar formulation in the absence of eddy events. The evolution of the reaction-diffusion terms is governed by a standard reaction-diffusion equation. The formulation used here is described elsewhere [Echekki et al., 2001, Hewson and Kerstein, 2000]. Diffusional transport and chemical source terms are evaluated using libraries from the Chemkin family of routines [Kee et al., 1983].

3. Extinction and reignition

Extinction in nonpremixed combustion may occur because heat losses exceed the rates of heat release. The rate of molecular transport, and thus heat loss, is typically characterized by a scalar dissipation rate, defined on the present one-dimensional domain as $\chi = 2D\,|dZ/dy|^2$, where D is the thermal diffusivity, and Z is the mixture fraction variable, a measure of the relative mass of fuel-derived species in a given mixture. The mixture fraction is bounded by zero and unity in the air and fuel streams, and the flame exists at a fixed value, Z_{st}, where reactants mix in stoichiometric proportions. There is a critical value of the scalar dissipation rate at Z_{st} in nonpremixed combustion above which the chemical reaction rates cannot keep up with the molecular transport rates [Peters, 1984]; this value is denoted χ_q in this paper. Steady flames cannot exist if $\chi > \chi_q$. The precise value for χ_q depends on the composition and enthalpy of the fuel and air mixtures; for the Syngas-air flames χ_q is 3000 s^{-1} based on the chemical mechanism used here. As $\chi \rightarrow \chi_q$ the peak temperature drops and leakage of reactants across the stoichiometric surface becomes significant. If the dissipation rate exceeds χ_q for a time of order $1/(\chi - \chi_q)$, the flame can be extinguished [Hewson and Kerstein, 2001].

Reignition can occur either by an autoignition mechanism in which slow exothermic reactions eventually reheat the mixture to the ignition temperature, or by heat transfer from adjacent burning regions to a non-burning region. For the ambient thermochemical conditions considered here, only the latter mechanism is sufficiently fast once the flame has cooled to the point of reactions being essentially quenched. The flame is considered to be essentially quenched when the temperature drops below the crossover temperature, which is slightly below 1000 K. Flames whose peak temperature remains above 1000 K are not considered to have extinguished for present purposes.

The heat-transfer reignition mechanism can be characterized by the direction of heat transfer with respect to the stoichiometric, but non-burning surface. Heat transfer that is primarily along the stoichiometric surface can be categorized as a triple flame or edge flame, which is a partially premixed flame propagating along that surface (cf. Chapter 4 of [Peters, 2000]). Heat transfer that is primarily normal to the stoichiometric surface is referred to here as flame-flame interaction (FFI), although in this case one 'flame' is non-burning.

Because the present model formulation is one-dimensional, it can only capture transport processes in which the gradients of all relevant properties are aligned with the computational domain. In this sense, triple flames are prohibited in the context of ODT because triple flames require non-aligned gradients of mixture fraction and enthalpy. ODT therefore captures only the FFI mode as defined here. Computed results suggest that this is not a severe limitation, at least for the cases considered here.

4. Piloted flame evolution

The typical evolution of the thermochemical state is described for both methane and Syngas flames here, followed in subsequent sections by more specific discussions of observations made for each series.

Both series of flames involve pilot flames between the fuel and coflowing air streams to stabilize the flames. Initialization involves a central fuel jet, a secondary jet of combustion products configured to represent the pilot, and coflowing air. For the Syngas flames, simple step function profiles for the velocity and species concentrations are adopted as initial conditions; species concentrations in the pilot are taken from a stoichiometric adiabatic premixed laminar flat flame calculation where the O-atom profile reaches a maximum. The methane-air flames take advantage of near-nozzle measurements [Barlow, 2000] for initialization to facilitate comparison with experimental results. Details of the dependence of ODT simulations on initial conditions are discussed elsewhere [Echekki et al., 2000, Hewson and Kerstein, 2000].

The fuel, oxidizer, and pilot are initially segregated, but they mix rapidly due to molecular processes that are augmented by frequent turbulent stirring events in the high shear regions near the nozzle. The mixing is sufficiently rapid to not only reduce the flame temperature due to finite-rate heat-release effects, but to also extinguish regions of

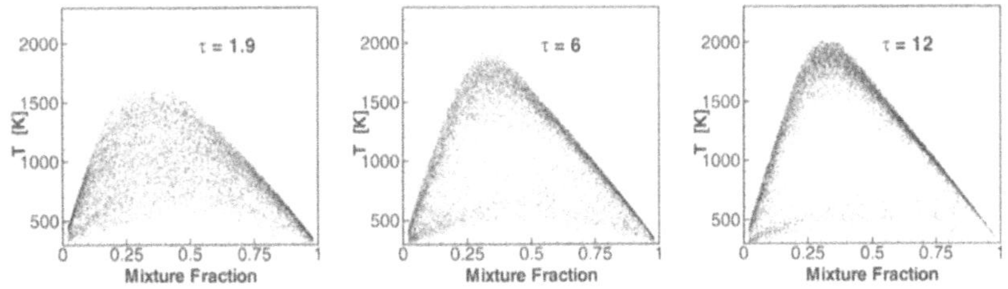

Figure 9.2. Scatter plots for the temperature in the Syngas flame E3 at three different evolution times. Flame E3 has the highest global mixing rate and consequently the highest degree of extinction among the Syngas flames.

the flame brush. In extinguished regions, relatively cool fuel and air may come together without burning. This local extinction is evident in scatter plots of the temperature versus the mixture fraction for Syngas flame E3 in Fig. 9.2; there, results are shown for three different evolution times, with $\tau = tU_0/d$ being a normalized evolution time. Samples near stoichiometric ($Z_{st} = 0.3$) with temperatures much below 1000 K are considered to be extinguished because, below these temperatures, the chemical rates become slow relative to the overall evolution time. Local extinction is significant early in the evolution when the dissipation rates are greatest. Turbulent entrainment and mixing of free-stream air increases the length scales and reduces the amplitudes of velocity and scalar fluctuations over the duration of the flow evolution, leading to lower dissipation rates. As the dissipation rates decline, the chance of extinction occurring becomes negligible. The reignition of the previously extinguished pockets occurs continuously and is evident in Fig. 9.2 through the reduction in the number of low temperature stoichiometric samples observed as the flow evolves. For all of the flames shown here, reignition is eventually complete and unburnt fuel is not emitted from these flames in significant quantities.

5. Methane flames

As discussed in Sec. 1, the ODT model has been validated through comparison with a number of extensively measured nonpremixed turbulent flames. Comparisons with measurements of flames D and F of the partially premixed methane-air series are shown here. These flames correspond to Re_{jet} values of 22,400 and 44,800, respectively; here $Re_{\text{jet}} = U_0 d/\nu$ with U_0 being the mean inlet velocity, d the width of the fuel inlet flow, and ν the kinematic viscosity. In the methane-air series, Re_{jet} was varied by varying the fuel inlet velocities while the noz-

Table 9.1. Initial conditions for simulated flames

	Fuel Stream			Pilot		Global Mixing
Flame	d [mm]	$\mathrm{Re_{jet}}$	U_0 [m/s] [a]	d_p [mm]	U_p [m/s]	Rate [s^{-1}] [b]
Methane D	7.2	22400	49.6	7.2	11.4	6.9×10^3
Methane F	7.2	44800	99.2	7.2	22.8	1.4×10^4
Syngas E1	6.25	15000	51	1.88	30	8.2×10^3
Syngas E2	5	15000	65	1.5	40	1.3×10^4
Syngas E3	4	15000	80	1.2	48	2.0×10^4

[a] $U_0 = Re_{\mathrm{jet}}\nu/d$.
[b] The global mixing rate is U_0/d.

zle and pilot sizes remain fixed. This results in simultaneous changes in the overall mixing rates (by a factor of two) and the dynamic range of scales, while the largest length scale remains fixed.

The ODT simulations are carried out in a planar-symmetric, temporally evolving jet configuration, and the domain is assumed to be convected downstream in a Lagrangian manner to simulate the spatially evolving laboratory flows. The convective velocity is obtained from the ratio of the momentum flux to the mass flux, both integrated across the ODT domain using the local, instantaneous velocity variable that is carried along the domain. Details of the fluid-mechanical representation of the jet flames considered here are the same as in a prior application of ODT to laboratory jet flames [Echekki et al., 2001, Hewson and Kerstein, 2000], including the choice of ODT parameter values. A twelve-step methane-air chemical mechanism is used [Sung et al., 1998].

Detailed comparisons of ODT results to these and other flames in the methane-air series are reported elsewhere [Echekki et al., 2000]. In Fig. 9.3, PDF's are shown for the temperature, conditionally sampled for $Z \in [0.3, 0.4]$. Local extinction is evident in the probability that the temperature is below 1000 K. Stoichiometric scalar dissipation rates are greatest between $x/d = 3$ and $x/d = 7.5$ in these flames [Pitsch and Steiner, 2000], and at $x/d = 7.5$ extinction is extensive in flame F and only slight in flame D. As the flow evolves downstream, extinguished flame segments are reignited while fewer new extinctions occur, and the PDF shifts towards higher temperatures.

It is seen that the simulations capture the development of extinguished flame elements and the subsequent reignition, as observed in the experiments. However, the rate of development in the ODT simulations differs somewhat from the measurements. Some of the differences can be attributed to the sensitivity of the ODT simulation to initial con-

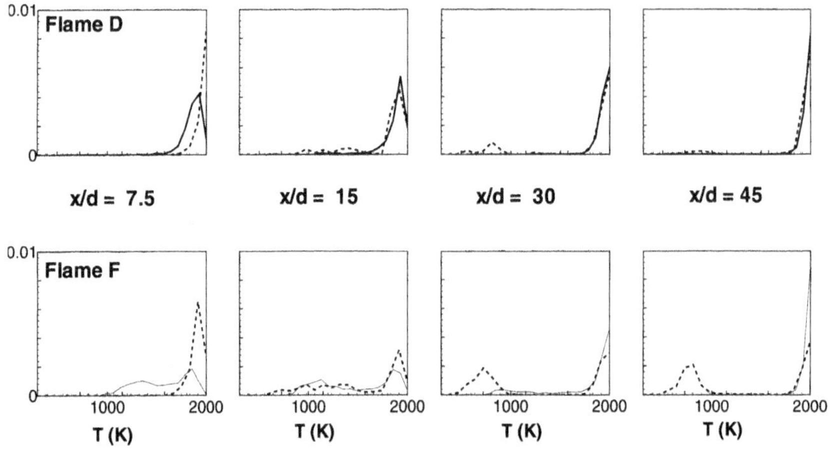

Figure 9.3. The conditional PDF of the temperature for methane-air flames D and F, conditionally sampled for $Z \in [0.3, 0.4]$. Predictions from the ODT simulations (dashed lines) are compared with measurements (solid lines) for x/d = 7.5, 15, 30 and 45 from left to right for each flame.

ditions [Echekki et al., 2000]. As in the experiments, the simulations indicate more local extinction for the higher-Re_{jet} case, which also has the higher mean scalar dissipation rates. While the mean mixing rate is roughly twice as great in flame F as in flame D, the probability of finding stoichiometric regions with temperatures below 1000 K increases by a larger factor, reflecting the nonlinear nature of the extinction and reignition process.

6. Syngas flames

For the Syngas flames simulated here, numerical experiments correspond to three different global mixing rates for a fixed Reynolds number $Re_{\text{jet}} = 15000$; these are the Syngas flames E1, E2 and E3 with the latter flames having the greater mixing rates. The mixing rate is varied by changing d as described in Table 9.1, while the range of length scales in the inertial range is kept constant. The conditions considered are similar to those studied in [Hewson and Kerstein, 2000], the primary difference being the use here of a small pilot flow (width d_p on each side of the fuel inlet) of hot combustion products to stabilize the flame.

In Fig. 9.4, PDF's of the scalar dissipation rate and temperature near stoichiometric are shown for various evolution times. For all Syngas flame simulations, conditional sampling is done for $0.875 Z_{st} < Z < 1.125 Z_{st}$ with $Z_{st} = 0.3$. The scalar dissipation rate takes on a roughly

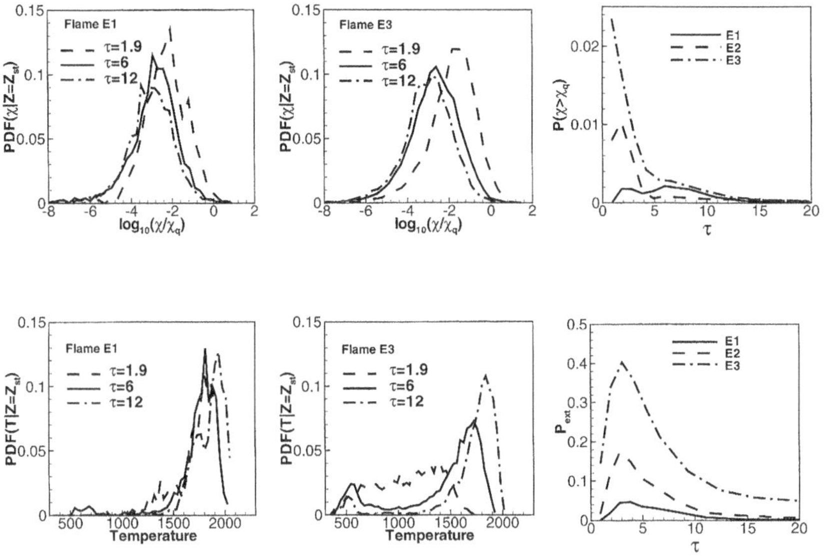

Figure 9.4. The conditional PDF of the scalar dissipation rate (upper series) and the temperature (lower series), conditionally sampled near stoichiometric, from the ODT simulations of Syngas flames E1 and E3 are shown in the left and center panes. The right pane shows the evolution of the probabilities that the $\chi > \chi_q$ and that $T < 1000$ K near Z_{st}. These are representive of the likelihood of extinction occurring and likelihood that extinction is observed, respectively.

lognormal distribution with a small chance of χ_q being exceeded. The probability that $\chi > \chi_q$ is also shown in Fig. 9.4 where the greater global mixing rates lead to significantly greater probabilities: while the global mixing rates change by only a factor of 2.5 from flame E1 to E3, the maximum probability that $\chi > \chi_q$ increases by roughly an order of magnitude.

The temperature PDF's in Fig. 9.4 are similar to those observed in the methane-air flames in Fig. 9.3, and the probability of finding low temperature stoichiometric regions is used to determine the probability of extinction. The quantitative criterion for extinction used here is that the temperature near stoichiometric is less than 1000 K, the temperature below which chemical reaction rates are significantly reduced; this defines the extinction probability, $P_{ext} \equiv P(T < 1000K \mid Z_{st})$. As with the probability that $\chi > \chi_q$, P_{ext} increases by roughly an order of magnitude from flame E1 to E3. These quantities differ primarily in that P_{ext} peaks slightly after $P(\chi > \chi_q)$ and is more than a magnitude greater than $P(\chi > \chi_q)$.

Of significant interest is the rate at which extinction and reignition occur; this is shown in terms of the rate of change of P_{ext} in Fig. 9.5.

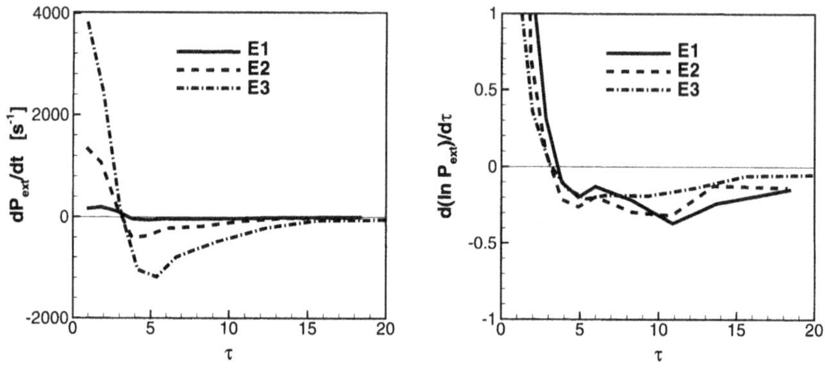

Figure 9.5. The rate of change in the probability of extinction and (right) its fractional rate of change scaled by the global mixing rate.

It can be observed, with respect to dP_{ext}/dt, that the rates of both extinction and reignition are much greater for flames with greater mixing rates. There are two sources for the greater dP_{ext}/dt: the greater global mixing rates and the greater overall degree of extinction. Differentiation of the fractional change in P_{ext} with respect to τ accounts for both of these effects, and this normalization collapses the evolution for all cases onto approximately a single line, as shown on the right-hand side of Fig. 9.5. While more rapid and more extensive extinction at higher mixing rates may be intuitive based on the PDF's of χ/χ_q shown in Fig. 9.4, more rapid reignition is also observed at higher mixing rates. This observation is interpreted as follows.

As noted in Sec. 3, the reignition of extinguished zones for the conditions of these simulations occurs by heat transfer between burning and non-burning regions. This process occurs over molecular mixing scales, among the smallest in these flames, yet they appear to be governed by a global time scale. The interpretation is that extinction events require a flame to be surrounded by cool reactants for the requisite heat loss to occur. For reignition to subsequently occur, the extinguished region needs to be brought into sufficient proximity with burning regions to enable the heat-transfer mechanism that causes reignition. The proximity that is required is of the order of the molecular mixing scales. The time scale for fluid to be mixed to this degree is the large scale mixing time scale, which scales with d/U_0 as was used to define τ. The significant result is that, despite the fact that molecular and chemical processes are involved in the reignition event, the rate of reignition appears to be governed by large scale mixing processes. In fact, reignition is accelerated for the same conditions that accelerate extinction. The rates of reignition and extinction are analyzed further in [Hewson and Kerstein, 2001].

7. Summary

In this paper, the ODT model has been used to study extinction and reignition processes in nonpremixed combustion. Certain aspects of the ODT model allow it to provide a unique perspective on these processes. First, within the ODT model statistical fluctuations in the local dissipation rate occur over a wide range of scales in a manner that creates, for the appropriate initial conditions, a small fraction of regions in which the flame will be locally extinguished. Reignition of these extinguished pockets requires molecular mixing processes to bring burning and non-burning flames into proximity. This is a process that appears to be well described by ODT. The time scale for ignition appears to scale with the global mixing time, d/U_0, even though the reignition involves diffusive and chemical processes. The scaling apparently arises because the reignition-limiting process is the bringing together of separated stoichiometric regions by turbulent advection. The subsequent molecular mixing occurring over much smaller time scales.

Acknowledgment

This research was supported by the Division of Chemical Sciences, Geosciences, & Biosciences, Office of Science, U.S. Department of Energy.

References

Barlow, R. S. (2000). International workshop on measurement and computation of turbulent nonpremixed flames. http://www.ca.sandia.gov/tdf/Workshop.html.

Bilger, R. W. (1993). Conditional moment closure for turbulent reacting flow. *Phys. Fluids A*, 5:436.

Echekki, T., Kerstein, A. R., and Chen, J.-Y. (submitted 2000). Computations of turbulent jet diffusion flames using the 'one-dimensional turbulence' model: Methane-air flames. *Combust. Flame.*

Echekki, T., Kerstein, A. R., Chen, J.-Y., and Dreeben, T. D. (2001). Computations of turbulent jet diffusion flames using the 'one-dimensional turbulence' model: Hydrogen-air flames. *Combust. Flame*, 125:1083–1105.

Hewson, J. C. and Kerstein, A. R. (submitted 2000). Stochastic simulation of transport and chemical kinetics in turbulent $CO/H_2/N_2$ flames. *Combust. Theory Modeling.*

Hewson, J. C. and Kerstein, A. R. (submitted 2001). Local extinction and reignition in nonpremixed turbulent $CO/H_2/N_2$ jet flames. *Combust. Sci. Tech.*

Kee, R. J., Warnatz, J., and Miller, J. A. (1983). Chemkin: A general-purpose, problem-independent, transportable, fortran chemical kinetics code package. Technical Report SAND83-8209, Sandia National Laboratory, Livermore, CA.

Kerstein, A. R. (1999). One-dimensional turbulence: Model formulation and application to homogenous turbulence, shear flows, and buoyant stratified flows. *J. Fluid Mech.*, 392:277–334.

Kerstein, A. R., Ashurst, W. T., Wunsch, S., and Nilsen, V. (2001). One-dimensional turbulence: vector formulation and application to free shear flows. *J. Fluid Mech.*, 447:85–109.

Klimenko, A. Y. and Bilger, R. W. (1999). Conditional moment closure for turbulent combustion. *Prog. Energy Combust. Sci.*, 25:595–687.

Peters, N. (1984). Laminar diffusion flamelet models in nonpremixed turbulent combustion. *Prog. Energy Combust. Sci.*, 10:319–339.

Peters, N. (2000). *Turbulent Combustion.* Cambridge University Press.

Pitsch, H. and Steiner, H. (2000). Scalar mixing and dissipation rate in large-eddy simulations of non-premixed turbulent combustion. *Proc. Combust. Inst.*, 28:41–49.

Pope, S. B. (1985). PDF methods for turbulent reacting flows. *Prog. Energy Combust. Sci.*, 11:119.

Sung, C. J., Law, C. K., and Chen, J.-Y. (1998). An augmented reduced mechanism for methane oxidation with comprehensive global parametric validation. *Proc. Combust. Inst.*, 27:295–304.

Xu, J. and Pope, S. B. (2000). PDF calculations of turbulent nonpremixed flames with local extinction. *Combust. Flame*, 123:281–307.

Discussion

A. Pollard, Queen's University "What would happen if you added in structure to this modelling attempt?"

Reply: I am not sure how you would add in structure. There is no structure here. In fact our present position is that the ODT basically works best as a model for these smaller scale structures coupled with for example large eddy simulation because we have no real way of capturing complex geometries or organized structures.

A. Pollard, Queen's University "So what is the next step?"

Reply: Well actually we have been working on implementing ODT as a boundary layer model within large eddy simulations and we are beginning to implement it as an actual sub rate scale for the reacting flows also. Although I guess maybe Suresh Menon will talk about a similar linear eddy model and he is also working in that area. I think he is going to talk about that. You will hear a lot more then.

Chapter 10

COMPRESSIBILITY EFFECTS ON THE SCALAR MIXING IN REACTING HOMOGENEOUS TURBULENCE

D. Livescu and C.K. Madnia
Department of Mechanical and Aerospace Engineering
State University of New York at Buffalo
Buffalo, NY 14260

Abstract

The compressibility and heat of reaction influence on the scalar mixing in decaying isotropic turbulence and homogeneous shear flow are examined via data generated by direct numerical simulations (DNS). The reaction is modeled as one-step, exothermic, irreversible and Arrhenius type. For the shear flow simulations, the scalar dissipation rate, as well as the time scale ratio of mechanical to scalar dissipation, are affected by compressibility and reaction. This effect is explained by considering the transport equation for the normalized mixture fraction gradient variance and the relative orientation between the mixture fraction gradient and the eigenvectors of the solenoidal strain rate tensor.

1. Introduction

Recent studies of scalar mixing in incompressible turbulence (e.g. Overholt & Pope 1996; Jaberi *et al.* 1996; Slessor, Bond & Dimotakis 1998; Livescu, Jaberi & Madnia 2000) have enhanced our understanding of the mixing mechanism, the behavior of the scalar PDF and the evolution of scalar moments. However, fundamental studies of scalar mixing in compressible turbulence are scarce. Blaisdell, Mansour & Reynolds (1994) show that the dilatational velocity has negligible contribution to the scalar flux in homogeneous turbulence. It is not clear, however, how the compressibility affects the scalar moments or the scalar dissipation. A better understanding of the compressibility effects is important in

A. Pollard and S. Candel (eds.), IUTAM Symposium on Turbulent Mixing and Combustion, 125-135.

the analysis of the combustion processes. The main objective of this work is to study the influence of compressibility and heat release on the evolution of the scalar dissipation rate in homogeneous turbulence.

2. Numerical methodology and DNS parameters

In order to assess the influence of compressibility on the mixing process, direct numerical simulations (DNS) of decaying isotropic and homogeneous sheared turbulence are performed under reacting (heat - releasing) and nonreacting conditions. In the nonreacting cases a passive scalar is simulated along with the turbulent field. The compressible form of continuity, momentum, energy and species mass fractions transport equations are solved using the spectral collocation method. In the reacting cases, the chemical reaction is modeled by a one-step irreversible Arrhenius-type reaction. In order to assess the influence of heat of reaction on the mixing process, the statistics pertaining to the mixture fraction Z are extracted from the two reacting scalar fields and compared to those obtained for a passive scalar in the nonreacting cases. The viscosity varies with temperature according to a power law and $Le = 1$ in all cases.

For the shear flow simulations, the velocity fluctuations are initialized as a random solenoidal, three-dimensional field with Gaussian spectral density function (with the peak at $k_{0v} = 10$) and unity rms. The initial pressure fluctuations are evaluated from a Poisson equation (except for the high Mach number case for which they are set to zero). For the decaying isotropic simulations the initialization proposed by Ristorcelli & Blaisdell (1997) is employed. The variables are then allowed to decay until they develop realistic turbulence fluctuations before the scalars are initialized. This corresponds to time $t = 0$ in all figures. The value of the Reynolds number at $t = 0$ is $Re_{\lambda_0} = 50$. The simulations are stopped at $t/\tau_t = 14.4$ such that the scalar variance becomes two orders of magnitude less than the initial value. For the shear flow cases $Re_{\lambda_0} = 21$ initially, and increases to values between $93 - 110$ in the direction of the mean velocity at the end of the simulations. The shear flow simulations are stopped between $18 < St < 20$, such that the integral scales remain small compared to box size and the Kolmogorov microscale is larger than the grid size. This corresponds to t/τ_t in the range 10.4 to 11.5.

The scalar fields are initialized as "random blobs", with double-delta PDFs (Overholt & Pope, 1996). The initial length-scale of the scalar field is controlled by changing the location, k_{0s}, of the peak of the Gaussian spectrum used to generate the scalar field. The variables are time advanced in physical space using a second order accurate Adams-Bashforth scheme. The code developed for this work is based on a fully parallel algorithm and uses the standard Message Passing Interface (MPI).

Compressibility effects on the scalar mixing are examined by defining a set of cases with different initial turbulent Mach number, M_{t_0}. The heat release effects are investigated by varying the values of the heat release parameter, Ce, computational Damköhler number, Da, and the initial lengthscale of the scalar field. In reacting cases the value of the Zeldovich number is $Ze = 8$. Table 10.1 summarizes the cases considered for this study. The isotropic turbulence cases are labeled with i and the shear flow cases with s. For the shear flow cases the value of the initial nondimensional mean shear rate, $S_0^* = \frac{S(2K)}{\epsilon}$, is in the range dominated by nonlinear effects, where K is the turbulent kinetic energy, ϵ is the viscous dissipation, and $S = \partial\tilde{u}_1/\partial x_2$ ($\tilde{}$ is the Favre average) is the mean shear rate.

Table 10.1. Parameters for the DNS cases.

Case #	M_{t_0}	S_0^*	Da	Ce	k_{0_s}
$i1$	0.2	0	0	0	4
$i2$	0.35	0	0	0	4
$i3$	0.5	0	0	0	4
$s1$	0.1	7.24	0	0	4
$s2$	0.2	7.24	0	0	4
$s3$	0.3	7.24	0	0	4
$s4$	0.4	7.24	0	0	4
$s5$	0.6	7.24	0	0	4
$s6$	0.3	7.24	1100	1.44	4
$s7$	0.3	7.24	1100	2.16	4
$s8$	0.3	7.24	1750	1.44	4
$s9$	0.3	7.24	1100	1.44	10

The reaction parameters chosen for the cases considered mimic the combustion of a typical hydrocarbon in air at low to moderate values of Reynolds number. For all reacting cases presented most of the reaction occurs between $2 < St < 8$ and the reaction rate peaks between $4 < St < 6$.

3. Results

In the absence of a mean scalar gradient, the transport equation for the mixture fraction variance, $\widetilde{Z''Z''}$, does not have a production term and $\widetilde{Z''Z''}$ decays continuously. Furthermore, if the transport equation for the mixture fraction variance is normalized by $\widetilde{Z''Z''}$ then the right hand side is proportional to the mixture fraction dissipation rate, $\epsilon_Z = \frac{1}{\widetilde{Z''Z''}} < \frac{\mu}{ReSc} \nabla Z'' \cdot \nabla Z'' >$ which is a key quantity in the modeling of both passive and reactive turbulent scalar fields. For all cases considered ϵ_Z increases initially as the turbulence breaks down the non-premixed "blobs" into smaller scalar structures, then it reaches a maximum and starts to decay. For the isotropic turbulence cases our results indicate a slight dependence of ϵ_Z on M_{t_0}. However, for the shear flow simulations, figure 10.1(a) shows that the values of ϵ_Z decrease as M_{t_0} increases. Since for the nonreacting cases the variation of the viscosity is small, the decrease in ϵ_Z with Mach number can be associated with the decrease in the normalized mixture fraction gradient variance, $\xi = < \nabla Z'' \cdot \nabla Z'' > / \widetilde{Z''Z''}$. This is in agreement with the results obtained for forced isotropic turbulence by Cai, O'Brien & Ladeinde (1998). In the presence of heat release, ϵ_Z increases its magnitude during the time when the reaction is significant (figure 10.1b). A further increase in the values of ϵ_Z is obtained by increasing Ce or Da. However, for the reacting cases, our results indicate that the values of ξ decrease due to the reaction. Therefore, the increase in ϵ_Z can be associated with the significant enhancement of the molecular transport properties due to the heat of reaction. By increasing the value of k_{0_s} (case $s9$) the initial length scale of the scalar field becomes smaller, which results in higher values of ϵ_Z compared to other cases.

The scalar dissipation rate is usually modeled by assuming that the ratio, $C_Z = \frac{2K/\epsilon}{1/\epsilon_Z}$, of the time scale of mechanical dissipation to that of the mixture fraction variance dissipation is constant, although there is much evidence that it does not take a universal value (Livescu, Jaberi & Madnia 2000). Figure 10.2 shows that for the isotropic turbulence cases C_Z varies in time initially and seems to asymptote to a constant at long time. Similar with the results obtained for the mixture fraction dissipation rate, C_Z is slightly affected by the change in M_{t_0} for the cases considered. The corresponding values of C_Z obtained using the model of Xu, Antonia & Rajagopalan (2000) are also shown for comparison. The model assumes local isotropy, and the prediction plotted in figure 10.2 does not take into account the intermittency of the small scales. The values of C_Z become close to the model prediction at late times.

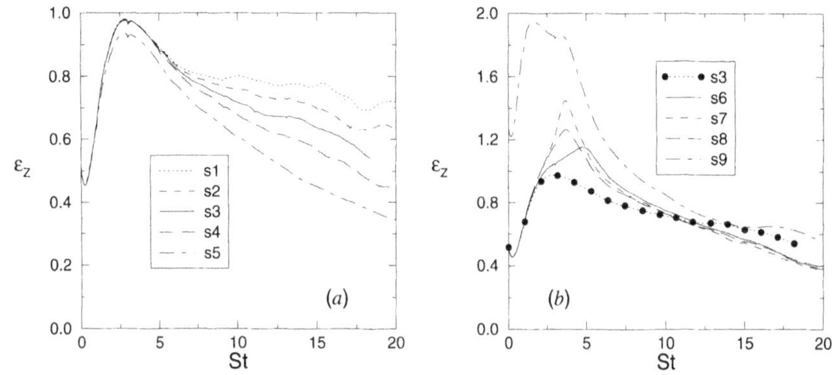

Figure 10.1. a) Mach number and b) heat release influence on the evolution of the mixture fraction dissipation rate.

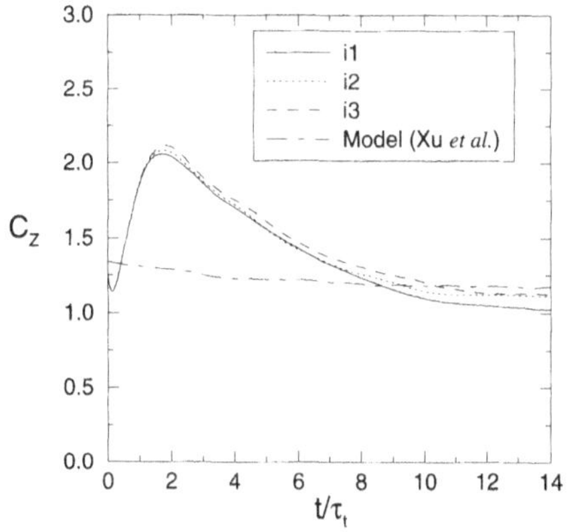

Figure 10.2. Time variation of C_Z for the isotropic turbulence cases.

Both M_{t_0} and heat of reaction influence the values of C_Z in the shear flow simulations (figure 10.3). For the nonreacting cases, C_Z decreases as M_{t_0} increases (figure 10.3a). At early times this is due to an amplification of the viscous dissipation rate, ϵ/K, and a decrease in ϵ_Z. At later times ϵ/K has lower values at higher M_{t_0}, which explains the change in the slope of C_Z observed in figure 10.3(a). For the reacting cases the viscous dissipation rate increases significantly during the time when reaction is important. This increase is larger than the increase in ϵ_Z, and C_Z becomes less than in the nonreacting case (figure 10.3b).

It is explained above that the normalized mixture fraction gradient, decreases at higher values of M_{t_0} or in the presence of heat release. This

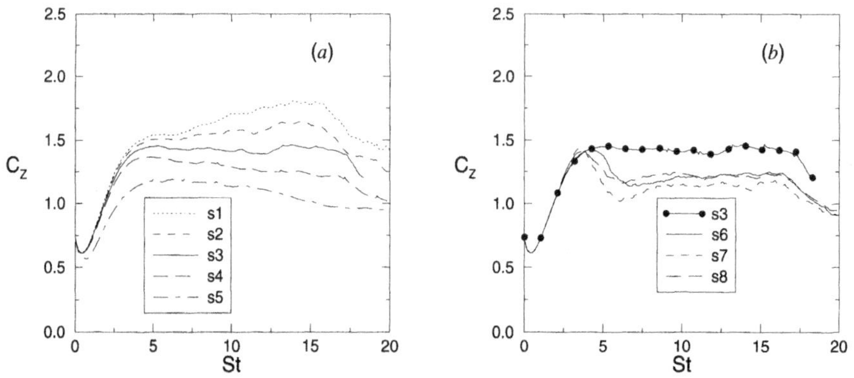

Figure 10.3. a) Mach number and b) heat release influence on the evolution of C_Z.

behavior is further examined by considering the transport equation for the normalized variance of the mixture fraction gradient:

$$\frac{d\xi}{dt} = I + II + III + IV + V \tag{10.1}$$

where

$$\begin{aligned} I &= -2 < \nabla Z'' \cdot \underline{\underline{S}} \cdot \nabla Z'' > /\widetilde{Z''Z''} \\ II &= -2 < \nabla Z'' \cdot \underline{\underline{s^s}} \cdot \nabla Z'' > /\widetilde{Z''Z''} \\ III &= -2 < \nabla Z'' \cdot \underline{\underline{s^d}} \cdot \nabla Z'' > /\widetilde{Z''Z''} \\ IV &= < \Delta(\nabla Z'' \cdot \nabla Z'') > /\widetilde{Z''Z''} \\ V &= -\left[\frac{2\tilde{\mu}}{ReSc < \rho >}\left\langle (\nabla^2 Z'')^2 - \frac{(\nabla Z'' \cdot \nabla Z'')^2}{\widetilde{Z''Z''}}\right\rangle\right] /\widetilde{Z''Z''} \end{aligned}$$

and the terms involving fluctuations of the viscosity are neglected. Here $s^d_{ij} = \frac{1}{2}\left(\frac{\partial u_i''^d}{\partial x_j} + \frac{\partial u_j''^d}{\partial x_i}\right)$ and $s^s_{ij} = \frac{1}{2}\left(\frac{\partial u_i''^s}{\partial x_j} + \frac{\partial u_j''^s}{\partial x_i}\right)$ are the dilatational and solenoidal strain rate tensors, respectively, and $\Delta = \frac{\partial u_i''}{\partial x_i}$ is the dilatation. The first two terms in equation 10.1 are production terms, due to the mean shear and solenoidal strain rate, respectively, terms III and IV are explicit dilatational terms and the last term is the molecular dissipation. For both reacting and nonreacting cases considered the explicit dilatational terms are found to be much smaller than the rest of the terms in equation 10.1, and will not be shown.

Figure 10.4 shows that for isotropic turbulence the two important terms in equation 10.1 (terms II and V), and therefore the rate of change of ξ, are slightly affected by the change in M_{t_0}.

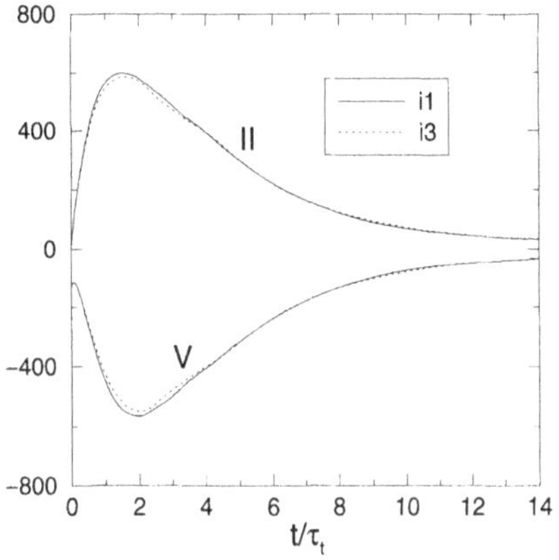

Figure 10.4. Important terms in equation 10.1 for isotropic turbulence cases.

For the shear flow simulations, term I, production due to the mean shear is also important (figure 10.5). However, for the value of S_0^* considered for this study, this term is smaller than term II. Unlike the isotropic cases, for the shear flow cases M_{t_0} has a strong influence on the evolution of the terms in equation 10.1 (figure 10.5a). The production terms and the dissipation term decrease their values significantly as M_{t_0} increases. However, the decrease in the viscous dissipation term leads to an increase in the values of ξ. Therefore, the reduction in ξ at higher M_{t_0} is due to a decrease in the production terms, primarily due to the reduction of term II. Similarly, all terms in equation 10.1 decrease their values for the reacting cases as compared to the nonreacting case (figure 10.5b). Nevertheless, during the time when the reaction is significant, the reduction in the production terms (mainly term II) is responsible for the decrease in ξ.

The results presented suggest that the reduction in the values of ξ (and hence ϵ_Z for the nonreacting cases) can be associated mainly to a decrease in the production due to the solenoidal strain. This term is dependent on the relative orientation between the mixture fraction gradient and the solenoidal strain rate tensor, the magnitude of the corresponding eigenvalues and the mixture fraction gradient variance. For compressible isotropic turbulence, Jaberi, Livescu & Madnia (2000) showed that the scalar gradient tends to align with the most compressive eigenvector of the strain rate tensor, similar with the behavior observed

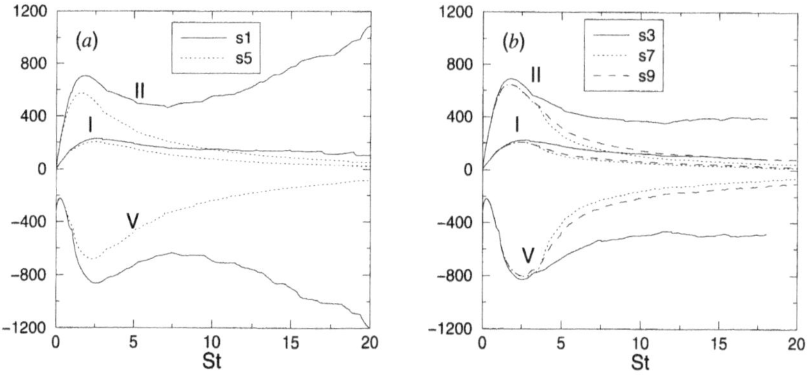

Figure 10.5. a) Mach number and b) heat release effect on the terms in equation 10.1.

earlier in incompressible turbulence (Ashurst *et al.* 1987). Our results indicate that this alignment is also obtained with the solenoidal part of the strain rate tensor and it does not change significantly with increasing M_{t_0}.

However, for the shear flow simulations the PDF of the cosine of the angle between $\nabla Z''$ and the γ-eigenvector of $\underline{\underline{s}}^s$ peaks at a value different than 1, indicating a most probable distribution towards an angle different from zero (figure 10.6). χ_1, χ_2, and χ_3, are angles between $\nabla Z''$ and the α-, β-, and γ-eigenvectors, respectively. These eigenvectors correspond to the eigenvalues labeled using the usual convention $\alpha > \beta > \gamma$. For a homogeneous flow, $< \alpha > + < \beta > + < \gamma >= 0$. Since $< \alpha >> 0$, it has a negative contribution to the magnitude of the production term (term II in equation 10.1), while $< \gamma >$ has a positive contribution. For all cases considered $< \beta >> 0$ and it is small compared to $< \alpha >$ and $< \gamma >$. As the initial turbulent Mach number increases, the peak of the PDF of $\cos \chi_3$ moves to smaller values, so the alignment worsens (figure 10.6a), contributing to a decrease in the production term. Furthermore, the peak of the PDF of $\cos \chi_1$ tends to occur at higher values as M_{t_0} increases, so that the alignment with the most dilatational eigenvalue improves, further decreasing the production term. Moreover, all three eigenvalues of the solenoidal strain rate tensor decrease their magnitude at higher M_{t_0} and it can be shown that this effect also contributes to the decrease in the production term.

For the reacting cases, near the time when the reaction rate peaks, the alignment between $\nabla Z''$ and the γ-eigenvector improves slightly (the peak of the PDF of $\cos \chi_3$ moves to higher values), which has a positive

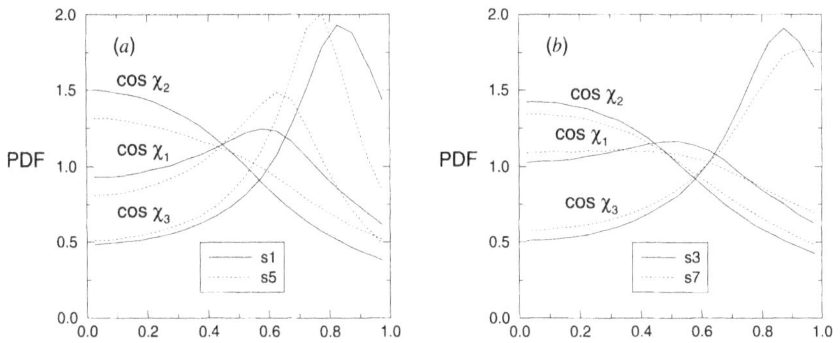

Figure 10.6. PDFs of the cosines of the angles between the eigenvectors of the solenoidal strain rate tensor and the mixture fraction gradient. a) Mach number effect ($St = 8$) and b) heat release effect ($St = 6$).

contribution to the production term (figure 10.6b). Later, the alignment becomes close to that obtained for the nonreacting cases. However, the decrease in the magnitudes of the eigenvalues is more significant and the production term has smaller values than in the nonreacting case.

4. Conclusion

DNS of compressible homogeneous turbulent shear flow and isotropic decaying turbulence are performed under reacting (heat releasing) and nonreacting conditions to examine the role of compressibility and heat release on the scalar mixing. For the reacting cases the chemical reaction is modeled as one step, irreversible, and Arrhenius type. The statistics pertaining to the mixture fraction are extracted from the reacting scalars and compared to those obtained for a passive scalar in the nonreacting cases.

For the range of M_{t_0} examined, it is found that the mixture fraction dissipation rate is less sensitive to the changes in M_{t_0} in isotropic turbulence than in shear flow. For the shear flow cases ϵ_Z is strongly affected by compressibility and it decreases as M_{t_0} increases. Although ϵ_Z increases for the reacting cases during the time when the reaction is important, this is primarily due to the increase in the viscosity. Similar with the Mach number effect, the mixture fraction gradient variance decreases for the reacting cases. The mechanical to scalar dissipation time scale ratio is also dependent on M_{t_0} in shear flow, unlike the isotropic turbulence where the dependence was found to be weak.

The transport equation for the normalized mixture fraction gradient variance was examined and was found that the explicit dilatational terms are much smaller than the other terms in the equation. For the shear flow simulations, the mixture fraction gradient variance decreases with Mach number primarily due to a misalignment between the mixture fraction gradient and the eigenvectors of the solenoidal strain rate tensor, and also a decrease in the magnitudes of the corresponding eigenvalues. However, for the reacting cases the normalized scalar gradient variance decreases mainly due to a reduction in the eigenvalues of the solenoidal strain rate tensor.

Acknowledgments

This work is sponsored by the American Chemical Society under Grant 35064-AC9 and by the National Science Foundation under Grant CTS-9623178. Computational resources were provided by the National Center for Supercomputer Applications at the University of Illinois at Urbana-Champaign, San Diego Supercomputer Center, and the Center for Computational Research at the State University of New York at Buffalo.

References

Ashurst, W. T., Kerstein, A. R., Kerr, R. M & Gibson, C. H. 1987 Alignment of vorticity and scalar gradient with strain rate in simulated Navier-Stokes turbulence. *Phys. Fluids* **30**, 2343-2353.

Blaisdell, G. A., Mansour, N. N. & Reynolds, W. C. 1994 Compressibility effects on the passive scalar flux within homogeneous turbulence. *Phys. Fluids* **6**, 3498-3500.

Cai, X. D., O'Brien, E. E. & Ladeinde, F. 1998 Advection of mass in forced, homogeneous, compressible turbulence. *Phys. Fluids* **10**, 2249-2259.

Jaberi, F. A., Miller, R. S., Madnia, C. K. & Givi, P. 1996 Non-Gaussian scalar statistics in homogeneous turbulence. *J. Fluid Mech.* **313**, 241-282.

Jaberi, F. A., Livescu, D. & Madnia, C. K. 2000 Characteristics of chemically reacting compressible homogeneous turbulence. *Physics of Fluids* **12**, 1189-1209.

Livescu, D., Jaberi, F. A. & Madnia, C. K. 2000 Passive scalar wake behind a line source in grid turbulence. *J. Fluid Mech.* **416**, 117-149.

Livescu, D., Jaberi, F. A. & Madnia, C. K. 2001 Heat release effects on the energy exchange in reacting turbulent shear flow. Accepted for publication in *J. Fluid Mech.*

Overholt, M. R. & Pope, S. B 1996 Direct numerical simulation of a passive scalar with imposed mean gradient in isotropic turbulence. *Phys. Fluids* **8**, 3128-3148.

Ristorcelli, J. R. & Blaisdell, G. A. 1997 Consistent initial conditions for the DNS of compressible turbulence. *Phys. Fluids* **9**, 4-6.

Slessor M. D., Bond C. L. & Dimotakis P. E. 1998 Turbulent shear-layer mixing at high Reynolds numbers: effects of inflow conditions. *J. Fluid Mech.* **376**, 115-138.

Xu, G., Antonia, R. A. & Rajagopalan, S. 2000 Scaling of mean temperature dissipation rate. *Phys. Fluids* **12**, 3090-3093.

Discussion

Stavros Tavoularis, University of Ottawa "What are the turbulence Reynolds numbers for the turbulent cases?"

Reply: For the decaying isotropic turbulent cases it is based on Taylor microscale it is 55 for the beginning of the simulation. For homogenous shear flow it is 21 but it increases during simulation and it reaches different in different directions. It reaches about 100 in each direction it averages around 60 at the end of the simulation.

Stavros Tavoularis, University of Ottawa "So could it be that this places a role in the differences you have? The shear flow is much more active."

Reply: Sure there is a dependence on the Reynolds number now the question is if it is during independence, it is also a dependence on the Mach number, yes the Reynolds number should play around this.

Thierry Poinsot, CNRS, Toulouse "How is temperature initialized for the reacting cases and how does combustion start? Do you have a hot side, one of the two reactants are hot, how does combustion start when you do the reacting cases?"

Reply: Because it is Arrhenius reaction action it starts.

Thierry Poinsot, CNRS, Toulouse "You still need a hot spot to do that, so how do you initialize temperature?"

Reply: I initialize the pressure and pressure fluctuations from the Poisson equation, and the density fluctuations are zero, temperature fluctuations are generated from the equation of state and temperature increases due to that because it is Arrhenius reaction.

Thierry Poinsot, CNRS, Toulouse "But one of the two sides has to be hot if you want this to happen otherwise it will never start. What is the maximum initial temperature?"

Reply: So that mean temperature is 1. Non dimensional.

Norman Peters, Sanford University "We found a similar behaviour between compressibility effects on heat release, do you have a physical explanation for that."

Reply: In heat release the dissipation effects are very important and the mechanism are very different.

Dan Ewing, McMaster University "How are you defining the Mach numbers, is this a convective Mach numbers?"

Reply: It is a turbulent Mach number based on velocity fluctuations.

Chapter 11

THE INTERACTION OF SCALAR MIXING AND HEAT RELEASE IN A REACTING SHEAR LAYER

Carlos Pantano, Sutanu Sarkar, and Forman Williams
Mechanical and Aerospace Engineering
University of California at San Diego
La Jolla, CA 92093

Abstract Direct numerical simulation of a reacting shear layer is performed to investigate the mixture fraction field and its dissipation. A systematic study of the influence of heat release is undertaken. Substantial changes, both qualitative and quantitative, in the mixing of an active scalar (with heat release) with respect to a passive scalar are observed.

Keywords: Turbulent combustion, scalar mixing, heat release effects

1. Introduction

The efficiency and environmental impact of nonpremixed combustion often depends on molecular mixing of fuel, oxidant and various intermediate chemical products in a background of turbulent flow. Therefore, a fundamental understanding of scalar mixing in turbulent reacting flows is required. In combustion, the scalars are active, that is, they impact the flow field through density and pressure changes associated with reaction. Some simplified models obviate the need to resolve the small length and time scales introduced by the chemical production terms by introducing a conserved scalar, ζ, an appropriately defined mixture fraction. Under the assumption of fast chemistry [Williams, 1985], a conserved scalar approach can be used and the overall reaction rate expressed [Bilger, 1980] in terms of the statistics of the scalar and a characteristic scalar mixing rate, $<\chi|\zeta=\zeta_s>$, the scalar dissipation, $\chi = 2D\nabla\zeta\cdot\nabla\zeta$, evaluated at stoichiometric mixture fraction, $\zeta=\zeta_s$. The laminar flamelet

A. Pollard and S. Candel (eds.), IUTAM Symposium on Turbulent Mixing and Combustion, 137-148.

approach [Peters, 1986, Pitsch, Chen and Peters, 1998] assumes that the flame corresponds to zones in physical space where the dominant balance is between reaction and diffusion. Consequently, an equation that determines species mass fractions, $Y_i(\zeta;\chi)$ is obtained with $\chi(\zeta)$ modeled using the counterflow one-dimensional profile for its shape and a reference value, $<\chi>(\zeta=\zeta_0)$, taken from the calculations with the turbulence model, to set its amplitude. The conditional moment closure (CMC) [Klimenko, 1990, Bilger, 1993] solves an equation for the conditional expectation of species mass fraction, $Q_i =< Y_i|\zeta >$. This equation requires the mixture fraction dissipation, $<\chi|\zeta>$, conditioned on the value of mixture fraction among other quantities, for closure. Although conditional averaging increases the dimensionality of the problem, the advantage [Bilger, 1993] is that, in shear flows, the transverse dependence of conditional statistics may be negligible.

Direct numerical simulation (DNS) is used to study the specific problem of non-premixed combustion in a temporally-evolving shear layer between methane and air. Previous DNS [McMurtry *et al.*, 1989, Miller, Madnia and Givi, 1994] and experiments [Mungal and Dimotakis, 1984, Hermanson and Dimotakis, 1989, Miller, Bowman and Mungal, 1998] in the reacting shear layer have given important information. The objective of the current work is to use DNS to investigate heat release effects on mixture fraction statistics relevant to the modeling of turbulence/chemistry interaction in a prototypical shear flow. Unlike other recent DNS studies concerned with models of turbulence/chemistry interaction, realistically large heat release with ratio of maximum to minimum density up to 8 is considered. Also, large computational grids are used so as to achieve Reynolds numbers sufficiently large to establish full-blown turbulent flow.

2. Formulation

A single-step, infinitely-fast, irreversible reaction is assumed,

$$CH_4 + 2O_2 \rightarrow CO_2 + 2H_2O\,. \tag{11.1}$$

While recognizing that finite rate effects may be important in practice, it is expected that such effects are not relevant to the objective of the current study ie; the investigation of heat release effects on scalar mixing. An equation of state corresponding to an ideal gas is assumed. Molecular transport coefficients of mass, momentum, and temperature are assumed constant and equal; $\rho D = \mu = \rho\alpha$. The specific heat is assumed to be a linear function of temperature.

The mixture fraction, ζ, is defined so that the value $\zeta = 0$ corresponds to the oxidizer stream while $\zeta = 1$ corresponds to the fuel stream. The value of the stoichiometric mixture fraction, $\zeta_s = 0.2$, is fixed by diluting the fuel stream with N_2. The species mass fractions are related to the mixture fraction by piecewise-linear functions, $Y_i^e(\zeta)$, as shown in Fig. 11.1(a). Instead of assuming that the density (temperature) is also determined solely by the mixture fraction, the full energy equation is solved. The scatterplot, Fig. 11.1(b), shows that there is some deviation with respect to the Burke-Schumann solution (curve around which the points are scattered). The Burke-Schumann solution is curved because the specific heat is not constant.

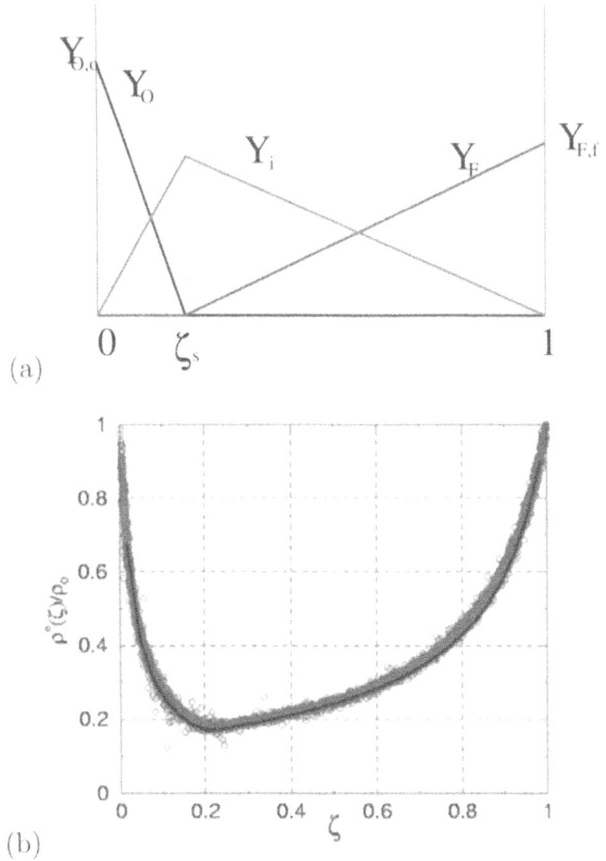

Figure 11.1. (a): Burke-Schumann solution for the mass fractions. (b): The dependence of density on mixture fraction

The three-dimensional, unsteady compressible Navier-Stokes equations are numerically solved using a sixth-order compact scheme for spatial derivatives, and a fourth-order, low storage Runge-Kutta scheme for the time advancement. Three simulations are performed with different values of the heat release parameter, $Ce \simeq T_{ad}/T_0 - 1$, where T_{ad} is the adiabatic flame temperature and T_0 is the temperature at the free streams. The choice $Ce = 7$ corresponds to conditions corresponding to the chosen methane/air stoichiometry, $Ce = 3.5$ corresponds to imposing partial heat release, and $Ce = 0$ to no heat release. The cases with heat release (active scalar mixing) are performed on a $768 \times 258 \times 192$ grid. The large grids allow high accuracy, $\Delta x \simeq 2\eta$, where η is the Kolmogorov scale as well as large evolution times required to achieve approximate selfsimilarity. The Reynolds number based on vorticity thickness evolves to approximately 10,000 and the microscale Reynolds number, $Re_\lambda \simeq 100$. Velocity and scalar fields exhibit broadband spectra, strong three-dimensionality and inviscid scaling of the turbulent dissipation rate, all indicators of full-blown turbulence.

For the fast-chemistry assumptions made in the current formulation, an expression for the average overall reaction rate has been derived [Bilger, 1980] as summarized below. The instantaneous reaction rate is

$$\dot{w}_i = -\rho D \nabla\zeta \cdot \nabla\zeta \frac{d^2 Y_i^e}{d\zeta^2} = -\frac{\rho\chi}{2}\frac{d^2 Y_i^e}{d\zeta^2}\,. \tag{11.2}$$

which, upon averaging in the selfsimilarly evolving shear layer, leads to

$$\overline{\dot{w}_i}(x_2) = -\frac{1}{2} Y_B < \rho\chi|\zeta_s; x_2) > P(\zeta_s; x_2) \tag{11.3}$$

where x_2 is the cross-stream coordinate normalized by the shear layer thickness and Y_B is a constant that depends on the stoichiometry. If ρ is a unique function of mixture fraction ie; $\rho = \rho^e(\zeta)$, exact for zero Mach number but only approximate in our case, a further simplification can be made,

$$\overline{\dot{w}_i}(x_2) = -\frac{1}{2}\overline{\rho} Y_B < \chi|\zeta_s; x_2 > \widetilde{P}(\zeta = \zeta_s; x_2) \tag{11.4}$$

where the Favre pdf is given by

$$\begin{aligned} \widetilde{P}(\zeta; x_2) &= \frac{1}{\overline{\rho}(x_2)} \int_0^\infty \rho P(\rho, \zeta; x_2)\, d\rho \\ &= \frac{\rho^e(\zeta)}{\overline{\rho}(x_2)} P(\zeta; x_2)\,. \end{aligned} \tag{11.5}$$

3. The density field

A longitudinal snapshot of the density field is shown in Fig. 11.2. The *mean* location of the flame sheet is displaced upwards to the air side at

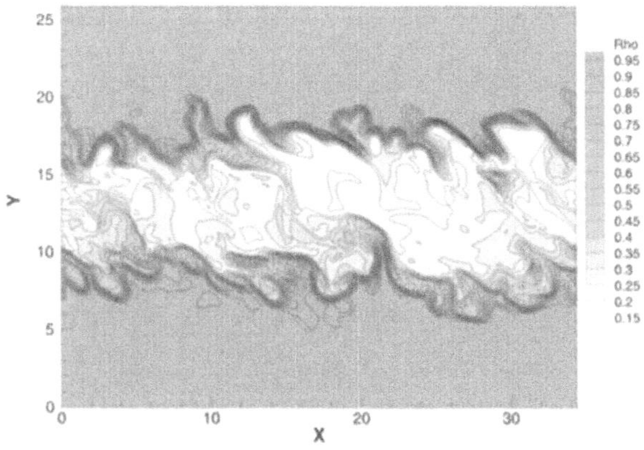

Figure 11.2. Instantaneous density visualization in a longitudinal plane

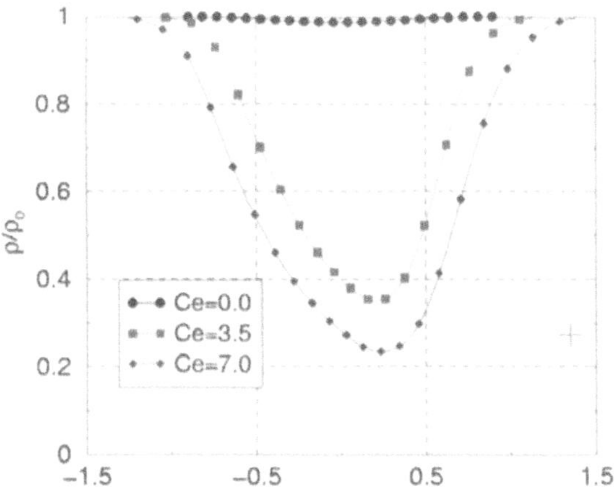

Figure 11.3. The mean density profile.

$y = 15.28 = 0.31\delta_\omega$ in the figure. However, the convective stirring of the flame sheet by the turbulent motion spans the entire width of the shear layer so that an instantaneous snapshot such as Fig. 11.2 shows a *wide* central core of hot, low-density fluid separated from the cold, high-density fluid on either side by thin regions with large values of the density gradient.

Figure 11.3 shows the average density for the different simulations as a function of the self-similar coordinate, η normalized by the vorticity thickness, $\delta_\omega = \Delta u / |\frac{\partial \tilde{u}_1}{\partial x_2}|_{max}$. The minimum density reaches 0.35

for $Ce = 3.5$ and 0.22 for $Ce = 7$. The mean density profile is somewhat flat near its minimum value corresponding to the central region of well-mixed, high-temperature fluid seen in the instantaneous snapshot, Fig. 11.2.

4. Conditional Scalar Dissipation

The conditional scalar dissipation which has units of inverse time characterizes the mixing rate at the flame surface and appears in the expression for overall reaction rate, Eq. (11.4).

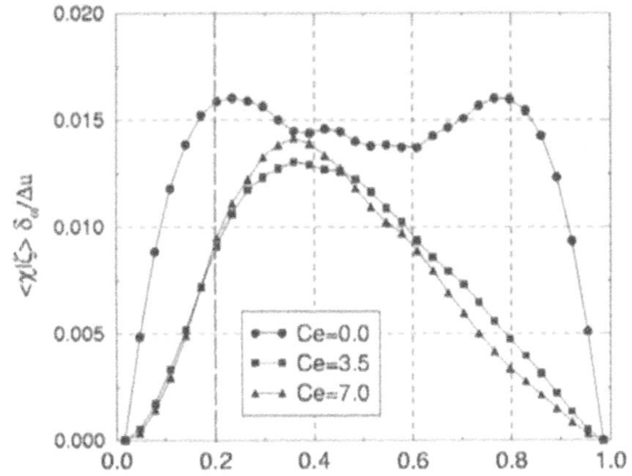

(a)

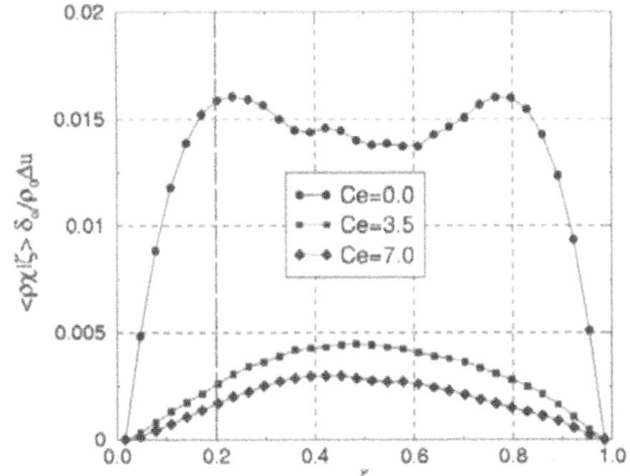

(b)

Figure 11.4. (a): Conditional mean of scalar dissipation, $< \chi|\zeta >$, which is an inverse time scale used to characterize mixing rate. (b): The conditional mean $< \rho\chi|\zeta >$.

It is observed in the DNS that the functional dependence of $<\chi|\zeta;x_2>$ on ζ does not vary much across the shear layer in the region $-\delta_\omega < x_2 < \delta_\omega$ and therefore, in order to improve statistical convergence, this entire region is used to calculate the conditional scalar dissipation, $<\chi|\zeta>$, now assumed independent of x_2. The observed insensitivity of conditional dissipation to the transverse coordinate is in accord with the expectation [Bilger, 1993] in shear flows. Figure 11.4(a) shows profiles, normalized using outer flow scales, of $<\chi|\zeta>$ in the approximately selfsimilar state. The peak value of the conditional scalar dissipation changes little with heat release. However, the shape is qualitatively different. The case with $Ce = 0$ exhibits values of $<\chi|\zeta>$ that are almost independent of the scalar value in the central portion of the curve. Such behavior is similar to that in homogeneous isotropic turbulence where the scalar gradient is found to be independent of the scalar value. The scalar dissipation decreases to zero at the limiting values of $\zeta = 0$ and $\zeta = 1$ which correspond to freestream conditions. In contrast to the case with $Ce = 0$, the cases with heat release exhibit a strong dependence on the scalar value and, instead of a central plateau, show a well-defined peak. The position of the peak shifts to a location between the center, $\zeta = 0.5$, and the stoichiometric value.

The quantity that directly appears in the reaction rate expression, Eq. (11.3), is $<\rho\chi|\zeta>$, shown in Fig. 11.4(b). The peak value$<\rho\chi>_p$ decreases with heat release; however, its location remains at $\zeta \simeq 0.5$.

5. Mixture fraction pdf

The mixture fraction pdf in the shear layer evolves to a self-similar profile, $P(\zeta;x_2/\delta_\omega)$ which varies strongly as a function of the cross-stream coordinate. Heat release is observed to cause qualitative differences in the pdfs.

Fig. 11.5 shows pdf's in the interior of the shear layer for a case with heat release parameter, $Ce = 7$. The variation of the pdf in the x_2 direction shows the so-called *marching* behavior ie; the value of ζ_p at which the pdf has a peak is coincident with the mean value, $\overline{\zeta}$, and therefore progressively marches over the range $0 < \zeta_p \simeq \overline{\zeta} < 1$ when the shear layer is traversed. Furthermore, the pdfs have a single peak without auxiliary peaks of free-stream fluid, suggesting classical small-scale mixing. In contrast, the pdfs in the nonreacting case (not shown) show dominant peaks of free-stream fluid in addition to a local peak of mixed fluid suggesting that large-scale coherent structures enable the penetration of freestream fluid into the core of the mixing layer. Also, the marching behavior of the mixed fluid peaks is less clear in the nonreacting case.

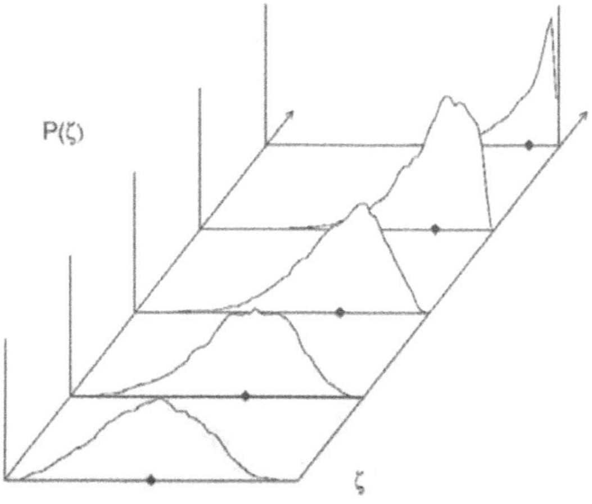

Figure 11.5. Mixture fraction pdfs at five cross-stream locations in a case with heat release parameter, $Ce = 7$. The locations, going from bottom to top in the figure, correspond to $\bar{\zeta} = 0.5$(centerline), $0.6, 0.7, 0.8$ and 0.9 (towards fuel freestream), respectively.

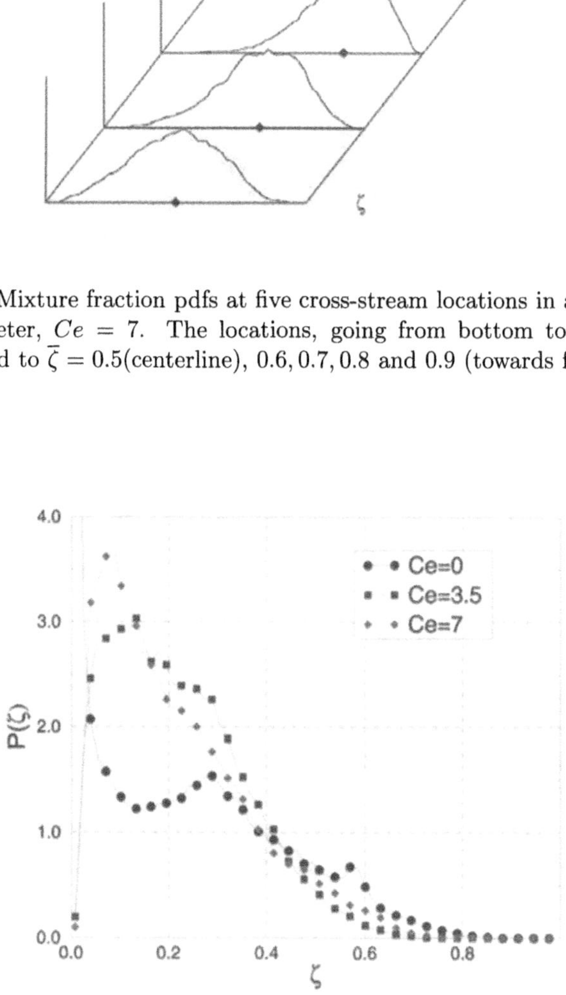

Figure 11.6. Scalar pdf at a cross stream location where the mean mass fraction is stoichiometric, $\bar{\zeta} = 0.2$.

The probability of stoichiometric mixture fraction, $P(\zeta = \zeta_s; x_2/\delta_\omega)$, is required to evaluate the overall reaction rate at any given cross-stream location. Because of the qualitative change in scalar pdfs due to heat release, values of $P(\zeta = \zeta_s; x_2)$ may be expected to show quantitative differences. Fig. 11.6 compares profiles of the scalar pdf in the three cases at the cross-stream location corresponding to stoichiometric mean value, $\overline{\zeta} = 0.2$. It is clear that the probability of stoichiometric composition, $P(\zeta = \zeta_s = 0.2)$ is much higher at that location in the cases with heat release.

6. Overall reaction rate

The overall reaction rate is calculated using Eq. (11.3). As discussed in the previous section, heat release affects both the conditional scalar dissipation and the scalar pdf and, therefore, the overall reaction rate may also be expected to be influenced by heat release.

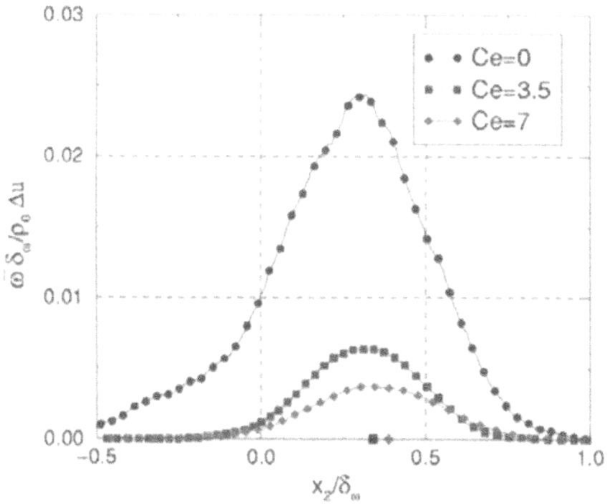

(a)

Fig. 11.7(a) shows that the reaction rate decreases strongly when heat release is present. Normalization by the mean density variation, $\overline{\rho}(x_2)$, reduces the differences between the three cases; nevertheless, such a normalization does not collapse the behavior of all the cases into a single curve.

7. Conclusions

Three-dimensional simulations of nonpremixed combustion in a shear layer between diluted methane and air have been performed to study the interaction between heat release and scalar (mixture fraction) mixing. Various values of heat release are considered in a range between zero and a large value corresponding to the ratio between maximum and minimum values of density being 8. Fast chemistry is assumed. The Reynolds number is sufficiently large, $Re_\lambda \simeq 100$, to obtain full-

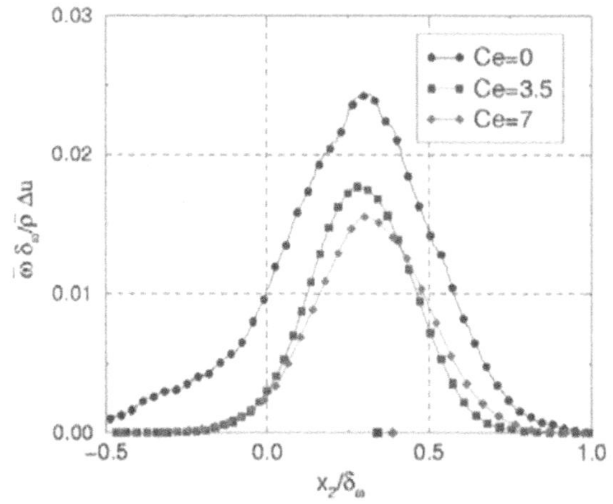

(b)

Figure 11.7. (a): The cross-stream variation of the average reaction rate. (b): The effect of normalizing the average reaction rate with the average density.

blown turbulence with approximately self-similar turbulence statistics. With increasing heat release, the pdf of the mixture fraction shows less evidence of external intermittency associated with coherent structures. The conditional scalar dissipation is also qualitatively different in the presence of heat release. Consequently, the overall reaction rate per unit mass in the combustion problem considered here is substantially less than thatobtained by ignoring the effect of density and temperature changes on the flow field and associated mixing.

Acknowledgments

The support of AFOSR through grant F49620-96-1-0106 is gratefully acknowledged. Computational time at the NAVO major shared resource center was provided through an HPCMP allocation.

References

Bilger, R. W., (1980) *Turbulent Reacting Flows*, P. A. Libby and F. A. Williams (eds.),65–113, Springer-Verlag.

Bilger, R. W., (1993) *Phys. Fluids A*, **5** 436–444.

Hermanson, J. C. and Dimotakis, P. E., (1989) *J. Fluid Mech.*, **199**, 333-375.

Klimenko, A. Y., (1990) *Fluid Dyn.*, **25**, 327.

(1989) McMurtry, P. A., Riley, J. J., and Metcalfe, R. W., *J. Fluid Mech.*, **199**, 297-332.

Miller, M. F., Bowman, C. T., and Mungal, M. G., (1994) *J. Fluid Mech.*, **356**, 25-64.
Miller, R. S., Madnia, C. K., Givi, P., (1994) *Combust. Sci. and Tech.*, 1994, **99**, 1-36.
Mungal, M. G., and Dimotakis, P. E, (1984) *J. Fluid Mech.*, **148**, 349-382.
Peters, N., (1986) *Twentyfirst Symp. (Intl.) on Combustion*, 1231-1250.
Pitsch, H., Chen, M., and Peters, N. (1998) *Twentyseventh Symposium (Intl.) on Combustion*, 1057-1064.
Williams, F. A. (1985) *Combustion Theory*, Addison Wesley.

Discussion

Paul Dimotakis, CalTech "In your simulations, the mixing seems to be relatively diffuse so you never find fluid from the two sides coming together and I am not surprised by the low values of scalar dissipation rates that you find, because it seems that your simulation does not admit the sort of frequent event, that I showed and happens quite often and is responsible for the very high values in the scalar dissipation rate. I was wondering if you compared the absolute value of the total mixed fluid to what has been measured under comparable conditions?"

Reply: No, I did not do that, but what we did find is that if you kept increasing the Reynolds number although the mean scalar dissipation remains the same, the largest values of the scalar dissipation increase and you can obtain them by fitting a log-normal PDF to the dissipation. If you can simulate and jack up the Reynolds number by a factor of 5 say, I don't think the average dissipation will change it will have the same inviscid scaling for Schmidt number of order 1; but larger values of maximum scalar dissipation but the probability of those events is small.

Paul Dimotakis, CalTech "My second question, we have done experiments with heat release that is as small as 3.5, we can do with hydrogen-fluorine. We do not find that overall rate of chemical reaction goes down and our explanation is that this is a process that is mixing limited for fast kinetics and therefore the fact that the kinetics have gone down is irrelevant since you are limited by entrainment and this is very weakly dependent on dilatation, so I was wondering if you compared your result with experiment?"

Reply: We compared the overall growth rate with your experiments and noting the fact that we are using a temporally evolving shear layer the agreement is fair. The overall production rate is this and the scalar dissipation rate is different, and there seems to be a heat release effect on the conditional scalar dissipation, and the PDF of the scalar seems to be different.

E. Villermaux, Marseilles "Generally I find your idea of computing L_r the thickness of the gradient close to the flame very intelligent and interesting and the result all the more interesting. The smallest value you compute the PDF, the Batchelor scale, the most probable value is much larger than this scale, as you showed, and it turns out to be of order of the Taylor microscale, but the Taylor microscale depends on the Reynolds number, so would you go as far as saying that this scale L_r the most probable one, is independent of Reynolds number or you don't have evidence of this..."

Reply: This is one of the nice things about DNS, in that we can't get to high enough Reynolds numbers so this is up to the experimentalists.

Norbert Peters, Stanford University "There was a strong effect of heat release on this linear region. By a factor of 2 maybe. Have you any explanation for that. I would have expected the other way around. I would have expected you to have gas expansion of the flame that the flame itself would generate a larger region in its vicinity where linear gradients would occur but you don't have any explanation for this."

Reply: We would have expected exactly what you expected and this is one of the reasons why we are redoing the calculations with the higher resolution just to make sure that things are fine.

Tadeo Takeno, Japan "I wonder if you have studied the effects of increase in transport coefficients due to temperature increase?"

Reply: In this case what we did was so as not to corrupt our simulations with low Reynolds number low Reynolds Schmidt, low Peclet numbers, we actually kept the viscosity constant. So the molecular coefficients are constant and not a function temperature.

A. Pollard, Queen's University "You have done a careful increase in the resolution of the simulation. Have you changed the size of the box?"

Reply: Yes we have. Essentially what happens is if you increase the lengths scale of the box so that you are always larger than some multiple of the integral lengths scale your statistics are invariant but if your length scale is smaller than that multiple of integral lengths scale then your statistics are not invariant.

I. Fedioun, Orleans, France "How can you reach 8000 Reynolds number with a resolution of 256?"

Reply: Delta X is approximately 4 times the Kolmogorov scale in this particular simulation. What we are trying to do in our new simulations is going to something like 2.5 x this scale. These are L/η increases with time because the Reynolds number is increasing.

Chapter 12

INFLUENCE OF DIFFERENTIAL DIFFUSION ON LOCAL EQUILIBRIUM AND SUPER-EQUILIBRIUM COMBUSTION IN TURBULENT NON-PREMIXED FLAMES

Olivier Gicquel
Ecole Centrale Paris/CNRS France
gicquel@em2c.ecp.fr

Renan Hilbert
Ecole Centrale Paris/CNRS France
hilbert@em2c.ecp.fr

Dominique Thévenin
Ecole Centrale Paris/CNRS France
thevenin@em2c.ecp.fr

Nasser Darabiha
Ecole Centrale Paris/CNRS France
darabiha@em2c.ecp.fr

Abstract Turbulent combustion can lead to very specific phenomena. One of them is called super-equilibrium combustion. In many applications, authors have indeed seen temperatures higher than those obtained by thermodynamic equilibrium. In the present work, this effect is studied for diluted hydrogen/air flames using DNS computations taking into account differential diffusion. The results show that transport phenomena lead to local accumulations of reactants or dilutants. These changes in the relative composition of the mixture induce large modifications of its thermodynamic properties. It becomes therefore possible to obtain tem-

A. Pollard and S. Candel (eds.), IUTAM Symposium on Turbulent Mixing and Combustion, 149-159.

peratures higher than those computed by neglecting these modifications of the local dilution rate.

Keywords: Detailed chemistry, DNS, Super-Equilibrium, Turbulence

Introduction

Although tremendous improvements have already been obtained in turbulent combustion modelling, the analysis of turbulent combustion phenomena either by experiments or numerical tools remains a major issue [Bilger, 2000]. Many authors have studied turbulent combustion using high-performance experimental techniques [Stepowski, 1992]. These results generally show a large scatter of the measured points when plotting the variation of the different parameters, such as species concentration or temperature, as a function of the mixture fraction Z. They also show that for a given mixture fraction of fuel and oxidizer, the local conditions of temperature and concentration of burned gases may reach values that can be beyond those that one may expect from the local value of Z. Authors generally call this condition super-adiabatic or super-equilibrium combustion [Vervisch and Poinsot, 1998,Barlow et al., 1990]. The objective of the present study is to analyze and understand the conditions for which such local super-equilibrium phenomena may take place. This analysis will put into evidence the physical process leading to this situation.

Description of the numerical simulation

In order to study the super-equilibrium combustion, results obtained from Direct Numerical Simulations [Thévenin et al., 1997] are used for diluted hydrogen/air non-premixed flames in a turbulent field taking into account detailed chemistry (9 chemical species and 37 reactions, [Warnatz, 1989]) and multicomponent transport velocities [Hirschfelder et al., 1954]. These simulations were performed with a diluted hydrogen mixture including 75% (in volume) of nitrogen at 300 K on one side and air at 300 K on the other side, leading to a global mixture ratio equal to 0.8. The Reynolds number of the initial turbulent field is equal to 200, based on the integral length-scale. The flame is ignited using a source term in the energy equation during the first time-steps of the computation. A typical result is shown in Figure 12.1 corresponding to the instantaneous field of OH after 2 eddy turn-over times [Hilbert and Thévenin, 2001].

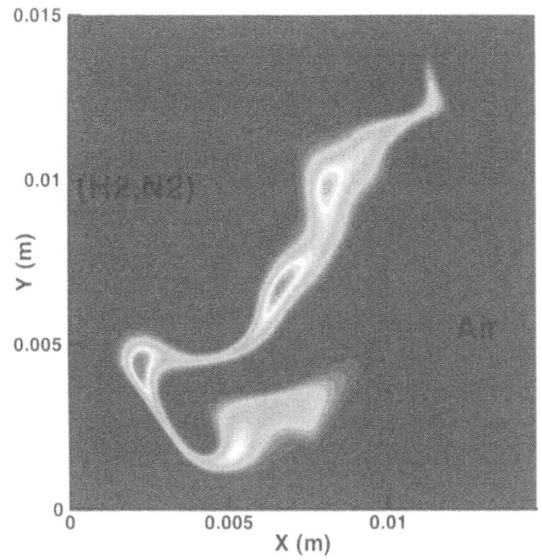

Figure 12.1. DNS results - instantaneous OH field (after 2 eddy turn-over times)

Global variable definitions

As DNS simulations give full access to the flame structure (species concentrations and temperature), it is necessary to introduce some global variables to post-process the results. Here, results are presented as a function of the mixture fraction Z. To take into account the decomposition of the main species into radicals, the definition of Z is based on element mass fractions of O and H as follows:

$$Z = \frac{\frac{\phi Y_H}{Y_H^\infty} - \frac{Y_O}{Y_O^\infty} + 1}{\phi + 1} \text{ with } \phi = \frac{sY_H^\infty}{Y_O^\infty} \tag{12.1}$$

where ϕ is the global mixture ratio, s is the mass stoichiometric coefficient and Y_H^∞ and Y_O^∞ are the element mass fractions of the fuel and oxidizer sides respectively.

The dilution of the mixture D, based on element mole numbers of O, H and N is introduced:

$$D = \frac{n_N}{n_N + n_O + n_H} \tag{12.2}$$

For instance, D is equal to 75% in diluted hydrogen and 79% in pure air. This parameter D quantifies differential diffusion, since it incorporates a direct comparison of diffusion velocities for all the elements in the system.

Definition of the equilibrium

In the literature, the scatter plots of the experimental results are associated with equilibrium curves [Barlow et al., 1990]. These curves represent the thermodynamic equilibrium based on the local value of Z, which is obtained assuming a perfect mixing between oxidizer and fuel. Here, this definition is referred to as *global equilibrium.*

As differential diffusion can lead to different element compositions for a same value of Z, the global equilibrium is not uniquely related to the local mixture. That is why the *local equilibrium* is introduced, which is based on the local composition and temperature of the mixture.

Results

In Fig 12.2, a scatter plot is given of the instantaneous local temperature as a function of the local value of the mixture fraction Z. For each value of Z, the global equilibrium temperature is plotted. The results

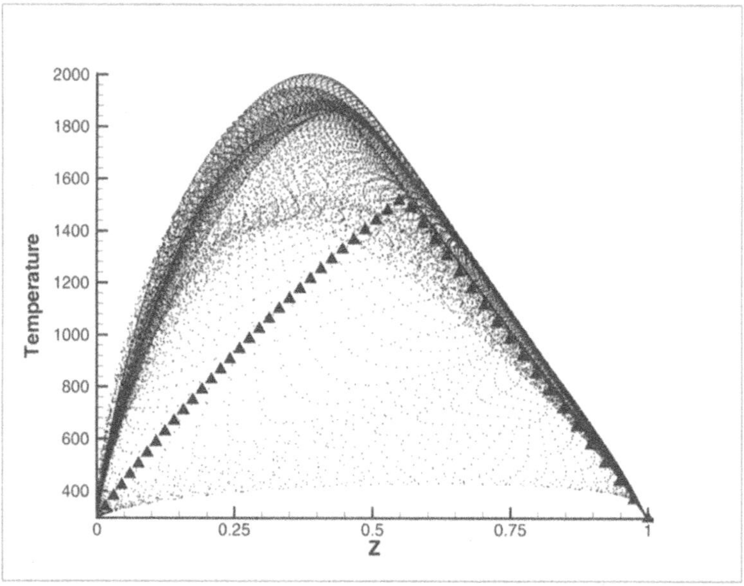

Figure 12.2. Scatter plot of temperature from DNS results. The triangular symbols represent global equilibrium temperature.

clearly show that the global thermodynamic equilibrium temperature is about 500 K lower than the temperatures obtained by DNS. This

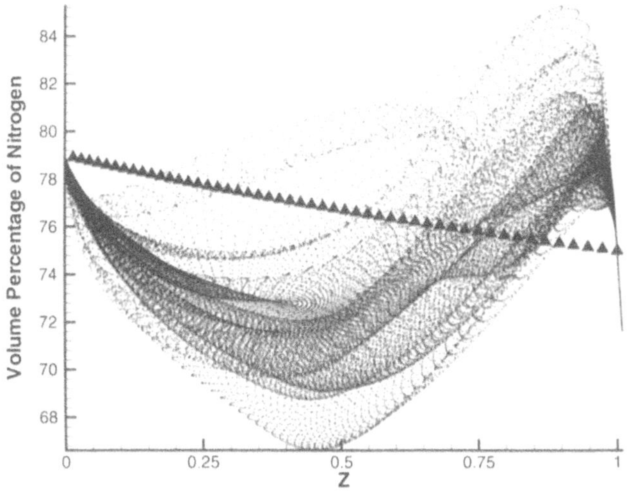

Figure 12.3. Scatter plot of the dilution D. The triangular symbols represent D as it would be obtained for perfect mixing.

phenomena is essentially due to differential diffusion of species, which results in a change in the local species distribution and consequently equivalence ratio and dilution.

In Fig 12.3, a scatter plot of the dilution D is given as a function of Z. The dilution is plotted, which corresponds to a mixing without differential diffusion (*perfect mixing*) between diluted hydrogen and pure air. It can be observed that values of D obtained with DNS are widely scattered as well above as below this last curve. In particular for low values of Z, the DNS results are always much below the perfect mixing curve. The minimum of this plot, corresponding to $Z = 0.45$ and $D = 66.5$, is associated with a point where differential diffusion has led to a considerable accumulation of reactive species in comparison with inert nitrogen N_2.

Considering these results, Fig 12.4 and 12.5 plots the turbulent flame temperature and the dilution D in the physical coordinates (X,Y). It is seen that high temperature zones are strongly correlated with the regions of low D. This is in particular due to the fact that in these regions, there is less inert species to be heated, while in the high D regions, a large part of the energy is absorbed to heat nitrogen. Simultaneously the heat released by the chemical reaction is locally reduced due to the lower available quantities of fuel and oxidizer in these last regions. As already mentioned these variations of dilution are due to differential diffusion.

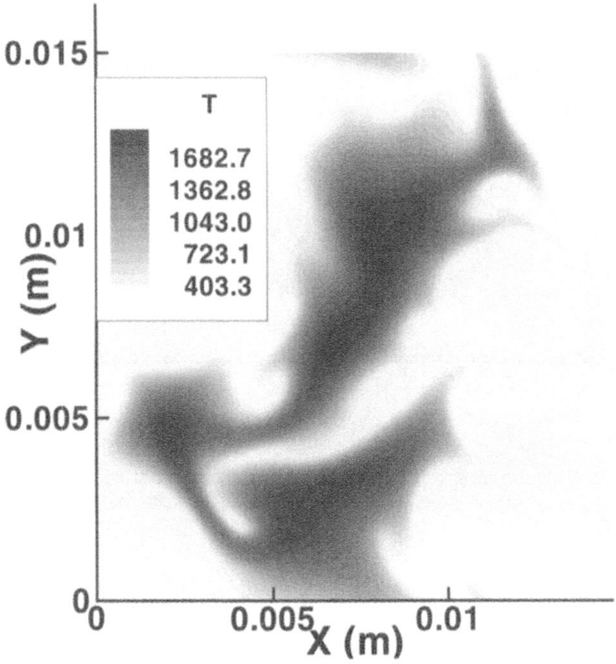

Figure 12.4. DNS results - instantaneous temperature field

The stoichiometric isoline, $Z = Z_{st}$ was extracted to check the correlation between the temperature and the dilution along the flame front. The results are presented in Fig. 12.6. The high values of temperature along the isoline are correlated to low values of the local dilution. Figure 12.7 confirms the strong correlation between the temperature and the dilution.

The exact composition of the local mixture is known since detailed chemistry is used. It is then possible to determine the local equilibrium associated to any point of the computational domain. Fig 12.8 provides the temperature along the stoichiometric isoline and the corresponding local equilibrium temperature. The local temperature is always lower than the equilibrium one. This clearly shows that the so-called super-equilibrium regime is just due in this case to an inappropriate picture of the phenomena, where differential diffusion has been neglected.

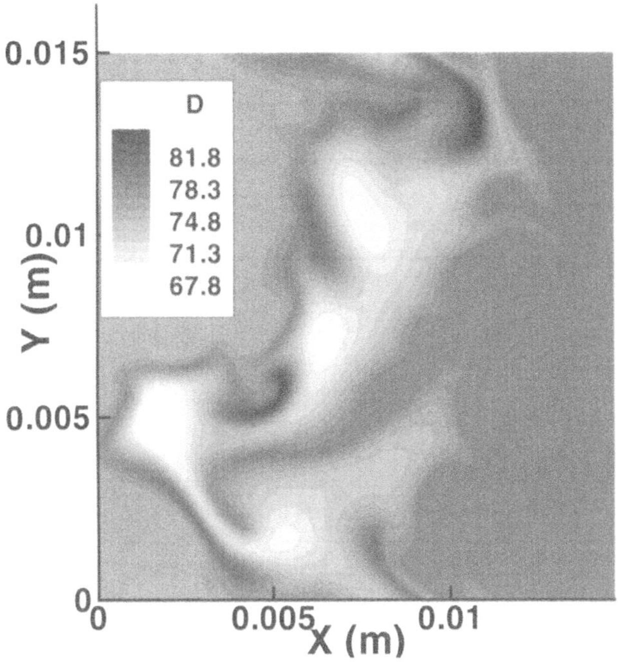

Figure 12.5. DNS results - instantaneous dilution D as defined by eq. (12.2)

Conclusion

Influence of differential diffusion on chemical equilibrium has been shown for a diluted hydrogen/air diffusion flame. The results show that differential diffusion induces a change in the local dilution compared to perfect mixing. This change leads to strong variations of the thermodynamic equilibrium associated to the local composition. Then temperatures are found, which are higher than the ones that could be obtained using a perfect mixing process. But these temperature always remain lower than the local equilibrium ones. Therefore the accuracy of the transport models is at least as important as employing detailed chemical mechanisms to describe correctly the turbulent flame structure. Super-equilibrium can, for this (H_2/N_2)/air flame, easily be understood if differential diffusion is taken into account by using an appropriate definition for Z and by introducing the dilution as a second parameter. Further investigations must be performed for more complex fuels and more turbulent flames to check the effects of differential diffusion in more realistic cases.

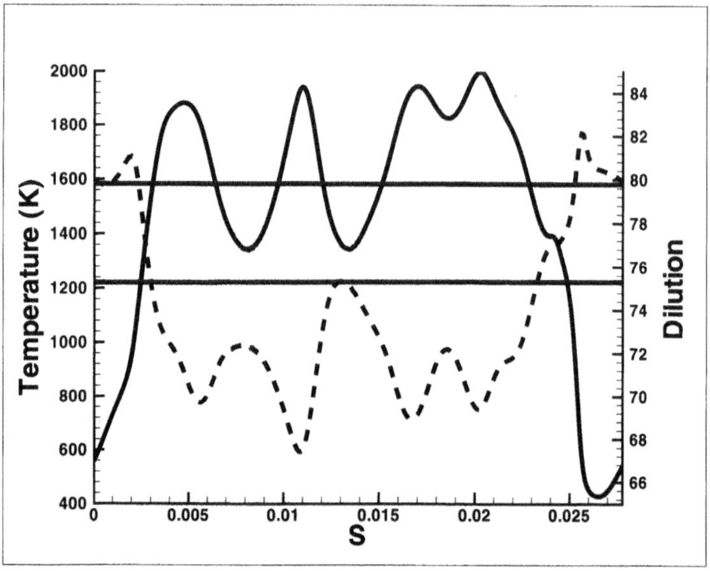

Figure 12.6. Evolution of temperature (solid line) and of dilution (dashed line) along the stoichiometric isoline. The two horizontal lines represent the dilution of the unburned reactants.

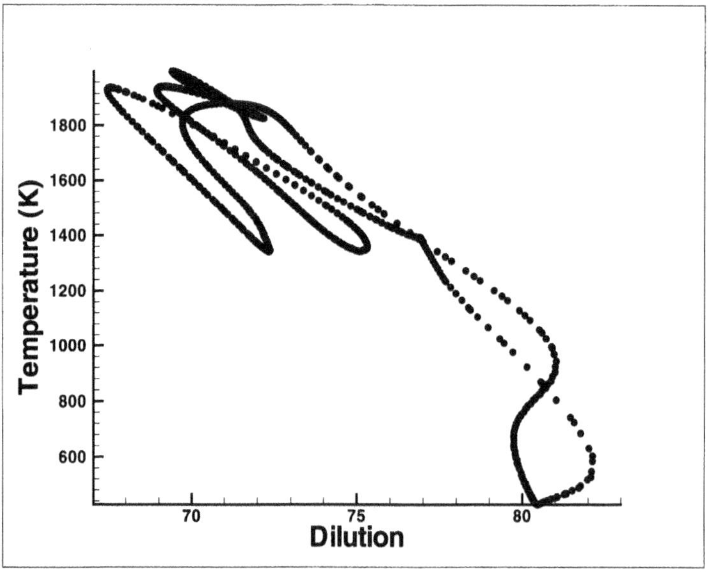

Figure 12.7. Correlation between temperature T and dilution D along the stoichiometric isoline.

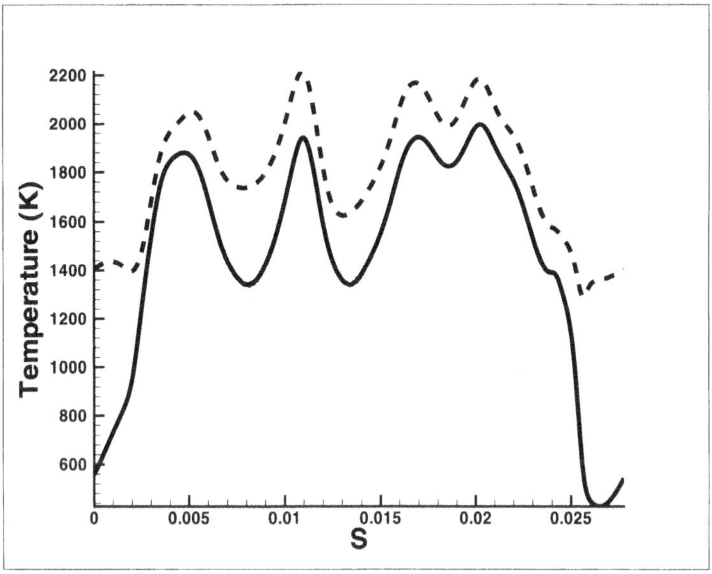

Figure 12.8. Comparison between the local temperature (solid line) and the local equilibrium temperature (dashed line) along the stoichiometric isoline.

References

Barlow, R. S., Dibble, R. W., Chen, J.-Y., and Lucht, R. P. (1990). Effect of Damköhler number on superequilibrium OH concentration in turbulent nonpremixed jet flames. *Combust. Flame*, 82(3):235–251.

Bilger, R. W. (2000). Future progress in turbulent combustion research. *Prog. Energ. Combust. Sci.*, 26:367–380.

Hilbert, R. and Thévenin, D. (2001). Nonpremixed turbulent flames investigated with direct numerical simulation using detailed chemistry. In *IUTAM symposium on turbulent mixing and combustion*, Eds: Pollard & Candel, pp. 251–258. Kingston, (Canada).

Hirschfelder, J. O., Curtis, C. F., and Bird, R. B. (1954). *Molecular Theory of Gases and Liquids*. John Wiley & Sons.

Stepowski, D. (1992). Laser measurements of scalars in turbulent diffusion flames. *Prog. Energ. Combust. Sci.*, 18:463–491.

Thévenin, D., van Kalmthout, E., and Candel, S. (1997). Two-dimensional direct numerical simulations of turbulent diffusion flames using detailed chemistry. In J.P.Chollet, Voke, P., and Kleiser, L., editors, *Direct and Large Eddy Simulation II*, pages 343–354. Kluwer Academic Publishers.

Vervisch, L. and Poinsot, T. (1998). Direct numerical simulation of non premixed turbulent flames. *Ann. Rev. Fluid Mech.*, 30:655–692.

Warnatz, J. (1989). Numerical simulation of ignition processes. In Dervieux, A. and Larrouturou, B., editors, *Numerical Combustion*, volume 351 of *Lecture Notes in Physics*. Springer-Verlag, New York, NY.

Discussion

Norbert Peters, Stanford University "In theory, differential diffusion effects are always associated with stretching and curvature effects and this is clear in premixed flames. We have recently done a similar theory for non-premixed flame and from that theory there is a possibility we have super-equilibrium temperatures, it just depends on the sign of the curvature. So the question that arises is what is, well the mean curvature is zero, of course, so what is the probability of larger or smaller curvatures in such a flame as you have?"

Reply: I don't know, we did the simulations and we wanted to develop a chemical model for turbulent combustion. We show that we don't have to compute super-adiabatic combustion maybe in some cases super-adiabatic combustion appears. We had a simulation in a methane-air flame and we saw some super-adiabatic combustion, you pick a point in the flame, and you compute the local equilibrium and the temperature but we found this in the rich part of the domain and locally we saw there is no chemical effect and it is only mixing of hot inert gases.

Josette Bellan, JPL, Pasadena "I am intrigued how you calculated the diffusion coefficients for a multicomponent mixture?"

Reply: It is a classical method...

Josette Bellan, JPL, Pasadena "No, there is no classical method, and in fact you can compute for a binary mixture, but not for a multi-component mixture. In fact you have to calculate what it is for a component into a mixture of several components and if you use a 13 moment theory that will not give you what is the diffusion of one component into a mixture of several other components, certainly not as a function of temperature and every thing else..."

Reply: **D. Thevenin (co-author of the paper, replies)** It is true this is a complex question, but what we use is basically transport (a computer package similar to Chemkin Transport, Sandia Labs) there is an option for multicomponent velocities are computed separately for each species... but you can do something else.

Tadeo Takino, Japan "I think you are using detailed kinetics then I think that it is not hydrogen concentration that govern the heat release, you have to take into account the dilution of OH or H to study..."

Reply: We take into account the local dilution, so we account for the local dilution of OH on the species... I don't see the point of your question.

Tadeo Takino, Japan "Because the detailed kinetics, it is not the hydrogen consumption that determines the heat release I think the effect of dilution has effect on the hydrogen atom OH concentration"

Reply: Yes, of course, when we look at the dilution, how many inert species do you have that can keep the energy to heat and thus you reduce the temperature of the domain so does energy come from OH or H it always goes to N2 so yes we account for all the species. When I speak about dilution, because if you want dilution based on mole fraction and for the mole fraction domain you have many species and if you have H or 2H so to be coherent in the mole domain to compare you have to come back to a simple case which is equal for all the points to come back to the fresh gases.

Everything is changing in the mole domain so it is not easy. So when I define dilution, I come back to O2/N2 mixture, but of course locally we have all the species. But this is to have a common point in the entire domain.

Chapter 13

DIFFUSION EDGE-FLAME QUENCHING

Joan Boulanger and Luc Vervisch
Institut National des Sciences Appliquées de Rouen
UMR-CNRS-6614 / CORIA
Campus du Madrillet - Avenue de l'Université - BP 8
76801 Saint Etienne du Rouvray Cedex, France
boulanger@coria.fr - vervisch@coria.fr

Abstract The quenching properties of the reaction zone located at the edge of a burning diffusion flame (edge-flame) are studied. For large activation energy, a relation is established between a geometrical aspect ratio of the edge-flame and a quenching Damköhler number. This number characterizes the transition from burning to quenching observed in diffusion edge-flames and is expressed as a function of the quenching Damköhler of a one-dimensional flamelet.

Keywords: Combustion, diffusion flame, quenching, edge-flame, partial premixing.

Introduction

In many combustion systems, fuel and oxidizer are injected separately and combustion occurs in a flow where the reactants are nonpremixed at the molecular level [Bray, 1996]. In these turbulent flames, reaction zone extremities are observed in the vicinity of the flame stabilization point or, eventually, when local quenching occurs. Then, along the surface where the mixtures are in stoichiometric proportion, a transition from burning to quenching appears.

These burning extremities may evolve into partially premixed fronts accompanied by a trailing diffusion flame leading to triple flame combustion [Phillips, 1965, Dold, 1989, Kioni et al., 1993, Ghosal and Vervisch, 2000]. However, when one of the burning extremity undergoes a high rate of strain, diffusion edge-flame quenching is observed [Buckmaster, 1996, Shay and Ronney, 1998, Favier and Vervisch, 2001].

A. Pollard and S. Candel (eds.), IUTAM Symposium on Turbulent Mixing and Combustion, 161-168.

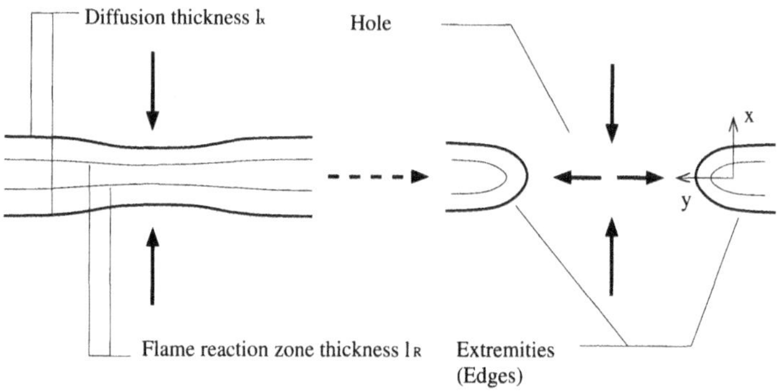

Figure 13.1. Schematic of the simplified problem. After straining a diffusion flame (left) a flame hole and edge-flames are formed (right). The quenching properties of the diffusion edge-flame are then studied (right).

Diffusion flames are usually parametrized in term of a Damköhler number, $Da = (\tau_\chi/\tau_c)$, ratio of a mechanical time τ_χ to a chemical time τ_c, where τ_χ is estimated from the scalar dissipation rate $\chi = (\lambda/\rho C_p)|\nabla Z|^2$ as $\tau_\chi = \chi^{-1}$. Z is the mixture fraction, a passive scalar with $Z = 0$ in the oxidizer stream and $Z = 1$ in the fuel stream. For a given chemistry and free stream conditions, Z and χ are the two control parameters of one-dimensional flamelets constructed in the direction normal to the stoichiometric surface [Peters, 2000].

It is intended in the following to derive a relation between the quenching Damköhler number of a diffusion edge-flame and a geometrical aspect ratio of the mixture fraction field bordering the reaction zone extremity.

1. Model Problem

Along a burning stoichiometric surface ($Z = Z_s$), we locate an extremity of a quenched diffusion flamelet at the origin of a coordinate system (x, y), x and y denote the normal and the tangential directions to the iso-Z_s (fig. 13.1). The primitive equations for temperature and species mass fraction may be written:

$$\rho C_p \left(\frac{\partial T}{\partial t} + \mathbf{u} \cdot \nabla T \right) = \nabla \cdot (\lambda \nabla T) + Q\rho\dot{\omega}_F \qquad (13.1)$$

$$\rho \left(\frac{\partial Y_i}{\partial t} + \mathbf{u} \cdot \nabla Y_i \right) = \nabla \cdot (\rho D_i \nabla Y_i) - \rho\dot{\omega}_i$$

where $i = F$ or $i = O$ denote either Fuel or Oxidizer, Q is the amount of heat released per unit of mass of fuel burnt, usual notations are otherwise adopted. A global single step chemistry, $\nu_F Y_F + \nu_O Y_O \longrightarrow \nu_P Y_P$ is assumed, with free stream conditions (Y_{F_0}, T_{F_0}) and (Y_{O_0}, T_{O_0}), ν_i are the molar stoichiometric coefficients. Close to the origin, the y axis is tangential to the stoichiometric surface that is not curved, then the stoichiometric scalar dissipation rate may be written $\chi_s = (\lambda/\rho C_p)|\nabla Z|_s^2 = (\lambda/\rho C_p)(\partial Z/\partial x)_s^2 = \chi_{xs}$. A normal to the flame characteristic length $l_x = (\partial Z/\partial x)^{-1} = \sqrt{(\lambda/\rho C_p)/\chi_{xs}}$ is introduced. The length l_x measures the local thickness of the mixing layer in the direction normal to the iso-Z_s surface. Accordingly, $l_y = f l_x$ is a characteristic thickness in the transverse direction y, where f is an edge-flame aspect ratio larger than unity.

From the characteristic thickness of the reaction zone l_R, a small parameter $\varepsilon = (l_R/l_x)$ may be defined and the flame solution is constructed from developments in the vicinity of the surface of stoichiometric mixture, where the quantity $(Z - Z_s)$ stays small. A reduced coordinate system is introduced, where $x^* = \varepsilon^{-1}(Z - Z_s)/(2Z_s(1 - Z_s))$ and $y^* = (y/\varepsilon l_y)$. Also, T^* is a normalized temperature so that, at $Z = Z_s$, $T^* = 1$ for infinitely fast chemistry and $T^* = 0$ for frozen flow mixing.

To study the transition from burning to quenching, the edge-flame is approached as a diffusion flame in the x direction to which additional fluxes of mass and heat are added in the y direction (fig. 13.1). In the developments discussed thereafter, T^* is a linear function of Z (and x^*) to which a small deviation of order ε is added. For $Z > Z_s$, the transverse diffusive budget may be approximated as $\left(\partial^2 T^*/\partial y^{*2}\right) \approx \varepsilon^2(T^*(Z) - T_o^*(Z)) \approx (1 - x^*(2Z_s\varepsilon))\left(\partial^2 T^*/\partial y^{*2}\right)_s$, where $T_o^*(Z) = (T_{F,o} - T_{O,o})Z + T_{O,o}$ denotes the frozen flow mixing temperature, the subscript $_s$ indicates a quantity measured at stoichiometry. Using boundary-layer derivation, the temperature balance equation may then be cast in:

$$\left(\frac{\partial^2 T^*}{\partial x^{*2}}\right) + \mathcal{F}^*(x^*) = -\left(\frac{gLe_F}{Y_{F,o}Z_s}\dot{\Omega}_F\right)\chi_x^{-1} \qquad (13.2)$$

Where $g = 4Z_s^2(1 - Z_s)^2$ and $\dot{\Omega}_F = \varepsilon^2 \dot{\omega}_F$ since the chemical source term is expected to be dominant. $\mathcal{F}^*(x^*)$ is a transverse (y) diffusive budget defined on the fuel side $(Z > Z_s)$ as:

$$\mathcal{F}^*(x^*) = gf^{-2}\left(\frac{\partial^2 T^*}{\partial y^{*2}}\right) \approx gf^{-2}(1 - x^*(2Z_s\varepsilon))\left(\frac{\partial^2 T^*}{\partial y^{*2}}\right)_s$$

The function $\mathcal{F}^*(x^*)$ is constructed so that it is maximum at the stoichiometric surface $(x^* = 0)$ and vanishes at the flame boundary conditions, especially when $Z \to 1$ then $x^* \to 1/2Z_s\varepsilon$. The solution of strained diffusion flames [Liñán, 1974] would correspond to the solution of Eq. 13.2 when $\mathcal{F}^* = 0$. Solution of Eq. 13.2 is now discussed for given values of f.

2. Solution

The general case of non-unity Lewis number is preserved using the generalized equivalence ratio [Liñán et al., 1994]:

$$\phi = \frac{\nu_O W_O Y_{F,o} Le_O}{\nu_F W_F Y_{O,o} Le_F} \tag{13.3}$$

W_i is the molar weight of the species i, Le_i its Lewis number. Then, the mixture fraction may be written:

$$Z = \frac{\phi Y_F / Y_{F,o} - Y_O / Y_{O,o} + 1}{1 + \phi} \tag{13.4}$$

with $Z_s = 1/(1+\phi)$.

In the limit of infinitely fast chemistry, piecewise relations exist between T, Y_F, Y_O and Z [Poinsot and Veynante, 2001]. In the flame featuring finite rate chemistry, the small departure from infinitely fast chemistry is measured using the small parameter $\varepsilon\Gamma$, where $\Gamma(x^*)$ is of the order of unity at $Z = Z_s$. Close to the stoichiometric condition and for $Z > Z_s$:

$$T^* = 1 - \varepsilon(\Gamma + (1+\gamma)x^*) \tag{13.5}$$

with $\gamma = 2(Z_s \Delta_{T_o} - 1)(1 - Z_s) + 1$, where one makes use of the normalized free stream temperature $\Delta_{T_o} = (T_{O,o} - T_{F,o})(Le_F C_p)/(Q Y_{F,o} Z_s)$. Eq. 13.5 also implies for the reactants correlated departures from their infinitely fast chemistry values. The burning rate may be written:

$$\dot{\omega}_F = Y_{F,o} B \left(Y_F^*\right)^{\nu_F} \left(Y_O^*\right)^{\nu_O} \exp\left(-T_a/T\right) \tag{13.6}$$

where $B = Y_{F,o} \nu_F \rho^{\nu_O + \nu_F - 1} \left(Y_{F,o}/W_F\right)^{\nu_F - 1} \left(Y_{O,o}/W_O\right)^{\nu_O} A$, in which A denotes the pre-exponential constant, T_a the activation temperature and $Y_i^* = (Y_i/Y_{i,o})$. $\beta = \alpha(T_a/T_s)$ is the Zel'dovitch number and $\alpha = (T_s - T_o(Z_s))/T_s$ a heat release factor. A Damköhler number may be expressed as [Liñán, 1974]:

$$Da = 4 Z_s^{\nu_F + \nu_O + 1} (1 - Z_s)^2 \phi^{\nu_O} Le_F B \exp(-\beta/\alpha)/\chi_x$$

Further analyzing the relations between temperature and reactants, and noting that the chemical source should be a leading order term, one finds well known relations:

$$\varepsilon^{\nu_O + \nu_F + 1} Da \approx 1 \quad \text{and} \quad \beta\varepsilon \approx 1 \tag{13.7}$$

Using Eq. 13.5 and the previous expressions, the reference balance equation Eq. 13.2 may be written on the rich side ($Z > Z_s$):

$$\frac{\partial^2 \Gamma}{\partial x^{*2}} = (\Gamma + 2x^*)^{\nu_F} (\Gamma)^{\nu_O} \exp\left(-\frac{\beta}{Da^{\frac{1}{\nu_F+\nu_O+1}}} ((1+\gamma)x^* + \Gamma) \right) + \frac{\mathcal{F}^*}{\varepsilon} \tag{13.8}$$

Eq. 13.8 is integrated over the reduced coordinate x^* on the rich (fuel) side. When $Z \to 1$, $x^* \approx 1/(2Z_s\varepsilon)$, thus for large Damköhler number $x^* \to \infty$, since from Eq. 13.7 when $Da \to \infty$, $\varepsilon \to 0$. Moreover $(\partial T^*/\partial x^*) \to -\varepsilon(1+\gamma)$. After integration, the balance equation becomes with $\mathcal{F} = \beta^2 g f^{-2} (\partial^2 T^* / \partial y^{*2})_s$:

$$(1+\gamma)\left(1 - \frac{\mathcal{F}}{4Z_s}\right) = \int_0^{+\infty} (\Gamma + 2x^*)^{\nu_F} (\Gamma)^{\nu_O} \exp(\psi(Da))\, dx^* \tag{13.9}$$

where $\psi(Da)$ is the exponential argument of Eq. 13.8. The integral in the RHS of Eq. 13.9 is a function of Da and we extract the minimum value of this Damköhler number for which the equality is verified. This value Da_q corresponds to quenching of the reaction zone. The edge-flame quenching Damköhler Da_q may then be written:

$$Da_q = Da_q^* \left(1 - \frac{\mathcal{F}}{4Z_s}\right) \tag{13.10}$$

In this relation, Da_q appears as a modification of Da_q^*, the value of the 1-D diffusion flamelet quenching Damköhler obtained when $\mathcal{F} = 0$. Da_q^* is a known function of the diffusion flame boundary conditions through γ [Liñán, 1974].

3. Discussion

In order to calibrate the impact of the mixture fraction field aspect ratio f on diffusion edge-flame quenching, we define the ratio:

$$\Delta_{Da} = \frac{Da_q - Da_q^*}{Da_q^*} = -\frac{\mathcal{F}}{4Z_s} \tag{13.11}$$

where one makes use of Eq. 13.10. The ratio Δ_{Da} is easily related to $\mathcal{R}$, the ratio of transverse and normal diffusive fluxes:

$$\mathcal{R} = \frac{(\lambda/\rho C_p)(\partial^2 T/\partial y^2)}{(\lambda/\rho C_p)(\partial^2 T/\partial x^2)} \approx \left(\frac{1}{f}\right)^2 \tag{13.12}$$

Using Eq. 13.8 and Eq. 13.7, the definition of T^*, x^* and Γ, one may write:

$$\frac{\lambda}{\rho C_p}\frac{\partial^2 T}{\partial x^2} = \chi_x \frac{\partial^2 T}{\partial Z^2} = -\left(\frac{Q Y_{F,o} Z_s}{Le_F C_p}\right) \frac{\beta}{4Z_s^2(1-Z_s)^2} \chi_x \frac{\partial^2 \Gamma}{\partial x^{*2}} \tag{13.13}$$

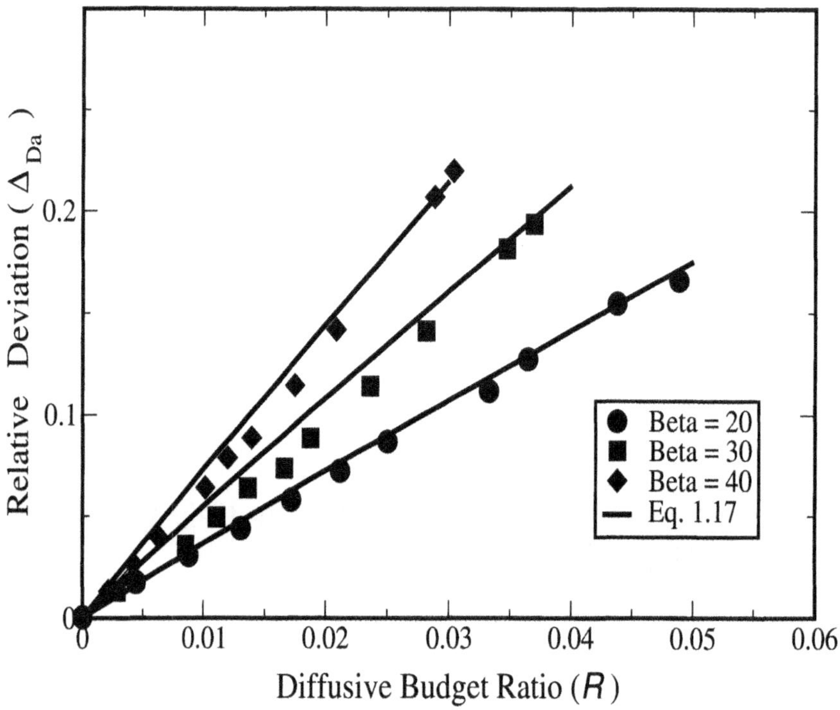

Figure 13.2. Relative Damköhler deviation Δ_{Da} versus $\mathcal{R}$. Symbols: Simulation of Eq. 13.2, line: analytical result (Eq. 13.17).

At the stoichiometric point ($x^* = 0$) and close to quenching (from Eq. 13.5 $T^* \approx 1 - \varepsilon$ and $\Gamma \approx 1$), using Eq. 13.8:

$$\frac{\partial^2 \Gamma}{\partial x^{*2}} = \beta \mathcal{F}^* + e^{-1} \tag{13.14}$$

Accordingly, using Eq. 13.2 at $x^* = 0$, the transverse budget is:

$$\frac{\lambda}{\rho C_p}\left(\frac{\partial^2 T}{\partial y^2}\right)_s = \left(\frac{Q Y_{F,o} Z_s}{Le_F C_p}\right)\frac{\lambda}{\rho C_p}\frac{1}{l_y^2 \varepsilon^2}\left(\frac{\partial^2 T^*}{\partial y^{*2}}\right)_s = \tag{13.15}$$

$$\left(\frac{Q Y_{F,o} Z_s}{Le_F C_p}\right)\chi_x \frac{\beta^2 \mathcal{F}^*}{4 Z_s^2 (1 - Z_s)^2}$$

Then,

$$\mathcal{R} = -\frac{\beta^2 \mathcal{F}^*}{\beta^2 \mathcal{F}^* + \beta e^{-1}} \tag{13.16}$$

Combining this relation with Eq. 13.11 provides:

$$\Delta_{Da} = \frac{\beta}{4 e Z_s (1 + \gamma)}\left(\frac{\mathcal{R}}{1 + \mathcal{R}}\right) \tag{13.17}$$

Assuming thin flamelets featuring normal fluxes much larger than their transverse counterpart, $f >> 1$, then $\mathcal{R} << 1$ and $\Delta_{Da} \approx (\beta/4eZ_s(1+\gamma))\mathcal{R}$. Typical values for hydrocarbon flames would lead to $\beta \approx 8$ and $Z_s \approx 0.05$, then $\Delta_{Da} \approx 30\mathcal{R}$. Because Zel'dovitch numbers are large, a transverse diffusive contribution of the order of 1% of the normal one (corresponding to $l_x \approx 0.1 l_y$) would thus lead to a modification of the Damköhler number of 30%. Confirming results from experiment [Shay and Ronney, 1998] and direct numerical simulation [Favier and Vervisch, 2001].

Eq. 13.10 describes quenching properties of an edge-flame, which therefore strongly differ from those of one-dimensional diffusion flame. To estimate the validity of these analytical developments, numerical simulations of Eq. 13.2 have been performed for $\gamma = 0$. Results are summarized in fig. 13.2, where the computed relative deviation Δ_{Da} is compared with the analytical result of Eq. 13.17. The simulations are carried out for several values of the Zel'dovitch number β, a good agreement is observed between the theory and the numerical simulations.

4. Conclusion

The quenching properties of a diffusion edge flame have been investigated using large activation energy. Assuming that the edge-flame may be approximated as a one-dimensional diffusion flame undergoing transverse fluxes of heat and mass, an expression for the diffusion edge-flame quenching Damköhler is derived. The geometrical aspect ratio of the mixture fraction field bordering the reaction zone enters this expression that may be useful in turbulent flamelet modeling. Because activation energies are large in flames, even when a small amount of heat and mass leak along the stoichiometric plane of the edge-flame, the quenching Damköhler strongly differs from the one expected from planar diffusion flame theory.

References

Bray, K. N. C. (1996). The challenge of turbulent combustion. In *Proceedings of the 26th Symp. (Int.) on Combustion.* The Combustion Institute, Pittsburgh, pp. 1-26.

Buckmaster, J. (1996). Edge-flames and their stability. *Combust. Sci. Tech.*, 115:41–68.

Dold, J. W. (1989). Flame propagation in a nonuniform mixture: analysis of a slowly varying triple flame. *Combust. Flame*, 76:71–88.

Favier, V. and Vervisch, L. (2001). Edge flames and partially premixed combustion in diffusion flame quenching. *Combust. Flame*, 125: 788–803.

Ghosal, S. and Vervisch, L. (2000). Theoretical and numerical investigation of symmetrical triple flame using a parabolic flame tip approximation. *J. Fluid Mech.*, 415:227–260.

Kioni, P. N., Rogg, B., Bray, K. N. C., and Liñán, A. (1993). Flame spread in laminar mixing layers: the triple flame. *Combust. Flame*, 95:276.

Liñán, A. (1974). The asymptotic structure of counterflow diffusion flames for large activation energies. *Acta Astronautica*, 1: 1007-1039.

Liñán, A., Orlandi, P., Verzicco, R., and Higuera, F. J. (1994). Effects of non-unity lewis numbers in diffusion flames. In *Studying turbulence using numerical databases - V*, pages 5–18. CTR, Stanford U.

Peters, N. (2000). *Turbulent Combustion.* Cambridge University Press.

Phillips, H. (1965). Flame in a buoyant methane layer. In *Proceedings of the 10th Symp. (Int.) on Combustion.* The Combustion Institute, Pittsburgh.

Poinsot, T. and Veynante, D. (2001). *Theoretical and Numerical Combustion.* R. T. Edwards, Inc.

Shay, M. L. and Ronney, P. D. (1998). Nonpremixed edge-flames in spatially varying straining flows. *Combust. and Flame*, 112(1/2):112–171.

Chapter 14

A MODEL DESCRIPTION OF THE EFFECTS OF VARIABLE FUEL-AIR MIXTURE COMPOSITION ON TURBULENT FLAME PROPAGATION

Jérôme Hélie
Institut Français du Pétrole
Division Techniques d'Applications Energétiques
Rueil-Malmaison, France

Arnaud Trouvé
University of Maryland
Department of Fire Protection Engineering
College Park, MD, USA
atrouve@eng.umd.edu

Abstract Stratified direct-injection spark-ignition engines feature a fuel-air mixing field characterized by large spatio-temporal variations of the mean mixture composition, $\widetilde{Z}(x_i, t)$, as well as regions with significant statistical fluctuations around $\widetilde{Z}$. Modifications to the Coherent Flame Model (CFM) are proposed in the present study to account for the effects of variable mixture strength on the primary premixed flame, as well as for the downstream formation of a secondary non-premixed reaction zone. Compared to previous work, the domain of validity of the present modifications is extended to the case of combustion with occurences of premixed flame extinction. The modeling strategy is based on previous results from direct numerical simulations as well as a theoretical analysis of a simplified problem due to Kolmogorov, Petrovskii and Piskunov (KPP). The KPP problem corresponds to a one-dimensional, turbulent flame propagating steadily into frozen turbulence and frozen fuel-air distribution, and provides a convenient framework to test the modified CFM model. In this simplified but somewhat generic configuration, the following trends are predicted: for small variations in mixture strength, *i.e.* in the absence of premixed flame extinction, partial fuel-air premix-

A. Pollard and S. Candel (eds.), IUTAM Symposium on Turbulent Mixing and Combustion, 169-179.

ing has a net negative (positive) impact on the turbulent flame speed, s_T, for mixtures close to stoichiometry (close to the flammability limits); for large variations in mixture strength, *i.e.* in the presence of premixed flame extinction, partial fuel-air premixing has a net negative impact on s_T and results in increased levels of secondary burning.

Keywords: Turbulent flame propagation, modeling, partially-premixed combustion, Internal Combustion engines, spark-ignition, direct-injection

1. Introduction

Direct-injection spark-ignition (DI-SI) engines represent one of the most promising technology to achieve further reductions in fuel consumption in automotive transport [Zhao *et al.*, 1997] [Takagi, 1998]. Under light- or part-load operation, these engines use fuel stratification to provide additional control on combustion performance and establish overall fuel lean operating conditions. In stratified DI-SI engines, the reactive mixture features temporal and spatial variations in the statistically-averaged fuel-air composition, with stoichiometric to slightly rich conditions ($\widetilde{Z} \approx Z_{st}$) achieved at the spark plug at the time of ignition, and very lean conditions ($\widetilde{Z} \ll Z_{st}$) sustained close to the cylinder walls. In addition, the reactive mixture also features regions with significant statistical fluctuations in the charge distribution, $(Z'/\widetilde{Z}) = O(1)$. Statistical fluctuations are produced by turbulent mixing in regions with strong fuel stratification as well as by the late vaporization of over-sized liquid fuel droplets. These spatio-temporal and statistical Z-variations correspond to inhomogeneities in mixture strength into which the spark-ignited turbulent flame propagates.

The propagation of turbulent flames into mixtures with variable fuel-air composition has been described recently using direct numerical simulations (DNS) in a series of related studies [Poinsot *et al.*, 1996] [Hélie & Trouvé, 1998] [Haworth *et al.*, 2000]. The numerical configurations considered in the DNS studies correspond to propane-air flames propagating into isotropic turbulent flow and into non-homogeneous reactants characterized by different levels of unmixedness. They also correspond to a regime of flame-flow conditions where the flow to flame velocity scale ratio is large, $(u'/s_L) > 1$, and the variations in mixture strength remain small, $(Z'/\widetilde{Z}) \ll 1$. The latter restriction is used as an intermediate step in the DNS studies, where the initial focus is on inflammable mixtures, *i.e.* on mixture compositions within the propane-air flammability limits and without the additional complication of premixed flame extinction.

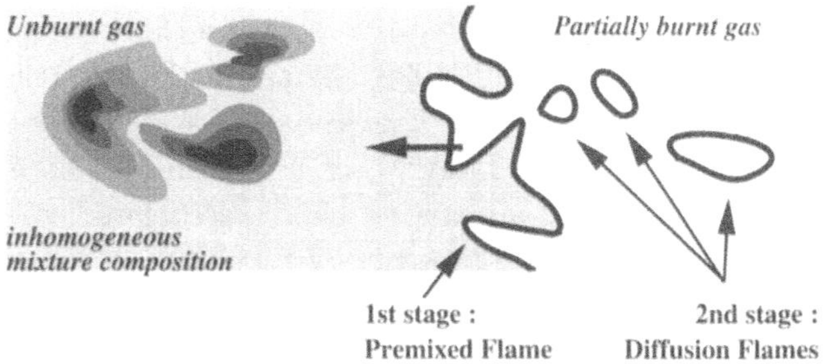

Figure 14.1. A representation of turbulent flame propagation into non-homogeneous reactants. The reaction zone can be described as a staged combustion system with a primary stage corresponding to a propagating premixed flame front followed by a secondary stage corresponding to multiple non-premixed flames. The primary premixed stage produces partially burnt gas, *i.e* a mixture composed of hot combustion products mixed with excess fuel fragments or with excess air. The excess reactants subsequently mix and burn in the secondary stage.

The DNS results show that : (1) under lean-rich conditions, the reaction zone can be described as a staged premixed/non-premixed combustion system (Fig. 1) ; (2) the first premixed stage can be conveniently described using classical laminar flamelet concepts; (3) for (small) variations in mixture strength around mean stoichiometric conditions, $\widetilde{Z} \approx Z_{st}$, unmixedness tends to have a net negative impact on the overall mean premixed reaction rate (*i.e.* on the turbulent flame speed); (4) this effect is related to changes in the mean flamelet structure, *i.e.* to a decrease of the mean flamelet mass burning rate per unit flame surface area with increasing values of Z'; (5) for $(u'/s_L) > 1$, the effect of unmixedness on the turbulent premixed flame surface area remains negligible.

The general objective of the present study is to examine how the DNS results above may be incorporated into current model descriptions of turbulent flame propagation in SI engines. A number of previous modeling efforts aimed at introducing the effects of fuel-air unmixedness into several classical turbulent combustion models may be found in the literature. Previous efforts include extensions of widely used flamelet models, such as the Bray-Moss-Libby (BML) model [Bray, 1990] [Bray & Libby, 1994], the Coherent Flame Model (CFM) [Candel *et al.*, 1990] [Fichot *et al.*, 1993], and the G-equation model [Peters, 1992] [Wirth & Peters, 1992]. For instance, an early extension of CFM to the case of combustion of mixtures with partially premixed reactants may be found in Veynante *et al.* (1991). This extension includes a description of the effects of mean fuel stratification as well as a description of secondary non-premixed burning. The effects of statistical variations in mixture strength are, however, neglected. Note that this modified version of CFM has been used recently to simulate stratified charge combustion in DI-SI engines [Baritaud *et al.*, 1996] [Duclos & Zolver, 1998]. Furthermore, a recent extension of BML to the case of partially premixed systems may also be found in Lahjaily *et al.* (1998). This extension includes a description of fuel-air mixing effects associated with both a spatially variable mean and the presence of turbulent fluctuations, but remains limited to the case of lean combustion without secondary burning. Finally, an extension of the G-equation model may be found in Müller *et al.* (1994). While this extension is originally proposed to describe flame propagation in the stabilization region of turbulent lifted jet diffusion flames, it may also be applied to the problem of flame propagation in stratified SI engines.

The present paper is a continuation of earlier work in Veynante *et al.*, Baritaud *et al.* and Duclos & Zolver aimed at adapting CFM to treat turbulent flame propagation in mixtures with variable fuel-air composition. It is also an extension of previous work in Hélie & Trouvé (2000) where a preliminary version of a modified CFM, was proposed using DNS for basic guidance and accounting for fluctuations in mixture strength as well as secondary burning. This preliminary version, however, was restricted to the case of inflammable fuel-air mixtures (the small-Z' regime). The case of mixtures characterized by large values of Z' with occurences of premixed flame extinction is considered in this chapter.

2. The modified Coherent Flame Model

2.1. Formulation of CFM-Z

Following the modeling strategy proposed in Veynante *et al.* (1991) and Haworth *et al.* (2000), we adopt a conditional two-fluid approach to describe the two combustion stages (Fig. 1). The conditional formulation allows to distinguish between fresh fuel in the unburnt gas, that is consumed by the primary premixed flame, and fuel fragments in the partially burnt gas, that are consumed by the secondary non-premixed flames. We accordingly use the following decomposition : $\widetilde{Y_F} = \widehat{Y_F}^0 + \widehat{Y_F}^1$, where $\widehat{Y_F}^0 \equiv (\overline{\rho Y_F I}/\overline{\rho})$, $\widehat{Y_F}^1 \equiv (\overline{\rho Y_F(1-I)}/\overline{\rho})$, with Y_F the fuel mass fraction and I a marker for the unburnt gas (see Hélie & Trouvé, 2000, for more details).

$\widehat{Y_F}^0$ and $\widehat{Y_F}^1$ are used as principal variables and are obtained as solutions of modeled balance equations:

$$\frac{\partial \overline{\rho}\widehat{Y_F}^0}{\partial t} + \frac{\partial}{\partial x_i}(\overline{\rho}\widetilde{u_i}\widehat{Y_F}^0) = \frac{\partial}{\partial x_i}(\overline{\rho}\frac{\nu_t}{\sigma_F}\frac{\partial \widehat{Y_F}^0}{\partial x_i}) - \overline{\dot{\omega}_{F,0\rightarrow 1}} - \overline{\dot{\omega}_{F,0\rightarrow 2}} \quad (14.1)$$

$$\frac{\partial \bar{\rho} \widehat{Y_F}^1}{\partial t} + \frac{\partial}{\partial x_i}(\bar{\rho} \widetilde{u_i} \widehat{Y_F}^1) = \frac{\partial}{\partial x_i}(\bar{\rho} \frac{\nu_t}{\sigma_F} \frac{\partial \widehat{Y_F}^1}{\partial x_i}) + \overline{\dot{\omega}_{F,0\to1}} - \overline{\dot{\omega}_{F,1\to2}} \quad (14.2)$$

where $\overline{\dot{\omega}_{F,0\to2}}$ and $\overline{\dot{\omega}_{F,1\to2}}$ are the mean fuel mass reaction rates associated with the first and second combustion stages, and where $\overline{\dot{\omega}_{F,0\to1}}$ is the mean fuel mass leakage term that represents transformation of fuel in the unburnt gas to excess fuel in the partially burnt gas. Closure models for these terms will be presented in the next section.

Note that to achieve our specific objective of capturing the effects of variations in mixture strength, the closure models will require a description of the mixing field, *i.e.* a description of the probability density function (Pdf) for mixture fraction. We adopt here a standard presumed β-Pdf for Z, where $\widetilde{p}(Z)$ is parametrized in terms of $\widetilde{Z}$ and $\widetilde{Z''^2}$.

Finally, in addition to $\widehat{Y_F}^0$, $\widehat{Y_F}^1$, $\widetilde{Z}$ and $\widetilde{Z''^2}$, the premixed flame surface density Σ is treated as a principal variable and is obtained from the following balance equation:

$$\frac{\partial \Sigma}{\partial t} + \frac{\partial \widetilde{u_i} \Sigma}{\partial x_i} = \frac{\partial}{\partial x_i}(\frac{\nu_t}{\sigma_\Sigma} \frac{\partial \Sigma}{\partial x_i}) + \alpha \kappa_t^* \Sigma - D_\Sigma \quad (14.3)$$

where κ_t^* and D_Σ are the modified turbulent flame stretch and Σ-dissipation term, and are described below.

2.2. Closure models

We now briefly summarize the closure models used to describe the source and sink terms present in Eqs. 1-3 and refer the reader to Hélie & Trouvé (2000) for a more detailed presentation. A new important feature in the present version of CFM-Z is the use of Z-conditioning to distinguish between inflammable ($Z_L < Z < Z_R$ with Z_L and Z_R the lean and rich fuel-air mixture flammability limits) and non-inflammable ($0 \leq Z \leq Z_L$ or $Z_R \leq Z \leq 1$) mixture compositions, that are respectively associated with burning and extinguished premixed flamelets.

For instance, primary burning is associated with flammable fuel-air conditions and is described using a modified flamelet expression:

$$\overline{\dot{\omega}_{F,0\to2}} = \langle \dot{m}_L \rangle_b \, \Sigma \quad (14.4)$$

where $\langle \dot{m}_L \rangle_b = (\int_{Z_L}^{Z_R} \dot{m}_L(Z) \widetilde{p}(Z) dZ)/(\int_{Z_L}^{Z_R} \widetilde{p}(Z) dZ)$ is the re-normalized mean fuel mass burning rate per unit premixed flame surface area.

Fuel mass transfer across the primary flame has contributions from both burning and extinguished premixed flamelets and is described using the following decomposition:

$$\overline{\dot{\omega}_{F,0\to1}} = \overline{\dot{\omega}_{F,0\to1,b}} + \overline{\dot{\omega}_{F,0\to1,e}} \quad (14.5)$$

We write:

$$\overline{\dot{\omega}_{F,0\to1,b}} = \frac{\overline{Y_{F,1}}^{b}}{\overline{Y_{F,0}}^{b} - \overline{Y_{F,1}}^{b}} \, \overline{\dot{\omega}_{F,0\to2}} \quad ; \quad \overline{\dot{\omega}_{F,0\to1,e}} = \frac{\overline{Y_{F,0}} - \overline{Y_{F,0}}^{b}}{\overline{Y_{F,0}}^{b} - \overline{Y_{F,1}}^{b}} \, \overline{\dot{\omega}_{F,0\to2}} \tag{14.6}$$

where $\overline{Y_{F,0}} = Y_F^\infty \widetilde{Z}$, $\overline{Y_{F,n}}^{b} = \int_{Z_L}^{Z_R} Y_{F,n}(Z)\widetilde{p}(Z)dZ$, $n = 0$ or 1, in which $Y_{F,0}(Z)$ and $Y_{F,1}(Z)$ respectively describe the mixture composition upstream and downstream of the primary flame and are estimated by the classical pure mixing and Burke-Schumann solutions.

Secondary burning is simply described using a classical, mixing controlled model proposed by Magnussen & Hjertager (1976):

$$\overline{\dot{\omega}_{F,1\to2}} = A \, \overline{\rho} \, \frac{\epsilon}{k} \, Min \left(\widehat{Y_F}^{1}, \frac{\widehat{Y_{O_2}}^{1}}{r_s} \right) \tag{14.7}$$

where classical coupling relations are used to relate the mean oxidizer mass fraction $\widehat{Y_{O_2}}^{1}$ to $\widehat{Y_F}^{0}$, $\widehat{Y_F}^{1}$, and $\widetilde{Z}$.

Finally, the CFM-Z modifications to the Σ-equation are based on the following Z-sensitive expressions of D_Σ and κ_t^*:

$$D_\Sigma = \beta \, \frac{\overline{Y_{F,0}}}{\overline{Y_{F,0}}^{b} - \overline{Y_{F,1}}^{b}} \, \frac{\langle \dot{m}_L \rangle_b \Sigma^2}{\overline{\rho} \widehat{Y_F}^{0}} \tag{14.8}$$

$$\kappa_t^* = \int_{Z_L}^{Z_R} \kappa_t(Z)\widetilde{p}(Z)dZ \tag{14.9}$$

In Eq. 9, the flame stretch $\kappa_t(Z)$ is estimated using the closure expression proposed by Meneveau & Poinsot (1991) where κ_t is related to the Kolmogorov time scale: $\kappa_t = (\epsilon/\nu_t)^{1/2}\gamma_k$, with γ_k a function of the relative flow to flame velocity and length scale ratios, (u'/s_L) and $(l_t/l_F) = (l_t/(D_{th}/s_L))$. We obtain: $\kappa_t^* = (\epsilon/\nu_t)^{1/2} \int_{Z_L}^{Z_R} \gamma_k(Z)\widetilde{p}(Z)dZ$. While this last expression may be used directly, the following approximation is adopted in order to avoid a costly integration of γ_k over the inflammable Z-domain:

$$\kappa_t^* = (\frac{\epsilon}{\nu_t})^{1/2} \, \gamma_k \left(\frac{u'}{\langle s_L \rangle_b}, \frac{l_t}{D_{th}/\langle s_L \rangle_b} \right) \left(\int_{Z_L}^{Z_R} \widetilde{p}(Z)dZ \right) \tag{14.10}$$

where $\langle s_L \rangle_b = (\int_{Z_L}^{Z_R} s_L(Z)\widetilde{p}(Z)dZ)/(\int_{Z_L}^{Z_R} \widetilde{p}(Z)dZ)$, and where κ_t^* depends on fuel-air mixture composition via $\langle s_L \rangle_b$ and $(\int_{Z_L}^{Z_R} \widetilde{p}(Z)dZ)$.

The present model follows closely the modeling strategy proposed by Hélie & Trouvé (2000) and accounts for premixed flame extinction due to

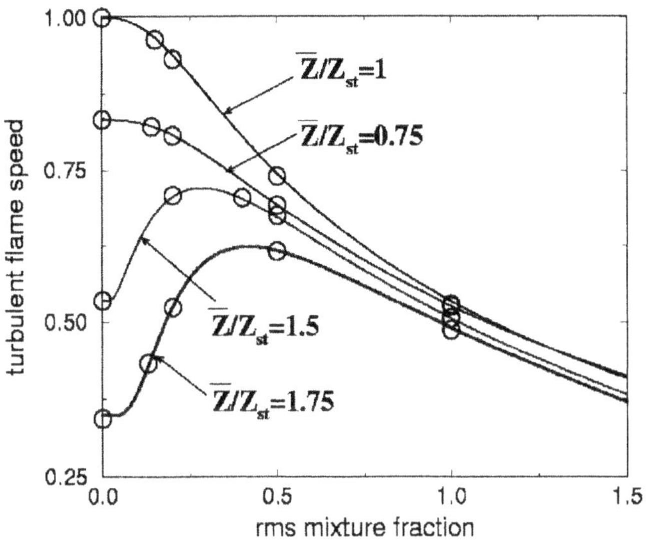

Figure 14.2. Variation of the turbulent flame speed, s_T, with the degree of unmixedness, $(Z'/\widetilde{Z})$, for different mean mixture compositions, $0.75 \leq (\widetilde{Z}/Z_{st}) \leq 1.75$. The turbulent flow field is the same in all cases : $u'/s_L(Z_{st}) = 10$ and $l_t/l_F(Z_{st}) = 100$. The variations $s_T(\widetilde{Z}, Z')$ are based on both a KPP analysis (lines) and a numerical evaluation (circles) of CFM-Z. $s_T(\widetilde{Z}, Z')$ is made non-dimensional with $s_T(Z_{st}, 0)$.

non-inflammable fuel-air compositions via the following two ingredients: an additional fuel mass leakage term across the primary burning zone, $\dot{\omega}_{F,0\rightarrow 1,e}$; and a modified turbulent flame stretch κ_t^*. Note that the last term on the right-hand side of Eq. 11 is dominant for large values of Z' and is responsible in CFM-Z for the reduced values of stretch, premixed flame surface area and turbulent flame speed obtained with large levels of fuel-air unmixedness.

3. Numerical evaluation of CFM-Z

Let us now consider the ideal problem due to Kolmogorov, Petrovskii and Piskunov (1937), referred to hereafter as the KPP problem, and corresponding to a statistically one-dimensional, plane, turbulent flame that propagates steadily into frozen turbulence and frozen fuel-air distribution. This KPP configuration is considered in several studies in the literature, because its simplicity allows the derivation of exact analytical expressions [Hakberg & Gosman, 1984] [Duclos *et al.*, 1993] [Fichot *et al.*, 1993]. For instance, the analytical methods presented in Fichot *et al.* (1993) and Duclos *et al.* (1993) show that the turbulent flame speed predicted by CFM-Z is equal to :

$$s_T = 2\,\left(\frac{\nu_t}{\sigma_\Sigma}\,\alpha\kappa_t^*\right)^{1/2} \tag{14.11}$$

Therefore, the effects of fuel-air unmixedness on s_T depend exclusively on their influence in the model description of κ_t^*.

The present KPP formulation uses Eqs. 1-3 with constant values of k, ϵ, $\widetilde{Z}$ and $\widetilde{Z''^2}$, coupled with the usual conservation equations for mass, momentum and energy. The CFM-Z closure models of §2 are combined with the β-Pdf description of $\widetilde{p}(Z)$ and a description of $\dot{m}_L(Z)$ from a flamelet library. Mixture fraction is defined using $Z_{st} \approx 0.06$, $r_s \approx 3.64$, $Y_F^\infty = 1$, $Y_{O_2}^\infty \approx 0.233$. The model constants are $\sigma_F = \sigma_\Sigma = 1$, $\alpha = 2.1$, $\beta = 1$, $A = 2$. In the energy equation, we use $\tau(Z_{st}) = 5$ and assume for simplicity that the heat of reaction per unit mass of fuel is independent of mixture strength.

The numerical solutions of the KPP problem are obtained using a high-order finite difference solver similar to that used in previous DNS studies [Hélie & Trouvé, 1998]. The solutions provide information on both the turbulent flame speed and the turbulent flame structure. For instance, figure 2 presents the estimated variations of s_T with Z', for different values of $\widetilde{Z}$. Consistent with previous results [Hélie & Trouvé,

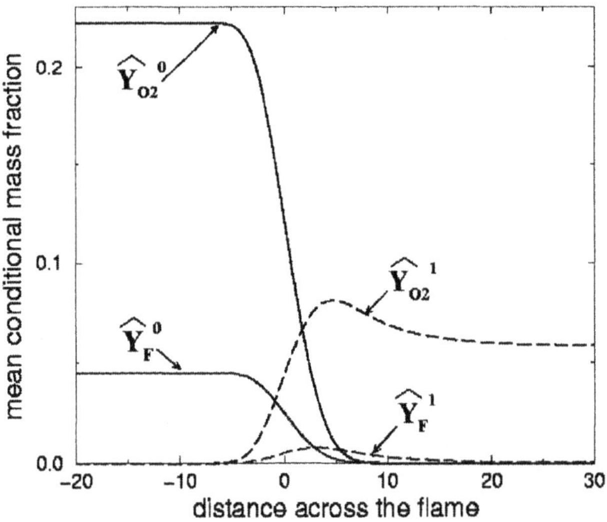

Figure 14.3. Spatial variations of the conditional mean fuel and oxidizer mass fractions across the flame. $(\widetilde{Z}/Z_{st}) = 0.75$, $(Z'/\widetilde{Z}) = 1.$, $u'/s_L(Z_{st}) = 10$ and $l_t/l_F(Z_{st}) = 100$. The distance x_1 is made non-dimensional with the turbulent integral length scale. The unburnt (burnt) gas side corresponds to $x_1 < 0$ ($x_1 > 0$).

2000], in the small-Z' regime, *i.e.* for $(Z'/\widetilde{Z}) < 0.2$, s_T is predicted to decrease with increasing values of Z' when the mixture compositions are close to stoichiometry, whereas an opposite trend is predicted when the mixture compositions are close to the flammability limits. In the large-Z' regime, *i.e.* for $(Z'/\widetilde{Z}) \geq 0.2$, s_T is affected primarily by occurences of premixed flame extinction and predicted to decrease with increasing values of Z'. In Fig. 2, the numerical results are also compared to the analytical expression in Eq. 11. The excellent agreement supports the extension of the KPP analysis to the case of partially premixed systems.

We now examine a typical CFM-Z flame structure. Figure 3 is of particular interest since it provides an illustration of the conditional approach perspective. In this perspective, the reactants present in the unburnt gas $(x_1 \to -\infty)$ are consumed by primary burning or leaked to the partially burnt gas, and $\widehat{Y_F}^0$ and $\widehat{Y_{O_2}}^0$ must vanish when crossing the flame. In addition, the excess reactants in the partially burnt gas are produced by leaking across the primary flame and consumed by secondary burning, and $\widehat{Y_F}^1$ and $\widehat{Y_{O_2}}^1$ increase or descrease, depending on the relative weights of these two processes. Secondary burning lasts until one of the reactant is depleted and accordingly in Fig. 3, $\widehat{Y_F}^1 \to 0$ when $x_1 \to +\infty$.

4. Conclusion

The Coherent Flame Model (CFM) is modified in the present study to treat turbulent flame propagation in mixtures with variable fuel-air composition. The modifications include: a description of the reaction zone as a staged premixed/non-premixed combustion system using a conditional two-fluid approach; a description of primary burning in the premixed stage using simple extensions of classical flamelet expressions; a description of secondary burning in the non-premixed stage using a mixing controlled model; a description of statistical fluctuations in mixture strength using a standard presumed β-Pdf approach for mixture fraction; a description of premixed flame extinction associated with non-inflammable fuel-air compositions via an additional fuel mass leakage term across the primary burning zone and a modified expression for the turbulent flame stretch.

The modified CFM model is tested numerically in a simplified configuration due to Kolmogorov, Petrovskii and Piskunov (KPP). The KPP problem corresponds to a plane, turbulent flame propagating steadily into frozen turbulence and frozen fuel-air distribution. The following trends are predicted from the KPP analysis: in the absence of significant premixed flame extinction, fuel-air unmixedness has a net negative impact on the turbulent flame speed for variations in mixture strength around mean stoichiometric conditions, and a net positive impact for variations close to the flammability limits; in the presence of significant premixed flame extinction, fuel-air unmixedness has a net negative impact on s_T and results in increased levels of secondary burning.

References

Baritaud, T. A., Duclos, J. M. and Fusco, A. (1996). *Proc. Combust. Inst.* 26:2627–2635.

Bray, K. N. C. (1990). *Proc. R. Soc. Lond. A* 431:315–335.

Bray, K. N. C. and Libby, P. A. (1994). In *Turbulent Reacting Flows* (P. A. Libby and F. A. Williams, Eds.), Academic Press, pp. 115–151.

Candel, S. M., Veynante, D., Lacas, F., Maistret, E., Darabiha, N. and Poinsot, T. (1990). In *Recent Advances in Combustion Modeling* (B. R. Larrouturou, Ed.), World Scientific, Singapore.

Duclos, J. M. and Zolver, M. (1998). In *Fourth International Symposium on Diagnostics and Modeling of Combustion in Internal Combustion Engines,* COMODIA, Japan Soc. Mechanical Engineers, pp. 335–340.

Duclos, J. M., Veynante, D. and Poinsot, T. (1993). *Combust. Flame* 95:101–117.

Fichot, F., Lacas, F., Veynante, D. and Candel, S. M. (1993). *Combust. Sci. and Tech.* 90:35–60.

Hakberg, B. and Gosman, A. D. (1984) *Proc. Combust. Inst.* 20:225–232.

Haworth, D., Cuenot, B., Poinsot, T. and Blint, R. (2000). *Combust. Flame* 121:395–417.

Hélie, J. and Trouvé, A. (1998). *Proc. Combust. Inst.* 27:891–898.

Hélie, J. and Trouvé, A. (2000). *Proc. Combust. Inst.* 28:193–201.

Kolmogorov, A. N., Petrovskii, I. G. and Piskunov, N. S. (1937). *Bjul. Moskovskovo Gos Univ.*, 1:1–72.

Lahjaily, H., Champion, M., Karmed, D. and Bruel, P. (1998). *Combust. Sci. and Tech.* 135:153–173.

Magnussen, B. F. and Hjertager, B. H. (1976). *Proc. Combust. Inst.* 16:719–729.

Meneveau, C. and Poinsot, T. (1991). *Combust. Flame* 86:311–332.

Müller, C., Breitbach, H. and Peters, N. (1994). *Proc. Combust. Inst.* 25:1099–1106.

Peters, N. (1992). *J. Fluid Mech.* 242:611–629.

Poinsot, T., Veynante, D., Trouvé, A. and Ruetsch, G. R. (1996). In *Proc. Summer Program,* Center for Turbulence Research, NASA Ames/Stanford University, pp. 111–136.

Takagi, Y. (1998). *Proc. Combust. Inst.* 27:2055–2068.
Veynante, D., Lacas, F., Maistret, E. and Candel, S. M. (1991). in *Seventh Turbulent Shear Flows*, Springer-Verlag, pp. 367–378.
Wirth, M. and Peters, N. (1992). *Proc. Combust. Inst.* 24:493–501.
Zhao, F.-Q., Lai, M.-C. and Harrington, D. (1997). *SAE Paper No. 970627.*

Chapter 15

CHALLENGES IN MODELING SCALARS IN TURBULENCE AND LES

Anisotropy, dynamic models, and scale separation

C. Meneveau[1,2], H. S. Kang[1], F. Charlette[1,3], J. Averill[1,4], O. Knio[1,2], & D. Veynante[5]
[1] Department of Mechanical Engineering, Johns Hopkins University, Baltimore, MD 21218, USA
[2] Center for Environmental & Applied Fluid Mechanics
[3] Institute Français du Pétrole, Rueil-Malmaison, France
[4] National Institute of Standards & Technology, Gaithersburg, USA
[5] Laboratoire EM2C, Ecole Centrale, Paris, France

Abstract This paper discusses three interrelated aspects of modeling scalar transport, mixing and combustion, in the context of large-eddy simulations (LES). In terms of passive scalar modeling, we find from heated wake measurements that whereas the kinetic energy dissipation tensor tends towards isotropy at small scales, in the presence of a mean scalar gradient the SGS scalar variance dissipation remains anisotropic independent of filter scale. The eddy-diffusion model predicts isotropic behavior, whereas the nonlinear model reproduces the correct trends, but overestimates the level of scalar dissipation anisotropy. The results provide some support for mixed models. Applications of dynamic models are also discussed. Initial tests on non-buoyant jet flows show that the constant-coefficient Smagorinsky model generates an entirely laminar jet, whereas the Lagrangian dynamic model yields very good results without the need to tune the model coefficient. In the context of applying the dynamic model to premixed combustion, or other physical processes with large scale-separations, we show that the dynamic determination of unknown scaling exponents is a more promising approach than dynamic evaluation of model coefficients.

Keywords: LES, passive scalar anisotropy, modeling, scale separation, dynamic model, premixed combustion

A. Pollard and S. Candel (eds.), IUTAM Symposium on Turbulent Mixing and Combustion, 181-199.

1. Introduction

This study addresses three distinct issues related to modeling of scalars in large-eddy simulations (LES) of turbulent flows. The first part addresses scalar mixing due to small scales from a fundamental point of view. Specifically, a major assumption of LES is the universality of subgrid-scale (small) motions and transport in turbulent flows. Also, one typically assumes a certain level of small-scale isotropy. In the past decade, careful measurements of passive scalar statistics have raised serious doubts about isotropy and universality of small-scale scalar statistics (Warhaft, 2000). However, the implications of these fundamental observations upon LES modeling are not well understood. We review results from experiments that were designed specifically to address these issues. The experiments are carried out in a moderately high Reynolds number wake generated by a heated cylinder in the Corrsin wind tunnel of the Johns Hopkins University. Measurements are performed using an array of four X-wire and four cold-wire probes. This array enables us to obtain two-dimensional filtered quantities in the cross-stream and streamwise directions by invoking Taylor's hypothesis (Cerutti & Meneveau, 2000). By spatially filtering the signals, we are able to measure the so-called subgrid-scale heat flux. This quantity represents how much heat flux is due to the small-scale motions of turbulence. Three different filter sizes are considered by varying the separation between the four probes. We compute several terms that comprise the subgrid dissipation tensor of kinetic energy and scalar variance and test for isotropic behavior, as a function of filter scale.

The above-mentioned experimental work raises questions about the fundamental structure of eddy-diffusion models. However, in practical applications, the eddy-diffusion models are likely to remain popular due to the numerical robustness and ease of their implementation. Regardless, models require the specification of model coefficients, and the predictions often depend heavily on these coefficients. The dynamic model (Germano *et al.*, 1991) has proven quite successful in providing reliable estimates for coefficients associated with models for unresolved momentum and scalar fluxes. In the second part of this paper, we briefly present recent results from an implementation of the Lagrangian dynamic model (Meneveau *et al.*, 1996) in a LES code used for fire predictions (McGrattan *et al.*, 1998).

In the third part of this work, we comment upon generalizations of the dynamic approach to physical processes with large scale-separations. An example is premixed combustion in which the reaction occurs in thin flames where the flame thickness δ_l^0 and the flame wrinkles occur typically at scales much below that of the LES mesh or filter. Traditional models for subgrid flame speed typically have adjustable model coefficients which, to leading order, appear as multiplying parameters. It is argued that determination of such multiplicative coefficients using the dynamic approach appears to be ill-posed asymptotically when the scale separation increases. Alternatively, it will be shown that dynamic determination of unknown scaling exponents is more promising, and initial applications to LES of premixed combustion will be presented.

2. Passive scalar anisotropy

The effects of passive scalar anisotropy on subgrid-scale (SGS) physics and modeling for large-eddy simulations were studied. In order to span a sizeable range of inertial-range filter scales, a sufficiently high Reynolds number had to be considered. Hence, this study was based on experimental data and was performed in a canonical shear flow, the heated cylinder wake. The main results of this study are presented in Kang & Meneveau (2001), but here we also present results for the mixed model.

Experiments were performed in the return type Corrsin Wind Tunnel. A heated smooth cylinder of diameter $D = 4.83$ cm was located horizontally at the centerline of the test section. The measurement location in the streamwise direction (x_1) was fixed at $x_1/D = 25$. To obtain the filtered and SGS quantities, an array of four custom-made miniature probes was used. Each probe was composed of one X-type hot-wire and one I-type cold-wire for the velocities in the $x_1 - x_2$ plane and the temperatures, respectively. Here x_2 is the "cross-wake" direction. The separation distance h between the probes in the cross-wake direction x_2 could be adjusted manually and three configurations ($h = 5$, 10 and 20 mm, with $\Delta = 2h$, where Δ is the filter-width) were used. The signals were low-pass filtered at a frequency of 20 kHz and sampled at 40 kHz during 60 seconds.

Table 15.1. Parameters of heated wake flow at $x_1/D = 25$.

U_{CL}	U_d	θ_d	ℓ	ϵ_{CL}	$\epsilon_{\theta CL}$	η	Re_λ
13.6 m/s	4.4 m/s	0.61 °C	0.08 m	87 m^2/s^3	0.65 °C^2/s	0.08 mm	350

Table 15.1 shows, the mean centerline velocity U_{CL}, the defect velocity $U_d = U_\infty - U_{CL}$, the defect temperature $\theta_d = \theta_{CL} - \theta_\infty$, the half-width of the wake ℓ, the molecular kinetic energy dissipation at the centerline ϵ_{CL}, the molecular scalar-variance dissipation at the centerline $\epsilon_{\theta CL}$, the Kolmogorov length scale $\eta = (\nu^3/\epsilon)^{1/4}$, and the Reynolds number based

on Taylor micro-scale Re_λ. The turbulence statistics across the wake agree quite well with those reported in Matsumura & Antonia (1993) and Kiya & Matsumura (1988).

To get the spatial quantities along the streamwise direction from the temporal data, Taylor's hypothesis is invoked. In the present study, to separate between large and small scales, the box filter is applied to the streamwise and cross-wake directions, and the trapezoidal rule is used for the spatial integrations. The filtering and error analysis is documented in Cerutti & Meneveau (2000) and Cerutti *et al.* (2000).

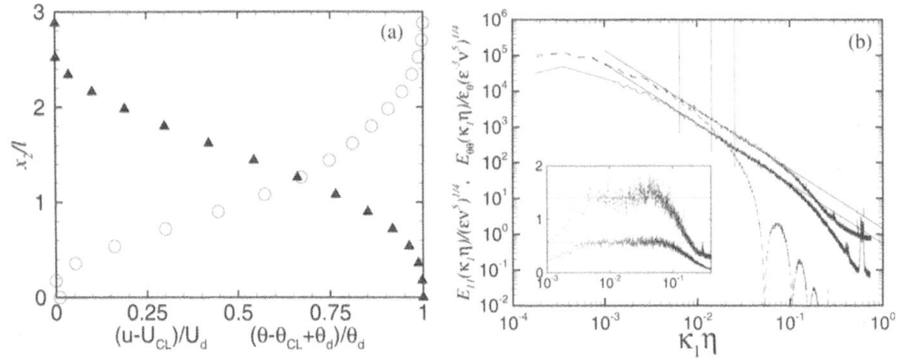

Figure 15.1. (a) Mean streamwise velocity and temperature profiles normalized with the defect velocity U_d and temperature θ_d, respectively, across the wake at $x_1/D = 25$: O, mean streamwise velocity; ▲, mean temperature; (b) the longitudinal spectra of the u_1 component (solid line), of the temperature (dashed line), and of the filtered temperature, filtered at $\Delta/\eta = 125$ (dashed line with lobes) at the centerline, as a function of the longitudinal wavenumber, κ_1. The insert shows the compensated spectra. The three vertical dotted lines are the wavenumbers corresponding to the filter sizes of $\Delta/\eta = 125$, 250 and 500. The slops of the straight solid lines are $-5/3$.

Figure 15.1(a) shows a comparison between the mean streamwise velocity and the mean temperature profiles normalized with the defect velocity U_d and temperature θ_d, respectively. The minimum mean streamwise velocity and the maximum mean temperature occur near at the centerline. Note that the location of the maximum temperature gradient $x_2|_{\frac{\partial\theta}{\partial x_2}=MAX}$ is $x_2/\ell \approx 1.5$. Figure 15.1(b) shows the longitudinal spectra of the u_1 component (solid line), of the temperature (dashed line), and of the filtered temperature, filtered at $\Delta/\eta = 125$ (dashed line with lobes), at the centerline. Here, κ_1 is the longitudinal wave number. The three vertical dashed lines correspond to the filter sizes of $\Delta/\eta = 125$ (10 mm), 250 (20 mm), 500 (40 mm). All the filter sizes are in the inertial range. As seen in the compensated spectra shown in the insert, there is a fairly clear inertial range with a $-5/3$ slope. The lobes in the filtered temperature spectrum at scales below the filter size are due to the streamwise box filter used in the present study. Spectra of filtered velocity (not shown) have similar shape. For a more detailed discussion of the spectral slope, see Kang & Meneveau (2001).

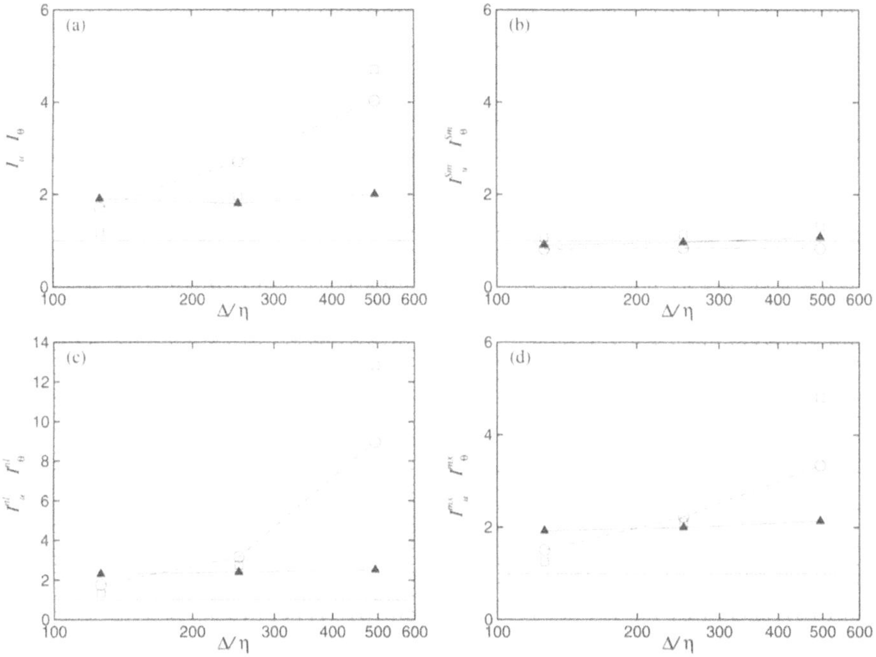

Figure 15.2. Isotropy ratios of SGS dissipations at $x_2/\ell = 1.44$, where the mean temperature gradient peaks: (a) real isotropy ratios ○, I_{u22}; □, I_{u12}; ▲, I_θ; (b) eddy-diffusion models: ○, I_{u22}^{Sm}; □, I_{u12}^{Sm}; ▲, I_θ^{Sm}; (c) nonlinear models: ○, I_{u22}^{nl}; □, I_{u12}^{nl}; ▲, I_θ^{nl}; (d) mixed models: ○, I_{u22}^{mx}; □, I_{u12}^{mx}; ▲, I_θ^{mx}.

We define the following 'isotropy ratios':

$$I_{u22} \equiv \frac{\left\langle \tau_{22}\widetilde{S}_{22}\right\rangle}{\left\langle \tau_{11}\widetilde{S}_{11}\right\rangle}, \quad I_{u12} \equiv \frac{\left\langle \tau_{12}\widetilde{S}_{12}\right\rangle}{\frac{3}{4}\left\langle \tau_{11}\widetilde{S}_{11}\right\rangle}, \quad I_\theta \equiv \frac{\left\langle q_2\widetilde{G}_2\right\rangle}{\left\langle q_1\widetilde{G}_1\right\rangle}, \tag{15.1}$$

where τ_{ij} is the SGS stress tensor ($\tau_{ij} \equiv \widetilde{u_i u_j} - \widetilde{u}_i\widetilde{u}_j$) and the q_j is the SGS heat flux ($q_j \equiv \widetilde{\theta u_j} - \widetilde{\theta}\widetilde{u}_j$) (a *tilde* represents filtering at scale Δ and θ is the passive scalar field). $\widetilde{S}_{ij} \equiv \frac{1}{2}(\partial\widetilde{u}_i/\partial x_j + \partial\widetilde{u}_j/\partial x_i)$ and $\widetilde{G}_j \equiv \partial\widetilde{\theta}/\partial x_j$ are the resolved strain-rate tensor and scalar gradient, respectively. The isotropy ratios I_{u22}, I_{u12} and I_θ should equal unity for isotropic statistics.

Figure 15.2(a) shows the isotropy ratios as a function of scale, for the transverse location $x_2/\ell = 1.44$ close to the peak mean temperature gradient. As is evident, the isotropy ratios of the kinetic energy dissipation decrease towards unity as the filter size decreases, whereas the isotropy ratio of the scalar-variance dissipation remains unchanged near $I_\theta \sim 2$ as the filter size is decreased. Near the centerline, where the mean shear and scalar gradient vanish, the SGS isotropy ratios are all near unity (not shown here). Therefore, we conclude that as far as mean SGS dissipation is concerned (which is the most important aspect of the resolved and SGS scale interactions), the scalar field maintains strong anisotropy in the presence of a mean scalar gradient.

Next, we quantify the isotropy level of model predictions, when the data are analyzed in an 'a-priori' sense, i.e. by replacing τ_{ij} and q_j above by model expressions. First, the standard eddy-diffusion model is considered, i.e.,

$$\tau_{ij}^{\mathrm{Sm}} - \frac{1}{3}\tau_{kk}^{\mathrm{Sm}}\delta_{ij} = -2(C_S\Delta)^2|\widetilde{S}|\widetilde{S}_{ij}, \qquad q_j^{\mathrm{Sm}} = -2Pr_{\mathrm{sgs}}^{-1}(C_S\Delta)^2|\widetilde{S}|\widetilde{G}_j, \tag{15.2}$$

where $|\widetilde{S}| = \left(2\widetilde{S}_{mn}\widetilde{S}_{mn}\right)^{1/2}$ is the modulus of the resolved strain rate, C_S is the Smagorinsky coefficient, and Pr_{sgs} is the SGS Prandtl number. Consequently, and independent of the model coefficients, the isotropy ratios from the eddy-diffusion model are:

$$I_{u22}^{\mathrm{Sm}} \equiv \frac{\left\langle|\widetilde{S}|\widetilde{S}_{22}\widetilde{S}_{22}\right\rangle}{\left\langle|\widetilde{S}|\widetilde{S}_{11}\widetilde{S}_{11}\right\rangle}, \quad I_{u12}^{\mathrm{Sm}} \equiv \frac{\left\langle|\widetilde{S}|\widetilde{S}_{12}\widetilde{S}_{12}\right\rangle}{\frac{3}{4}\left\langle|\widetilde{S}|\widetilde{S}_{11}\widetilde{S}_{11}\right\rangle}, \quad I_{\theta}^{\mathrm{Sm}} \equiv \frac{\left\langle|\widetilde{S}|\widetilde{G}_2\widetilde{G}_2\right\rangle}{\left\langle|\widetilde{S}|\widetilde{G}_1\widetilde{G}_1\right\rangle}. \tag{15.3}$$

In computing the filtered strain-rate magnitude from the data, the following approximation is used (this is a 2D extension of the 1D approach of O'Neil & Meneveau 1997, and was also used in Liu, Katz & Meneveau 1999): $|\widetilde{S}| \approx \left[2\left(2\widetilde{S}_{11}\widetilde{S}_{11} + \widetilde{S}_{22}\widetilde{S}_{22} + 6\widetilde{S}_{12}\widetilde{S}_{12}\right)\right]^{1/2}$. The isotropy ratios of the SGS dissipations from the eddy-diffusion model are shown in figure 15.2(b). Results are near unity almost independently of scale, including the scalar variance dissipation. Profiles across the wake (not shown) also are near unity. This is mainly due to the isotropy of the square gradient tensors (not shown here, the only difference is the additional $|\widetilde{S}|$ factor). Hence, using the eddy-diffusion model one would (incorrectly) predict isotropy of SGS dissipation since the resolved second-order moments are isotropic.

The second model class considered here is the 'nonlinear model' or 'tensor eddy-diffusion model' (Leonard 1974; Clark *et al.* 1979; Liu,

Meneveau & Katz 1994; Leonard 1997; Borue & Orszag 1998; Meneveau & Katz 2000). The nonlinear model reads:

$$\tau_{ij}^{\mathrm{nl}} = C_{nl}\Delta^2 \frac{\partial \widetilde{u}_i}{\partial x_k}\frac{\partial \widetilde{u}_j}{\partial x_k}, \qquad q_j^{\mathrm{nl}} = C_{nl_\theta}\Delta^2 \frac{\partial \widetilde{\theta}}{\partial x_k}\frac{\partial \widetilde{u}_j}{\partial x_k}, \tag{15.4}$$

where C_{nl} and C_{nl_θ} are the nonlinear model coefficients. Therefore, the corresponding isotropy ratios from the nonlinear model are:

$$I_{u22}^{\mathrm{nl}} \equiv \frac{\left\langle \frac{\partial \widetilde{u}_2}{\partial x_k}\frac{\partial \widetilde{u}_2}{\partial x_k}\widetilde{S}_{22}\right\rangle}{\left\langle \frac{\partial \widetilde{u}_1}{\partial x_k}\frac{\partial \widetilde{u}_1}{\partial x_k}\widetilde{S}_{11}\right\rangle}, \quad I_{u12}^{\mathrm{nl}} \equiv \frac{\left\langle \frac{\partial \widetilde{u}_1}{\partial x_k}\frac{\partial \widetilde{u}_2}{\partial x_k}\widetilde{S}_{12}\right\rangle}{\frac{3}{4}\left\langle \frac{\partial \widetilde{u}_1}{\partial x_k}\frac{\partial \widetilde{u}_1}{\partial x_k}\widetilde{S}_{11}\right\rangle}, \quad I_{\theta}^{\mathrm{nl}} \equiv \frac{\left\langle \frac{\partial \widetilde{\theta}}{\partial x_k}\frac{\partial \widetilde{u}_2}{\partial x_k}\widetilde{G}_{2}\right\rangle}{\left\langle \frac{\partial \widetilde{\theta}}{\partial x_k}\frac{\partial \widetilde{u}_1}{\partial x_k}\widetilde{G}_{1}\right\rangle}, \tag{15.5}$$

where $k = 1, 2$ in the analysis of our 2D data. The isotropy ratios of the SGS dissipations from the nonlinear model are shown in figure 15.2(c). As is apparent in comparing with figure 15.2(a), the main features of anisotropy and filter-size dependence are correctly reproduced qualitatively. Quantitatively, the levels of anisotropy are overestimated.

The third modeling class is the 'mixed model' (Bardina *et al.* , 1980) which is composed of the linear combinations of the eddy-diffusion and nonlinear models. So, the corresponding isotropy ratios from the mixed model are defined as follows:

$$I_{u22}^{\mathrm{mx}} \equiv \frac{\beta\left\langle -2(C_S\Delta)^2|\widetilde{S}|\widetilde{S}_{22}\widetilde{S}_{22}\right\rangle + (1-\beta)\left\langle C_{nl}\Delta^2 \frac{\partial \widetilde{u}_2}{\partial x_k}\frac{\partial \widetilde{u}_2}{\partial x_k}\widetilde{S}_{22}\right\rangle}{\beta\left\langle -2(C_S\Delta)^2|\widetilde{S}|\widetilde{S}_{11}\widetilde{S}_{11}\right\rangle + (1-\beta)\left\langle C_{nl}\Delta^2 \frac{\partial \widetilde{u}_1}{\partial x_k}\frac{\partial \widetilde{u}_1}{\partial x_k}\widetilde{S}_{11}\right\rangle},$$

$$I_{u12}^{\mathrm{mx}} \equiv \frac{\beta\left\langle -2(C_S\Delta)^2|\widetilde{S}|\widetilde{S}_{12}\widetilde{S}_{12}\right\rangle + (1-\beta)\left\langle C_{nl}\Delta^2 \frac{\partial \widetilde{u}_1}{\partial x_k}\frac{\partial \widetilde{u}_2}{\partial x_k}\widetilde{S}_{12}\right\rangle}{\beta\frac{3}{4}\left\langle -2(C_S\Delta)^2|\widetilde{S}|\widetilde{S}_{11}\widetilde{S}_{11}\right\rangle + (1-\beta)\frac{3}{4}\left\langle C_{nl}\Delta^2 \frac{\partial \widetilde{u}_1}{\partial x_k}\frac{\partial \widetilde{u}_1}{\partial x_k}\widetilde{S}_{11}\right\rangle},$$

$$I_{\theta}^{\mathrm{mx}} \equiv \frac{\beta\left\langle -2Pr_{\mathrm{sgs}}^{-1}(C_S\Delta)^2|\widetilde{S}|\widetilde{G}_{2}\widetilde{G}_{2}\right\rangle + (1-\beta)\left\langle C_{nl_\theta}\Delta^2 \frac{\partial \widetilde{\theta}}{\partial x_k}\frac{\partial \widetilde{u}_2}{\partial x_k}\widetilde{G}_{2}\right\rangle}{\beta\left\langle -2Pr_{\mathrm{sgs}}^{-1}(C_S\Delta)^2|\widetilde{S}|\widetilde{G}_{1}\widetilde{G}_{1}\right\rangle + (1-\beta)\left\langle C_{nl_\theta}\Delta^2 \frac{\partial \widetilde{\theta}}{\partial x_k}\frac{\partial \widetilde{u}_1}{\partial x_k}\widetilde{G}_{1}\right\rangle}, \tag{15.6}$$

where, again, $k = 1, 2$ and the model coefficients C_S, C_{nl}, Pr_{sgs} and C_{nl_θ} are 0.09, 0.21, 0.56 and 0.21, respectively. These values are set so each model produces the correct SGS dissipation individually. The weighting factor β is set to $\beta = 0.2$, to obtain the isotropy ratios of the SGS dissipations from the mixed model shown in figure 15.2(d). The overestimations of anisotropy from the nonlinear model are reduced, and the overall trends predicted by mixed model are more realistic than those of the nonlinear or Smagorinsky terms separately. Some discrepancies still exist for the 22 term, but overall, we may conclude that the results provide added support for the mixed model.

3. Dynamic modeling applied to mixing

The preceding observations suggest that mixed models are able to reproduce better the fundamental structure of scalar SGS dissipation in turbulence. However, in applied situations, mixed models have not yet been applied extensively mainly due to the additional complexities and computational cost. Significant advances, however, have already been made even in the context of the simple Smagorinsky model by using the dynamic approach. This model has been described extensively in the literature (Germano *et al.* , 1991; Meneveau & Katz, 2000; Pope, 2000). For practical applications, however, there still are questions as to its value due to the fact that it also requires significant additional computer resources.

In this chapter we briefly summarize recent experience in implementing the dynamic model in a computer code used to model buoyant fire flows. The LES scheme developed at National Institute of Standards and Technology's Building and Fire Research Laboratory (Mell, McGrattan & Baum, 1995; Mell, Johnson, McGrattan & Baum, 1995; Baum, McGrattan & Rehm, 1996; McGrattan, Baum & Rehm, 1998), implements a low-Mach number formulation of the compressible conservation equations for an ideal, Newtonian gas. In the current version of the method, unresolved scales of motion are accounted for by a Smagorinsky (1963) eddy diffusivity scheme. Despite the spatially homogeneous model constant, the LES scheme provides reasonable predictions of stationary buoyant plumes with the model parameter that has been tuned appropriately. It is well-known, however, that the Smagorinsky scheme with a standard coefficient causes excessive damping in the non-turbulent regions, delaying transition and turbulent production (Piomelli & Zang, 1991). The dynamic model (Germano *et al.*, 1991) evaluates the closure parameter based on the turbulent stresses generated in the resolved velocity field. However, it requires averaging over directions of statistical homogeneity, which typically do not exist in complex geometry flows. The Lagrangian, dynamic model (Meneveau *et al.*, 1996), however, accumulates statistics over fluid-particle trajectories, which are independent of both boundary conditions and directions of statistical homogeneity. This approach allows spatial variations to occur without causing excessive fluctuations. Also, the model is constructed according to basic principles of Galilean invariance. Details about the model can be found in Meneveau *et al.* (1996) and Pope (2000).

Here we present initial test results from LES of an isothermal round air jet. Results of the dynamic approach are compared with the classical Smagorinsky model. The jet diameter was 20 cm. The round nozzle was approximated by 12 square cells, essentially a 4×4 square with each corner cell missing. The physical domain was 1.5 m wide, 1.5 m deep, and 3.5 m high. The computational grid was 30 by 30 by 72 cells in the y, z, and x directions, respectively. The inlet jet velocity was set to 0.3 m/s, giving a jet Reynolds number of $Re_D = 4000$. The simulation reached a steady-state after 500 seconds (in physical time) and was run for 2000 secs.

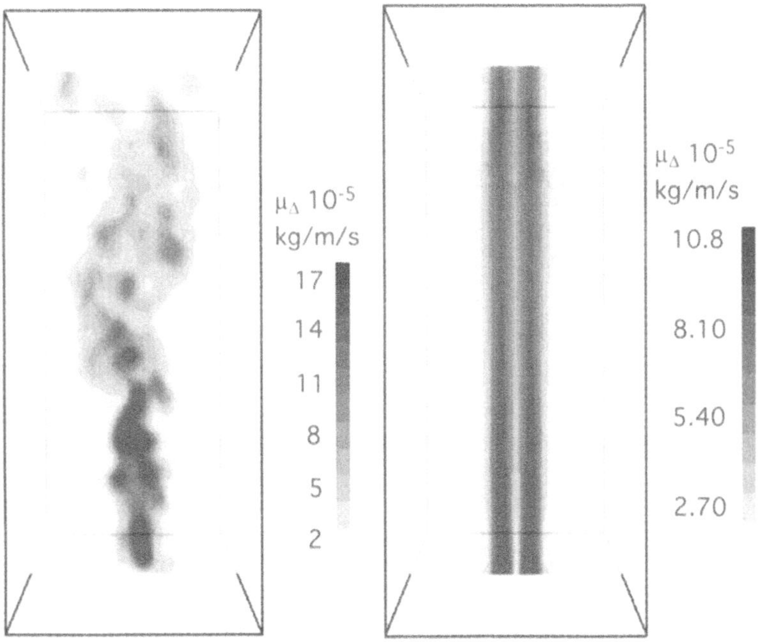

Figure 15.3. Contour maps of eddy-viscosity ($\mu_\Delta = \rho c_s^2 \Delta^2 |\tilde{S}|$) across the jet center-plane for a representative snapshot during the simulation. Left: $c_s(\mathbf{x}, t)$ is given by the Lagrangian dynamic procedure. Right: $c_s = 0.14$ is kept constant and prevents turbulence from developing.

Data were accumulated every 10 s from 500 s to 2000 s for statistical analysis. Figure 15.3 shows contour plots of the viscosity at an axial slice across the midplane of the domain for both the Lagrangian dynamic and classical models. Velocity measurements were recorded every 0.2 m from 1.0 m to 3.4 m, yielding results from 5 to 17 jet diameters downstream. Comparing the dynamic model to the classical model reveals that the classical model fails to transition to turbulence, resulting in negligible velocity decay. The spreading rate (jet half-width as function of down-stream distance) using the dynamic model is plotted in Figure 15.4(a). The spreading rate of the non-dynamic Smagorinsky model was nearly zero, whereas it was $S \sim 0.0985$ for the Lagrangian dynamic model, which compares favorably with a value of $S \sim 0.0985$ from Wygnansky & Fiedler (1969) and of $S \sim 0.094 - 0.102$ from Hussein, *et al.* (1994). Figure 15.4(b) shows the normalized velocity profile at several downstream distances as a function of normalized radial distance. Thus, the results of the dynamic model compare favorably with the experimental measurements. In summary, the Lagrangian dynamic model shows promise for accurately capturing turbulent behavior in computational fire modeling.

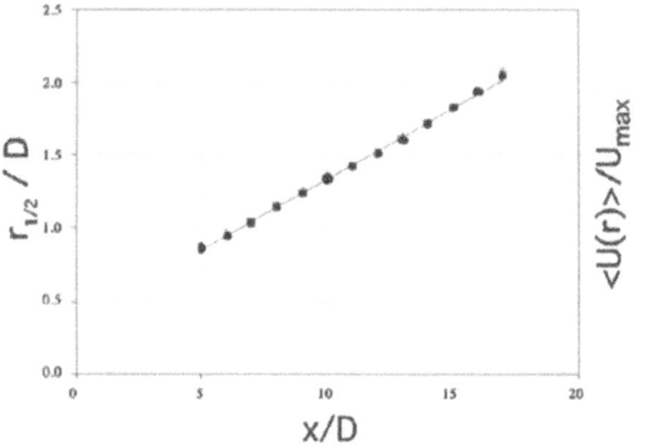

(a)

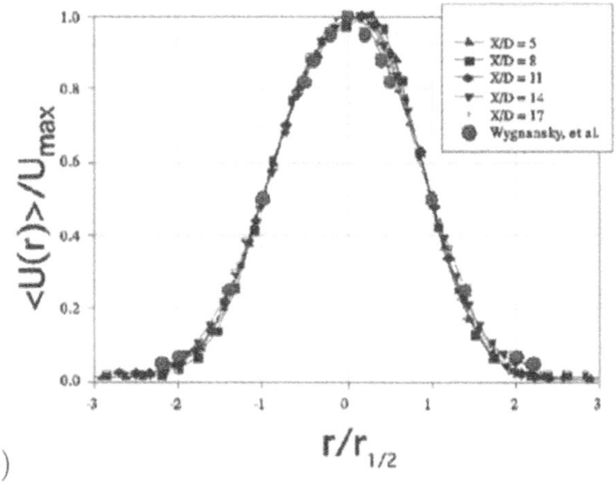

(b)

Figure 15.4. (a) Spreading rate (jet half-width as function of downstream distance) from coarse LES of round jet using the Lagrangian dynamic model. (b) Normalized mean axial velocity from LES using Lagrangian Dynamic Model and comparison with experimental data.

4. Dynamic modeling applied to premixed turbulent combustion

The previous section described the dynamic model as applied to modeling the subgrid transport of momentum. It is well known (see Meneveau & Katz, 2000) that the SGS fluxes of momentum and heat are dominated by local interactions, i.e. the dominant contributions to the SGS fluxes come from length-scales that are only marginally smaller than the filter size (e.g. the largest octave of unresolved scales). However, in recent years LES has also been applied increasingly for modeling reacting flows. In LES of premixed combustion, the typically thin reaction zone of the real flame front cannot be resolved on the computational grid and can be much smaller than the filter size Δ. This is an application where the dominant physics occurs at a scale which can be much smaller than the filter size. As will be shown below, this scale disparity can cause significant problems for the dynamic model when it is formulated in traditional fashion.

We restrict the discussion to modeling unresolved flame wrinkling in the context of the thickened flame model (TF-LES) as described in detail in Colin *et al.* (2000). More details of the dynamic model applied to this problem are given in Charlette *et al.* (2001b). In TF-LES the filtered reaction rate $\overline{\dot{\omega}}$ is modeled according to

$$\overline{\dot{\omega}} \; = \; \Xi_\Delta \, \frac{A}{F} \, \overline{\rho} \, \widetilde{Y_F} \, \widetilde{Y_O} \exp\left(-\frac{E_a}{R\widetilde{T}} \right), \qquad (15.7)$$

where F is the thickening factor defined as

$$F = \frac{\Delta}{\delta_l^0}, \qquad (15.8)$$

δ_l^0 is the laminar flame thickness, and

$$\Xi_\Delta \; = \; \frac{s_{T\Delta}}{s_l^0} \qquad (15.9)$$

is the wrinkling factor which accounts for increased reaction rate due to the increase in flame surface. $s_{T\Delta}$ is the SGS turbulent flame propagation speed. In addition to Eq. (15.7), the diffusion term is also renormalized (by a factor $F \cdot \Xi_\Delta$), but this need not be considered here since it does not affect the mean reaction rate.

A classical model (see e.g. Peters 2000) for turbulent flame speed or flame wrinkling factor can be written as follows

$$\frac{s_{T\Delta}}{s_l^0} \; \equiv \; \Xi_\Delta \; = \; \left(1 \, + \, \alpha \, \frac{u'_\Delta}{s_l^0} \right), \qquad (15.10)$$

where u'_Δ is the characteristic fluid SGS velocity, and α is an unknown model coefficient of order unity. A possible application of the dynamic model then is for the dynamic determination of the coefficient α.

The Germano identity for the reaction rate reads trivially as follows $\widehat{\dot{\bar{\omega}}} = \widehat{\dot{\bar{\omega}}}$, where an *over-brace* denotes a test-filter operation at scale $\gamma\Delta$ with $\gamma > 1$. The l.h.s. of this equation is interpreted as the reaction rate at scale Δ which has, subsequently, been test-filtered at scale $\gamma\Delta$. The r.h.s. is interpreted in the context of performing LES at scale $\gamma\Delta$. Replacing the model expressions (15.7) and (15.10) at the two scales leads to

$$\left\langle \overbrace{\left(1 + \alpha \frac{u'_\Delta}{s_l^0}\right) \frac{W_\Delta(\tilde{\mathbf{q}})}{\Delta}} \right\rangle = \left\langle \left(1 + \alpha \frac{u'_{\gamma\Delta}}{s_l^0}\right) \frac{W_{\gamma\Delta}(\widehat{\tilde{\mathbf{q}}})}{\gamma\Delta} \right\rangle, \quad (15.11)$$

where the term

$$W_\Delta(\tilde{\mathbf{q}}) = A\,\delta_l^0\,\bar{\rho}\,\widetilde{Y_F}\,\widetilde{Y_O}\exp\left(-\frac{E_a}{R\widetilde{T}}\right) \quad (15.12)$$

arises in the TF-LES expression for reaction rate in terms of the resolved variables $\tilde{\mathbf{q}} = (\widetilde{T}, \widetilde{Y_F}, \widetilde{Y_O}, \bar{\rho})$. $\widetilde{T}$, $\widetilde{Y_F}$ and $\widetilde{Y_O}$ are Favre-filtered temperatures and mass fractions, respectively, and $\bar{\rho}$ is filtered density. E_a and R are the activation energy and universal gas constant, respectively.

It is important to examine the asymptotic behavior in the limit $\alpha u'_\Delta / s_l^0 >> 1$. In this limit the fraction by which the turbulence increases the flame speed greatly exceeds unity. Then it becomes immediately clear that α cancels from each side of Eq. (1.11) and the Germano identity becomes useless in helping us determine α.

Instead, in Charlette *et al.* (2001a) a power-law model was proposed in terms of an inner cutoff scale η_c (defined as the average inverse radius of curvature of the flame) and an unknown scaling exponent β. The basic model for flame wrinkling factor is

$$\Xi_\Delta = \left(1 + \frac{\Delta}{\eta_c}\right)^\beta . \quad (15.13)$$

In principle the exponent β could depend on relevant parameters such as turbulence level, Reynolds number, proximity to walls, etc. As argued in Charlette *et al.* (2001a), this expression covers several possibilities: if the flame is not wrinkled, one has $\eta_c \to \infty$, leading to $\Xi_\Delta = 1$ and $s_{T\Delta} = s_l^0$. The same effect is obtained for $\beta = 0$. If β is assumed to be independent on scale (Δ or η_c), and if $0 < \beta < 1$, one recovers the fractal model with $\beta = D - 2$, where D is the fractal dimension of the flame. In Charlette *et al.* (2001a), following the procedures of Colin *et al.* (2000) and Meneveau & Poinsot (1991), a model was proposed for the inner cutoff scale η_c. We will employ this model and consider η_c as a prescribed quantity in the derivations that follow. However, this still leaves the exponent β undetermined.

In Charlette *et al.* (2001b), we have proposed applying the dynamic approach to find this exponent. The resulting power-law dynamic model consists in determining β according to

$$\beta = \log\left(\gamma \frac{\langle \ \widehat{W_\Delta(\tilde{\mathbf{q}})} \ \rangle}{\left\langle W_{\gamma\Delta}(\ \widehat{\tilde{\mathbf{q}}}\ )\right\rangle}\right) \Big/ \log\left(\frac{1+\frac{\gamma\Delta}{\eta_c}}{1+\frac{\Delta}{\eta_c}}\right). \tag{15.14}$$

This expression is obtained after replacing Eq. (15.13) in the Germano identity for reaction rate. We note that in this approach, no cancellations occurred.

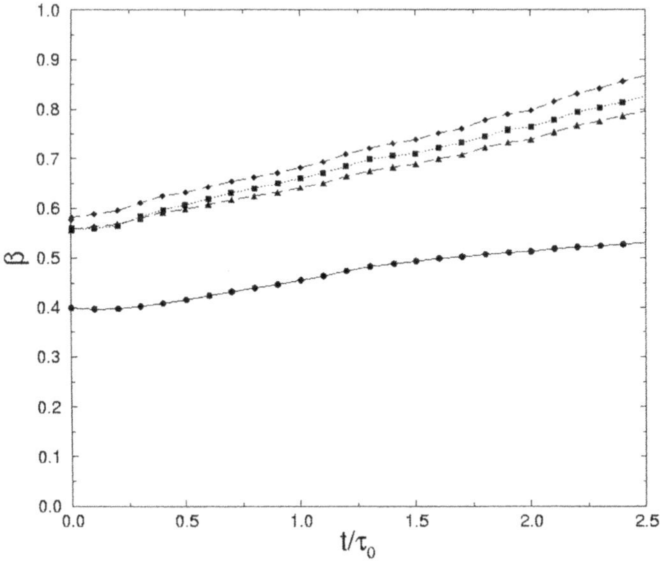

Figure 15.5. Time evolution of the dynamic exponent β in LES of flame in decaying isotropic turbulence. •: TF-1 ($F = 6.6$); ■: TF-2 ($F = 4.0$); ♦: TF-3a ($F = 2.8$, $\gamma = 2$); ▲: TF-3b ($F = 2.8$, $\gamma = 3$).

We have tested this approach in LES using a fully compressible DNS/LES code (NTMIX); see Stoessel (1995). The flow simulated is a premixed flame placed in 3D decaying isotropic turbulence. We perform DNS using 200^3 nodes, and perform several LES runs using various reso-

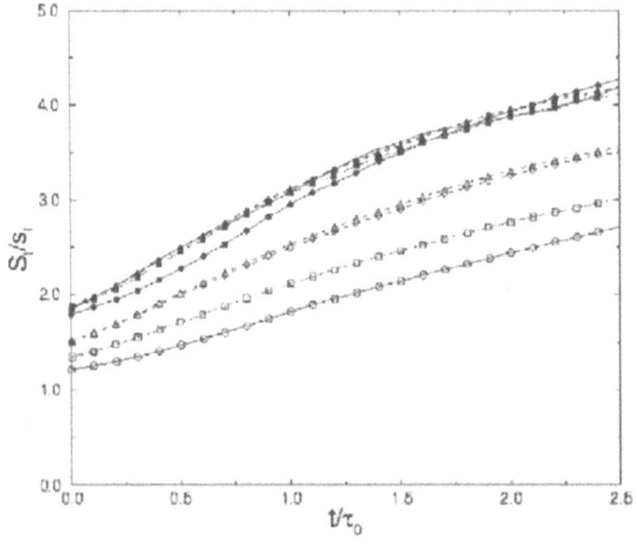

Figure 15.6. Time evolution of the turbulent flame speed in LES of flame in decaying isotropic turbulence using the power-law dynamic model. τ_0 is the turbulent eddy turn-over-time. Thick solid line: DNS; •, ○: TF-1 ($F = 6.6$); ■, □: TF-2 ($F = 4.0$); ♦, ◇: TF-3a ($F = 2.8$, $\gamma = 2$). ▲, △: TF-3b ($F = 2.8$, $\gamma = 3$). The empty symbols are for the resolved turbulent speed (computing the reaction rate without wrinkling factor), plain symbols are for the total turbulent speed (with wrinkling factor).

lutions (30^3, 50^3 and 70^3 with $F = 6.7$, $F = 4$ and $F = 2.8$, respectively). Figure 15.5 shows the exponents computed dynamically as function of time for the various cases. Clearly the results are fairly independent of F and γ except for the very coarse LES case. Considering that in this case the test-filter scale $\gamma\Delta$ corresponds to almost 1/3 of the entire simulation domain, it is not surprising that the result is less reliable. Nevertheless, the dynamically computed β begins at about $\beta = 0.55$, and rises steadily as function of time while the turbulence decays. Clearly the exponent is not a universal value. Physically, this time dependence can be interpreted as follows: while the turbulence decays (i.e. u'_Δ decreases) the flame wrinkling still is increasing due to non-equilibrium conditions in the dynamics of flame surface density. This is reflected in the smallest resolved scales of the LES and thus β increases.

In figure 15.6, the symbols denote the total and resolved part of the turbulent flame speed (or burning rate) from the LES, made non-dimensional by the laminar flame speed s_l^0. The figure shows results from different LES runs using the power-law dynamic model. They correspond to using different basic LES resolutions (from fine LES with $F = 2.8$ to very coarse LES with $F = 6$ using only 30^3 grid-points) with the same test-filter ratio $\gamma = 2$. Also, we show results for a larger test-filter ratio $\gamma = 3$. As expected, increasing the thickening factor F (or the parameter Δ/δ_l^0) reduces the contribution from the resolved part, but then the sub-grid portion increases. As shown in figure 15.6 compared to the DNS result, the overall turbulent speed is estimated fairly accurately by the various LES runs, including its temporal growth rate. Initially the very coarse LES underestimates the DNS reaction rate, but by no more than about 7%. More details can be found in Charlette *et al.* (2001b).

5. Conclusion

We have addressed several aspects of turbulent mixing and combustion associated with LES. We have identified particular challenges associated with scalar anisotropy, we have shown how the Lagrangian dynamic model can be deployed in predicting unresolved momentum fluxes in cases where the non-dynamic model fails, and we have presented a new dynamic model for premixed combustion. We have shown that the standard dynamic model which aims at determining multiplicative model coefficient is ill-posed in the context of finding parameters for modeling turbulent flame wrinkling in premixed combustion. The cause of the difficulties was argued to be the large scale disparity in premixed combustion, which differs from the closeness of the interactions that occur when modeling turbulent fluxes. It was shown that when seeking scaling exponents, however, the dynamic procedure leads to asymptotically well-posed expressions and procedures. We proposed a new dynamic method, the power-law dynamic model, to deal with such scale-disparate problems. Applications to flame propagation in decaying isotropic turbulence showed promising results.

Acknowledgments

We thank Professors Z. Warhaft and L. Mydlarski for useful comments and suggestions about the cold-wire calibration. This work was supported by the National Science Foundation (grant CTS-9803385) and the National Institute for Standards and Technology. C.M. gratefully acknowledges the hospitality and financial support from Ecole Centrale de Paris (Chatenay-Malabry, France) where much of the work on the power-law dynamic model was performed.

References

Bardina, J., Ferziger, J. H. and Reynolds, W. C. (1980) "Improved subgrid scale models for large eddy simulation," *AIAA Pap.* **25**, 80–1357.

Baum, H., McGrattan, K. and Rehm, R. (1996) "Large eddy simulations of smoke movement in three dimensions," *Proceedings of the Thirteenth Meeting of the UJNR Panel on Fire Research and Safety* edited by K. Beall, (NIST, Gaithersburg, MD), 249–256.

Borue, V. and Orszag, S. (1998) "Local energy flux and subgrid-scale statistics in three-dimensional turbulence," *J. Fluid Mech.* **366**, 1–31.

Cerutti, S. and Meneveau, C. (2000) "Statistics of filtered velocity in grid and wake turbulence," *Phys. Fluids* **12**, 1143–1165.

Cerutti, S., Meneveau, C. and Knio O.M. (2000) "Spectral and hyper eddy viscosity in high-Reynolds-number turbulence," *J. Fluid Mech* **421**, 307–338.

Charlette, F., Meneveau, C. and Veynante, D. (2001a) "Flame-wrinkling model and application in thickened-flame LES of premixed turbulent combustion," *Comb. Flame* (in preparation for submission).

Charlette, F., Meneveau, C. and Veynante, D. (2001b) "A power-law dynamic procedure for LES of turbulent premixed combustion," *Comb. Flame* (in preparation for submission).

Clark, R. G., Ferziger, J. H. and Reynolds, W. C. (1979) "Evaluation of subgrid models using an accurately simulated turbulent flow," *J. Fluid Mech.* **91**, 1–16.

Colin, O, Ducros, F., Veynante, D. and Poinsot T. (2000) "A thickened flame model for large eddy simulations of turbulent premixed combustion," *Phys. Fluids* **12**, 1843–1863.

Germano, M., Piomelli, U. Moin, P. and Cabot, W. (1991) "A dynamic subgrid-scale eddy viscosity model," *Phys. Fluids A* **3**, 1760–1765.

Hussein, H. J., Capp, S. and George, W. K. (1994) "Velocity measurements in a high-Reynolds-number, momentum-conserving, axisymmetric, turbulent jet," *J. Fluid Mech.* **258**, 31–75.

Im, H. G., Lund, T. S. and Ferziger J. H. (1997) "Large eddy simulation of turbulent front propagation with dynamic subgrid models," *Phys. Fluids* **9**, 3826–3833.

Kang, H. S. and Meneveau, C. (2001) "Passive scalar anisotropy in a heated turbulent wake: new observations and implications for large-eddy simulations," *J. Fluid Mech.* **442**, 161–170.

Kiya, M. and Matsumura, M. (1988) "Incoherent turbulent structure in the near wake of a normal plate," *J. Fluid Mech.* **190**, 343–356.

Leonard, A. (1974) "Energy cascade in large-eddy simulations of turbulent fluid flows," *Adv. Geophys* **18**, 237–248.

Leonard, A. (1997) "Large-eddy simulation of chaotic convection and beyond," *Am. Inst. Aeronaut. Astronaut. Pap. 97-0204: 1–8.*

Liu, S., Katz, J. and Meneveau, C. (1999) "Evolution and modelling of subgrid scales during rapid straining of turbulence," *J. Fluid Mech.* **387**, 281–320.

Liu, S., Meneveau, C. and Katz, J. (1994) "On the properties of similarity subgrid-scale models as deduced from measurements in a turbulent jet," *J. Fluid Mech.* **275**, 83–119.

Matsumura, M. and Antonia, R. A. (1993) "Momentum and heat transport in the turbulent intermediate wake of a circular cylinder," *J. Fluid Mech.* **250**, 651–668.

McGrattan, K., Baum, H. and Rehm, R. (1998) "Large Eddy Simulations of Smoke Movement," *Fire Science Journal* **30**, 161–178.

Mell, W., Johnson, A., McGrattan, K. and Baum, H. (1995) "Large eddy simulations of buoyant plumes," *Proceedings, Eastern States Section of Combustion Institute, Fall Technical Meeting* **HTD 304**, 187–190.

Mell, W., McGrattan, K. and Baum, H. (1995) "Large eddy simulations of fire driven flows," *National Heat Transfer Conference* **HTD 304**, 73–77.

Meneveau, C. and Poinsot, T. (1991) "Stretching and quenching of flamelets in premixed turbulent combustion," *Comb. Flame* **86**, 311–332.

Meneveau, C. and Katz, J. (2000) "Scale-invariance and turbulence models for large-eddy simulation," *Annu. Rev. Fluid Mech.* **32**, 1–32.

Meneveau, C., Lund, T. and Cabot, W. (1996) "A Lagrangian dynamic subgrid-scale model of turbulence," *J. Fluid Mech.* **319**, 353–386.

O'Neil, J. and Meneveau, C. (1997) "Subgrid-scale stresses and their modelling in a turbulent plane wake," *J. Fluid Mech.* **349**, 253–293.

Peters (2000) "*Turbulent combustion*," Cambridge University Press, Cambridge.

Piomelli, U. and Zang, T. (1991) "Large-eddy-simulation of transitional channel flow," *Comp. Phys. Comm.* 65–224.

Pope, S. B. (2000) "*Turbulent flow*," Cambridge University Press, Cambridge.

Smagorinsky, J. (1963) "General circulation experiments with the primitive equations. i. The basic experiment," *Mon. Weather. Rev.* 91–99.

Stoessel, A., (1995) "An efficient tool for the study of 3D turbulent combustion phenomena on MPP computers," *Proc. of the HPCN 95 Conference*, Milan (Italy), Springer-Verlag, 306–311.

Warhaft, Z. (2000) "Passive scalars in turbulent flows," *Annu. Rev. Fluid Mech.* **32**, 203–240.

Wygnanski, I. and Fiedler, H. (1969) "Some measurements in the self-preserving jet," *J. Fluid Mech.* **38**, 577–612.

Discussion

Emmanuel Villermaux, Marseille "There are indications both numerically and experimentally that the wrinkles on the flame front are very weakly dependent on the structure of the turbulence producing the wrinkles; in other words, the fractal dimension of the front if you make it fractal is very weakly dependent on the exponent of the spectrum of the turbulence. Do you agree with this statement? Is it consistent with what you showed and what is the meaning of the seven third that you showed at the end?"

Reply: I haven't looked at this long enough to give you a concise answer. What I don't believe is that there is a universal exponent, let's say of seven thirds, which would serve our purpose in fact which is exactly the point. I think the strongest dependency you can see there is essentially a time dependence; so, for example, relating that exponent to a spectral exponent of the velocity may make sense if you are in a hugely asymptotic regime, if the spectrum does not change in time. I believe this approach, which is a little bit of what you have, I've seen papers of yours that have exponents that depend on time. Certainly this is exactly what seems to be happening in this case. The one assumption that we have made here that we don't feel very secure about is to assume that this exponent is independent of scale. That is a strong assumption but in this case it seemed to work. I believe the process can be generalized if you do test filtering at various scales you can get an idea of how this beta depends on scale; for example, this might be a way of generalizing this.

Jeff Lienau, Pratt and Whitney, East Hartford, Conn. "Just wondering how well you think or just from your gut feeling or any experience you have, how well this wrinkle flamelet modeling will apply over different combustion regimes going from the wrinkle flamelet right up into the other regimes. How universal is it?"

Reply: From what I have seen here probably not too universal I probably am not the best person to answer that particular question in terms of combustion I have mainly look at this particular problem solving. The question you are asking is probably not the best one for me to answer. There are plenty of people in the room that will give you a better answer. A more philosophical answer. To bridge different regimes with a single model I believe this approach is probably better then other because we are at least letting the coefficients or the parameters of the model change gradually as the physics of the resolved scales changes so in terms of going in this direction is more hopeful in this direction then this is my model in this regime and the coefficients are this much and then somebody comes with a different regime and I have to change the coefficient this may allow you to gradually bridge different regimes.

Steve Pope, Cornell University "I have a question on the middle part of your talk on the coarse LES with the use of the dynamic model and there is a criticism of the dynamic model by Moser and Jimenez (LES: Where are we and what can we expect, AIAA paper 98-2891, 1998, ed.) where they say that success of the model does not seem to be at all connected to the physics on which it is based. When you presented this you presented an equation for the Germano identity; and in fact it had two equal signs: the first part was an identity but the second part involved the Smagorinsky constant and involves the assumption of scale similarity and in the sketch you used to explain this you showed a high Reynolds number spectrum with a filter delta and then twice the filter of two delta both in the initial sub range. Then you apply this model to this coarse LES where for sure the filter is not in the inertial sub range and there is no reason to suppose this scale similarity. So would you accept that this is another example of a dynamic model working for reasons that we don't understand?"

Reply: I agree with the middle part of your assessment. I think that is probably too harsh in some sense the basic issue that we are using. In this case the simulation worked because we started with the laminar jet, we apply the Germano identity and we find that the resolved scales are not turbulent and so the coefficient is small and almost zero. As you know it is not exactly zero but it is small enough to allow the turbulence to become transitional. So the question now is that 0 or .01 we are getting in the non turbulent region, is that really physical? I agree with you probably not exactly. There are large errors involved because the filters are not in the inertial range but the 0th order. Whether we are getting a very small or finite coefficient, which in this case made all the difference. That is, physically the large scales are laminar when we started the simulation and the coefficient is small that is not some accident that we were just lucky, that is real physics in my view. Whereas the precise value of the coefficient that we are getting may be wrong by forty percent because the filter is not really in the inertial range for example.

Jim Riley, University of Washington "When you went from the laminar to the turbulent like case did you need to force that at all or does it develop on its own?"

Reply: No it just developed on its own in this case.

Santanu Sarkar, San Diego "I had a question on your simulation of the jet and you applied your Lagrangian model. Did it have the same coarse grid as the grid that you used for the regular Smagorinsky model or did you use a fine grid?"

Reply: No. No, we used exactly the same grid.

Santanu Sarkar, San Diego "So you got a turbulent jet even without resolving the shear layers at the edge of the jet?"

Reply: Across the jet diameter we had four grid points; this is very, very coarse. We put a profile which is like a thick shear layer type profile. So it starts very, very thick right from the start in fact the virtual origin we get for our profiles is right at the nozzle exit so we don't get any shear layers developing on the sides or anything like that.

Santanu Sarkar, San Diego "If you use the regular dynamic model without the Lagrangian approach would you still get the shear layer?"

Reply: We have not.

Santanu Sarkar, San Diego "The reason I ask is that we tried that and we basically set a coarse grid and were not able to get a turbulent jet using either a classical Smagorinsky or a dynamics model. Of course a dynamics model worked much better with a grid that is not as coarse as four points across."

Reply: The problem then is essentially that I believe that one has to do some averaging in this case you have a circumferential direction but there are not enough grid points to really get any statistics from one circle around.

Santanu Sarkar, San Diego "The trick was the time thing..."

Reply:You have to do a lot of clipping in the standard non-Lagrangian in this case and probably has some effect where as we don't need to do any clipping really.

Santanu Sarkar, San Diego "Okay, thank you.

Chapter 16

PDF OF TEMPERATURE FLUCTUATIONS IN UNIFORMLY SHEARED TURBULENCE

M. Ferchichi
Department of Mechanical Engineering, Lakehead University
Thunder Bay, Ontario, Canada P7B 5E1
mohsen.ferchichi@lakeheadu.ca

S. Tavoularis
Department of Mechanical Engineering, University of Ottawa
Ottawa, Ontario, Canada K1N 6N5
tav@eng.uottawa.ca

Abstract Statistics of temperature fluctuations were measured in uniformly sheared turbulence with a uniform mean temperature gradient. The scalar pdf was Gaussian and the scalar-velocity joint pdf was jointly Gaussian, with the conditional expectations of the velocity fluctuations linearly dependent on the scalar value. The scalar dissipation rate was statistically independent from the scalar value. A Gaussian scalar pdf was also predicted analytically.

Keywords: Turbulence, shear, passive scalar, pdf, Gaussian

1. Introduction

Compared to moment-based formulations, the probability density function (pdf) formulation of scalar transport and mixing has the advantage that it does not require "closure" schemes, other than those used for the velocity field. The shape of the scalar pdf in homogeneous flows, and specifically whether it is Gaussian or not, remains a controversial issue. A number of experimental and analytical/numerical studies (Tavoularis and Corrsin, 1981; Thoroddsen and Van Atta, 1992; Mydlarski and

A. Pollard and S. Candel (eds.), IUTAM Symposium on Turbulent Mixing and Combustion, 201-209.

Warhaft, 1998, among others) confirmed Gaussianity, while others (Gollub et al., 1991; Jayesh and Warhaft, 1992) concluded that this pdf had tails broader than Gaussian (e.g. exponential). The scalar pdf equation contains expectations of the velocity and the scalar "dissipation rate" conditioned upon the scalar value. Much effort has been devoted to the study of a possible relationship between the shape of the scalar pdf and the conditional expectation of its dissipation rate (Eswaran and Pope, 1988; Jayesh and Warhaft, 1992; Kailasnath et al., 1993; Anselmet et al., 1994); such a relationship has also been examined in the context of local isotropy of the scalar (Mi et al., 1995). On the other hand, little attention has been paid to the effects of the conditional expectations of the velocity fluctuations on the scalar pdf, although it is known that such effects may be significant (Overholt and Pope, 1996).

The present paper contains an analysis of the scalar pdf equation and pdf measurements in uniformly sheared turbulence with a uniform mean scalar gradient, which was also the configuration used by Tavoularis and Corrsin (1981). It is hoped that the present results will help resolve the controversies surrounding the issue of scalar pdf. More details are given by Ferchichi (2000) and Ferchichi and Tavoularis (2001).

2. An Analytical Prediction

The pdf of a scalar random process $\theta(x_i,t)$ is defined as $p(\psi,x_i,t) = \langle \delta\left[\theta(x_i,t)-\psi\right]\rangle$, where δ is Dirac's delta function and the brackets $\langle\rangle$ indicate ensemble averaging. An equation for the evolution of the pdf may be obtained as

$$\begin{aligned}\frac{\partial p}{\partial t} &+ \langle U_j\rangle \frac{\partial p}{\partial x_j} = -\frac{\partial\left(\langle u_j|\theta=\psi\rangle p\right)}{\partial x_j} - \frac{\partial\langle \theta u_j\rangle}{\partial x_j}\frac{\partial p}{\partial \psi}\\ &+ \frac{\partial\langle T\rangle}{\partial x_j}\frac{\partial\left(\langle u_j|\theta=\psi\rangle p\right)}{\partial \psi} - \frac{\partial^2\left(\langle \epsilon_\theta|\theta=\psi\rangle p\right)}{\partial \psi^2} + \gamma\frac{\partial^2 p}{\partial x_j \partial x_j}\end{aligned} \tag{16.1}$$

In this equation, $\langle u_j|\theta=\psi\rangle$ and $\langle \epsilon_\theta|\theta=\psi\rangle$ denote, respectively, the expectations of the velocity fluctuations and of the fluctuating scalar destruction rate, $\epsilon_\theta = \gamma(\frac{\partial\theta}{\partial x_i}\frac{\partial\theta}{\partial x_i})$, conditioned upon the scalar value.

For stationary turbulent velocity and scalar fields, having uniform mean transverse gradients and transverse homogeneity, and omitting terms which experimental evidence proves to be negligible, equation (16.1) may be simplified to

$$\overline{U}_1\frac{\partial p}{\partial x_1} = \frac{d\overline{T}}{dx_2}\frac{\partial\left(\langle u_2|\theta=\psi\rangle p\right)}{\partial \psi} - \frac{\partial^2\left(\langle \epsilon_\theta|\theta=\psi\rangle p\right)}{\partial \psi^2}. \tag{16.2}$$

This equation is analogous to the equation for the scalar variance. However, due to the conditional expectations appearing in it, it cannot be solved for the pdf without further assumptions. The desired solution must be self-similar, which implies that $p(\psi, x_1) = f(\eta)/\theta'(x_1)$, where $\eta = \psi/\theta'(x_1)$. Once more resorting to the experimental evidence, one may consider that the conditional expectations of u_1 and u_2 are linearly related to the value of the scalar, namely that $\langle u_i|\theta = \psi\rangle = \rho_{i\theta} u_i' \eta$, and that the scalar dissipation rate is independent of the value of the scalar, such that $\langle \epsilon_\theta|\theta = \psi\rangle = \langle \epsilon_\theta\rangle \equiv \chi$.

Then, it is easy to transform equation (16.2) into the form $\frac{d^2 f}{d\eta^2} + \eta \frac{df}{d\eta} + f = 0$, which has the Gaussian function as its solution.

3. Experimental Conditions

The experiments were conducted in a wind-tunnel, having a test section with a height of h = 305 mm, a width of 457 mm and a length of 5180 mm. The mean shear was produced by a variable-obstruction-type shear generator, followed by a set of parallel channels to create uniform length scales (flow separator). The heating system, consisting of 47 heating ribbons, 0.8 mm wide, 0.08 mm thick and 6.3 mm apart, was inserted normal to the flow at a position 792 mm (2.6h) downstream from the exit of the flow separator, which was taken as the origin of the x_1 axis. The velocity fluctuations were measured using a sub-miniature cross-wire probe, while temperature fluctuations were measured with cold wire probes, with sensors made of 1 μm diameter, 100% Platinum wire, and having a typical length of about 0.5 mm. Streamwise derivatives were evaluated using Taylor "frozen flow" approximation. The scalar dissipation rate was evaluated from its streamwise component (Kailasnath et al., 1993; Jayesh and Warhaft, 1992). Typical values of the Taylor and Corrsin microscales were λ_{11} = 5.2 mm and $\lambda_{\theta 1}$ = 4.6 mm, throughout the measuring range, while the Kolmogorov microscale η was found to decrease from about 0.20 mm to about 0.17 mm, along the measuring section. Re_λ varied from 184 to 253.

The flow speed on the centerline of the tunnel was U_c = 6.6 m/s, while the centerline mean temperature rise was ΔT_c = 1.37 K. Transverse profiles of the mean velocity and mean temperature displayed good linearity and downstream constancy in the core of the test section. The mean shear was $d\overline{U}_1/dx_2 \approx 40$ s^{-1} and the mean temperature gradient was $d\overline{T}/dx_2 \approx 7.8$ K/m in the core of the wind tunnel.

4. Results and Discussion

Figure 1 shows that the pdf of the velocity and scalar fluctuations were essentially Gaussian, in agreement with the Tavoularis and Corrsin (1981) measurements. In particular, the scalar skewness and flatness were, respectively, -0.19±0.06 and 3.15±0.08 over the range of Re_λ considered, only slightly deviating from the Gaussian values.

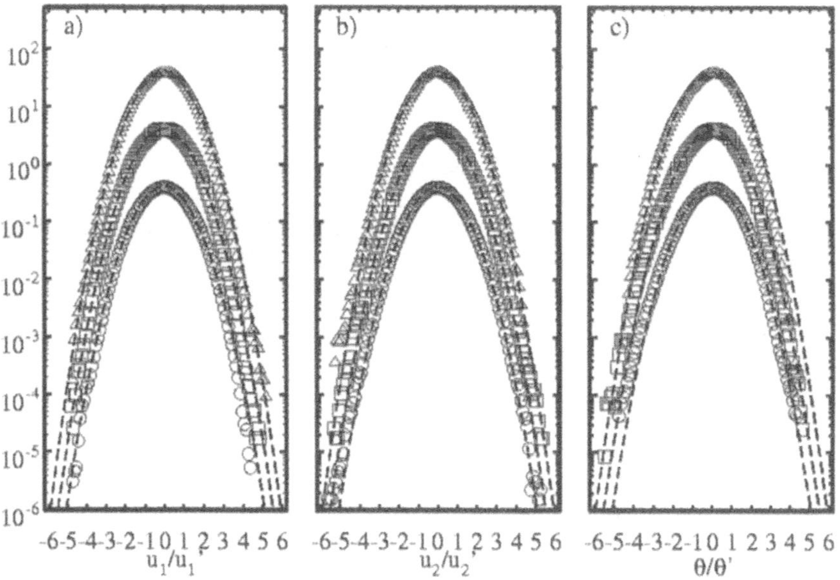

Figure 16.1. Pdf of the velocity and temperature fluctuations: a) $p(u_1/u_1')$, b) $p(u_2'/u_2')$ and c) $p(\theta/\theta')$; Re_λ = 184 (○), 200 (∇ , values multiplied by 10) and 253 (△, values multiplied by 100); dashed lines indicate Gaussian pdf.

As noted in the introduction, there has been some controversy about the shape of the tails of the scalar pdf. It has been suggested that there exists a transitional Reynolds number, above which the tails of the scalar pdf would change from Gaussian to exponential and Jayesh and Warhaft (1992) recommended the value of $Re_L = 70$, based on the integral length scale. In both the present and the Tavoularis and Corrsin (1981) experiments, Re_L was much greater than this value (for example, for Re_λ= 253, Re_L = 2380), without any hint of exponential tails. Furthermore, the lack of exponential tails in the scalar pdf cannot be attributed to the limited temperature range across the tunnel, because the present mean scalar profile extended by ± 10 scalar standard deviations on either side of the measuring point; for comparison, the mean scalar profile in the

Jayesh and Warhaft (1991) experiments extended by about ± 6 scalar standard deviations from their centerline. Another argument (Warhaft, 2000), that exponential tails of the scalar pdf would only occur if the wind tunnel cross section extended by at least 8 integral length scales, also seems to be contradicted by the present experiments, in which the ratio $h/L_{11,1}$ varied from 14 to 6, while the ratio $h/L_{\theta,1}$ varied from 30 to 8. The measurements presented here correspond to the range $8 < h/L_{\theta,1} < 12$.

Isocontours of the joint pdf of the temperature and velocity fluctuations, shown in figure 16.2, displayed only minor differences from the corresponding contours of the jointly normal pdf, in agreement with Tavoularis and Corrsin (1981).

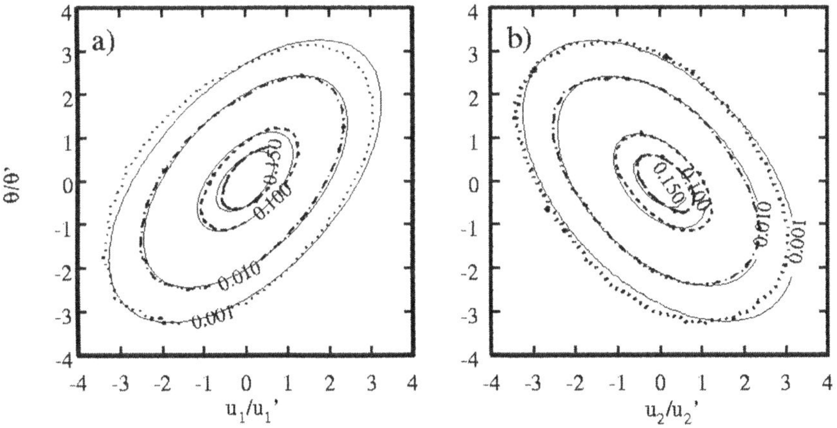

Figure 16.2. Isocontours of the joint pdf of the scalar and a) the streamwise and b) the transverse velocity fluctuations; solid lines indicate the corresponding contours of jointly Gaussian pdf with the same correlation coefficients as the measured ones.

The measured expectations of the velocity fluctuations conditioned upon the scalar are shown in figure 16.3a,b. The same figure shows straight lines with slopes equal to the corresponding correlation coefficients, representing jointly Gaussian random processes. This figure clearly demonstrates that the measured values of both conditional expectations adhered closely to the Gaussian lines, except possibly at large deviations of the scalar fluctuations, for which the experimental uncertainty is also increased. Nearly linear conditional expectations have also been observed by Tong and Warhaft (1995) in a heated jet and by Overholt and Pope (1996) in a numerical simulation.

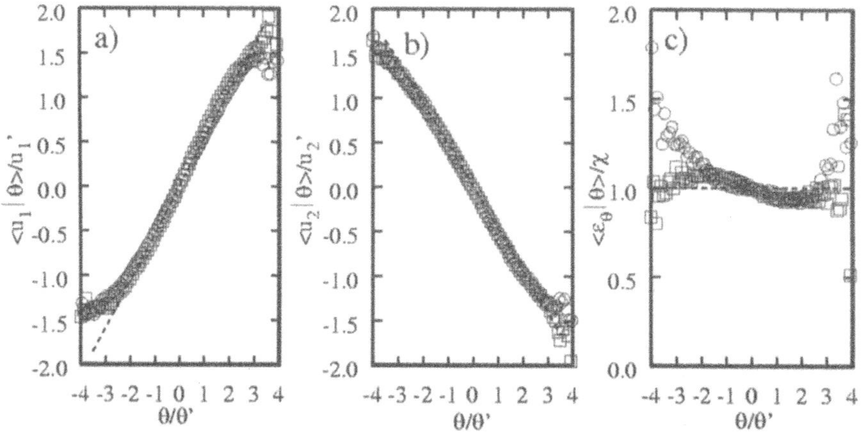

Figure 16.3. Normalized expectations of the a) streamwise velocity, b) transverse velocity and c) scalar dissipation rate, conditioned upon the scalar values; $Re_\lambda = 200$ (○), 253 (∇).

Local isotropy requires that the fine structure of the scalar field should be independent of the motions that contribute mostly to the scalar variance. A relevant indicator is the correlation coefficient between the scalar fluctuations and their dissipation rate (Eswaran and Pope, 1988). For all cases examined, the measured correlation coefficient was, in magnitude, less than 0.02, suggesting that the scalar and its destruction rate were essentially statistically independent. Measurements of the expectation of the scalar dissipation rate, conditioned upon the scalar and normalized by the average dissipation rate, are shown in figure 16.3c. The values were 1.0±0.1 for $|\theta/\theta'| < 2$, although displaying substantial scatter for $|\theta/\theta'| > 2$. The scatter of this parameter seems to decrease as Re_λ increases, which reflects an improved transverse homogeneity with increasing distance from the heating elements.

A stronger test of the statistical independence of θ and ϵ_θ may be made by considering their joint pdf. Statistical independence requires that $p(\theta, \epsilon_\theta) = p(\theta)p(\epsilon_\theta)$. Iso-probability contours of $p(\theta, \epsilon_\theta)$ are displayed in figure 16.4. These contours are nearly symmetrical with respect to the mean of the scalar fluctuations, suggesting that the fluid is well mixed and that the scalar dissipation gets nearly equal contributions from motions with negative and positive scalar fluctuations. Contours of the product $p(\theta)p(\epsilon_\theta)$, also plotted in figure 4, essentially collapsed on the corresponding contours of $p(\theta, \epsilon_\theta)$.

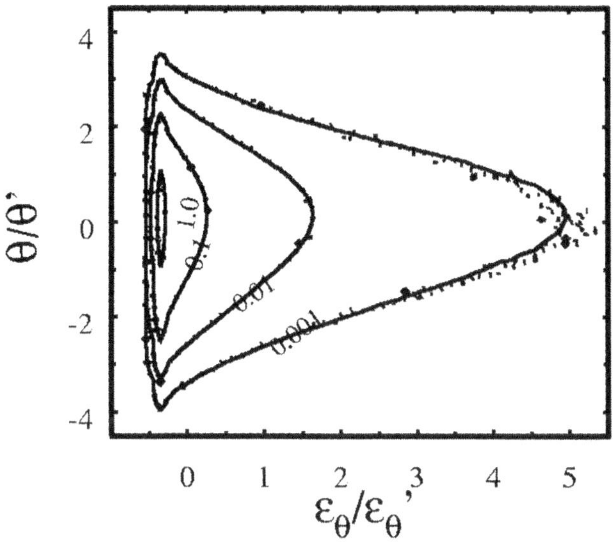

Figure 16.4. Contours of the joint pdf of the scalar and its dissipation rate (dots) and corresponding contours of the product of the pdf of the scalar and the pdf of the scalar dissipation rate (solid lines).

Based on the above tests, it seems reasonable to conclude that the conditional expectation of the scalar dissipation rate in uniformly sheared turbulence is independent of the scalar value. Such independence has also been observed, both experimentally and computationally, in other flows in which the scalar pdf was found to be Gaussian (e.g. Anselmet et al., 1994; Mi et al., 1995; and Overholt and Pope, 1996). In contrast, the conditional measurements of ϵ_θ by Jayesh and Warhaft (1992), in a flow in which the scalar pdf was non-Gaussian, displayed a V-shape.

5. Conclusions

The objective of the present study was to examine the pdf of a passive scalar field in uniformly sheared turbulent flow. An analysis of the scalar pdf equation, taking into consideration some experimental evidence, has concluded that the scalar pdf must be Gaussian. Measurements of the scalar pdf also clearly indicate the same. The joint pdf of the scalar and the velocity fluctuations were jointly Gaussian, with the conditional expectations of the velocity fluctuations linearly dependent on the scalar value. Finally, the scalar dissipation rate was found to be statistically independent from the scalar itself.

References

Anselmet, F., Djeridi, H. & Fulachier, L. (1994). Joint statistics between a passive scalar and its dissipation in turbulent flows. J. Fluid Mech. 280, 173.

Eswaran, V. & Pope, S.B. (1988). Direct numerical simulations of the turbulent mixing of a passive scalar. Phys. Fluids 31, 506.

Ferchichi, M. (2000). *Measurements of Small-Scale Statistics and Probability Density Functions in Passively Heated Shear Flow.* Ph.D. Dissertation, Department of Mechanical Engineering, University of Ottawa, Ottawa, Canada.

Ferchichi, M. & Tavoularis S. (2001). Scalar pdf and fine structure in uniformly sheared turbulence. J. Fluid Mech. (submitted).

Gollub, J.P., Clarke, J., Gharib, M., Lane, B. & Mesquita, O.N. (1991). Fluctuations and transport in a stirred fluid with a mean gradient. Phys. Rev. Lett. 67, 3507.

Jayesh & Warhaft, Z. (1992). Probability distribution, conditional dissipation, and transport of passive temperature fluctuations in grid-generated turbulence. Phys. Fluids A 4, 2292.

Kailasnath, P., Sreenivasan, K.R. & Saylor, J.R. (1993). Conditional scalar dissipation rates in turbulent wakes, jets, and boundary layers. Phys. Fluids A 5, 3207.

MI, J., Antonia R.A. & Anselmet, F. (1995). Joint statistics between temperature and its dissipation rate components in a round jet. Phys. Fluids 7, 1665.

Mydlarski, L. & Warhaft, Z. (1998). Passive scalar statistics in high-Péclet-number grid turbulence. J. Fluid Mech. 358, 135.

Overholt, M.R. & Pope, S.B. (1996). Direct numerical simulation of a passive scalar with imposed mean gradient in isotropic turbulence. Phys. Fluids 8, 3128.

Tavoularis, S. & Corrsin, S. (1981). Experiments in nearly homogenous turbulent shear flow with a uniform mean temperature gradient. J. Fluid Mech. 104, 311.

Thoroddsen, S.T. & Van Atta, C.W. (1992). Exponential tails and skewness of density-gradient probability density functions in stably stratified turbulence. J. Fluid Mech. 244, 547.

Tong, C. & Warhaft, Z. (1995). Passive scalar dispersion and mixing in a turbulent jet. J. Fluid Mech. 292, 1.

Warhaft, Z. (2000). Passive scalars in turbulent flows. Ann. Rev. Fluid Mech. 32, 203.

Discussion

James Riley, University of Washington "There are several ways to generate temperature fluctuations and one is with a uniform profile but another is if you have a jet with a fuel or something coming out and where you have 1 and 0 and I'm wondering if you would expect these types of results to hold in the second case."

Reply: There has been a lot of DNS studies where they use two delta functions I think one by Eswaran and Pope (Direct numerical simulations of the turbulent mixing of a passive scalar," Phys. of Fluids 31, 506 (1988), ed.) and the longtime PDF was basically a Gaussian PDF. It might be extended to do some experimental work in there too, I assume.

Chapter 17

A MATHEMATICAL PROTOTYPE TO VALIDATE LES STRATEGIES FOR TURBULENT FLAMES

A.Bourlioux, B.Khouider, & A.J.Majda

Département de Mathématiques et Statistique, Université de Montréal
& Courant Institute, New York University

Anne.Bourlioux@Umontreal.ca

Abstract An idealized model for turbulent flames is introduced. It consists of a scalar advection-reaction-diffusion equation that describes the interaction of a thin flame with a turbulent-like flow field acting on two separate scales. Rigorous asymptotic results as well as affordable and reliable direct numerical simulations are available to describe the effective large scale behavior of the idealized front; a strategy is proposed to use this framework to validate closure models for more realistic turbulent flames. Examples are given to illustrate the strategy as applied to premixed flames.

Keywords: turbulent burning velocity, large eddy simulation, direct numerical simulation, bending.

1. Overall strategy

The idealized model describes the evolution of the reactive scalar T which can be viewed as either the temperature field or a reactant mass fraction. The scalar obeys an advection-diffusion-reaction equation of the form:

$$\frac{\partial T}{\partial t} + (\overline{\mathbf{v}}(\mathbf{x},t) + \mathbf{v}'(\mathbf{x}/\epsilon^{\alpha}, t/\epsilon^{\alpha})).\nabla T = \epsilon \Delta T + \frac{1}{\epsilon} f(T) \qquad (17.1)$$

with $\epsilon > 0$ and $0 < \alpha < 1$ given. The nonlinear reaction rate function $f(T)$ must also be given; an example of reaction rate corresponding to a nonpremixed flame can be found in [Bourlioux and Majda, 2000] while an example for a premixed flame is given in the next section. The scaling in the equation is such that, in the absence of turbulent advection, the

A. Pollard and S. Candel (eds.), IUTAM Symposium on Turbulent Mixing and Combustion, 211-219.

combination of slow diffusion (order ϵ, with ϵ small) and fast reaction (reaction rate $f(T)/\epsilon$) leads to a very thin flame whose width scales like ϵ.

The synthetic velocity field on the left hand side must also be specified a priori. It has two components, both assumed to be incompressible: a large scale flow $\overline{\mathbf{v}}$ and a perturbation $\mathbf{v}'$ that acts at the scale ϵ^α intermediate between the flame thickness ϵ and the large scale (normalized here to be of order 1). An example of such flow is given in the next section. Turbulent mixing by the intermediate scale velocity always leads to an enhancement of the effective burning rate. There are two related point-of-views as far as characterizing that enhancement, and more generally describing the solution for T at large scales:

1 **mathematical statement of the problem:** what is the solution for T in the limit of $\epsilon, \epsilon^\alpha \to 0$, i.e. what is the effective behavior of the thin flame front at large scales, taking into account the effects of its distortion at the intermediate scale by the velocity perturbation $\mathbf{v}'$?

2 **practical statement of the problem**: what is an efficient procedure to compute numerically the solution for T at large scales when both the very small flame thickness ϵ and the intermediate scale ϵ^α associated with the turbulent flow $\mathbf{v}'$ are smaller than the discretization grid size ?

The practical formulation of the problem in 2 is the classical closure problem that one must tackle in large eddy simulations: the challenge is to account properly for the subgrid effects that contribute to the solution at large scales. There are two such effects - the first one is advection, which corresponds to the classical non-reactive turbulence closure problem, in the non-reactive case it is dealt with by specifying an enhanced diffusivity. The second one is specific to reactive turbulence and is associated with the nonlinear reaction rate function $f(T)$. There have been many propositions for ad-hoc closure in the reactive cases that capitalize on the fact that, at least in the flamelet regime, the reaction zone is very thin compared to all turbulent length scales. The strategy proposed here is to exploit the rigorous results from the mathematical approach as stated in 1 to offer a fresh perspective on the practical closure issue as stated in 2. This asymptotic approach to closure validation has two key advantages, both direct consequences of the relative simplicity of Eq.17.1 compared with the full reactive Navier-Stokes equations:

- Rigorous asymptotic results are available to provide useful information regarding the effective behavior at large scales, for example: scaling of subgrid models, upper bounds on effective transport coefficients, etc.;

- Eq.17.1 is orders of magnitude easier to solve numerically than the full reactive Navier-Stokes equations so that cheap, extensive, reliable numerical databases can be generated.

This overall strategy has been used very successfully in the past in the case of passive scalar turbulent advection - see [Majda and Kramer, 1999] for a self-contained introduction to that topic, with results that are directly relevant to the validation of flamelet models for large eddy simulations of nonpremixed flames [Bourlioux and Majda, 2000]. Below, the strategy is applied to an idealized turbulent premixed flame. A short introduction of the relevant homogenization results is given in Section 2. A combination of asymptotic predictions and numerical simulations is used in Section 3 to revisit two practical questions relevant to LES of premixed flames: how to identify a priori an infinitely thin front from direct numerical simulations and how to scale the subgrid burning speed enhancement.

2. Specific set-up for premixed flames

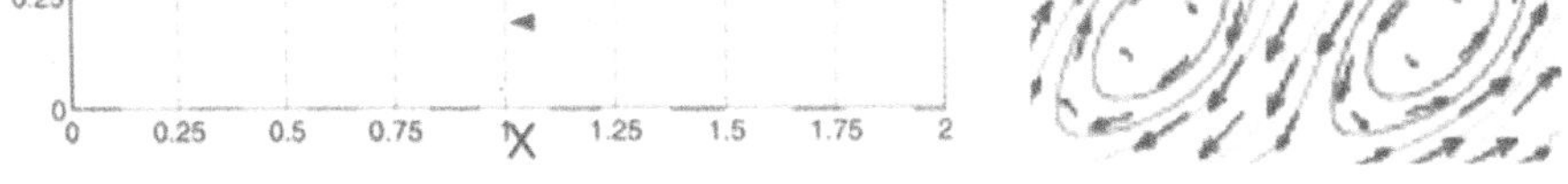

Figure 17.1. On the left: set-up for the premixed flame test-case with $\epsilon^\alpha = 1/4$. Each ϵ^α square tile corresponds to the streamlines on the right.

A premixed flame test case is set-up by assuming a KPP-type reaction rate function: $f(T) = T(1 - T)$ where the temperature T has been normalized so that the unburnt gas temperature corresponds to $T = 0$ and the burnt temperature to $T = 1$. Eq.17.1 is solved in a two-dimensional domain with periodic conditions in y. The initial conditions correspond to unburnt gas on the left side of the domain and burnt gas on the other side, so that overall the flame front is moving from right to left. The large scale velocity field is a simple horizontal shear

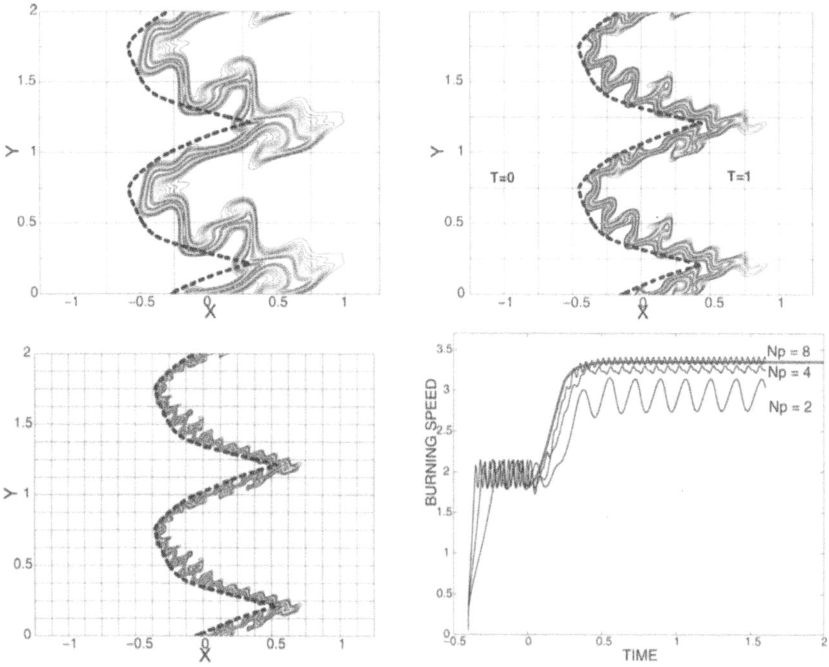

Figure 17.2. Direct numerical simulation results. Top left: $\epsilon = 1/4$, top right: $\epsilon = 1/16$ and bottom left: $\epsilon = 1/64$ - finite thickness reaction rate contours along with the prediction for the asymptotic front (thick dash-line). Bottom right picture shows the effective burning speed history for all three direct numerical simulations along with the asymptotic prediction (in that figure, $N_p = 1/\epsilon^\alpha$).

$\overline{\mathbf{v}} = (\sin(2\pi y), 0)$ and the intermediate scale flow field is obtained from the Childress-Soward stream function $F(x', y') = \sin(\pi x'/2) sin(\pi y'/2) + 0.5\cos(\pi x'/2)\cos(\pi y'/2)$, with $(x', y') = (x/\epsilon^\alpha, y/\epsilon^\alpha)$. This corresponds to a mix of shears and eddies, with the streamlines shown in Fig.17.1 along with the initial set-up for the simulations.

2.1. Direct numerical simulations

Results from the direct numerical simulation are reported in Fig.17.2. Eq.17.1 was solved with a standard second order finite difference scheme, with $\alpha = 1/2$ and for three different values for $\epsilon = 1/4, 1/16, 1/64$. The purpose of that sequence of simulations is to illustrate the asymptotic convergence of the solution as $\epsilon \rightarrow 0$. The top two plots in Fig.17.2 as well as the left bottom plot show a snapshot of the reaction rate contour for each of the three values of ϵ along with the asymptotic flame front predicted by the theory (see below), with very good agreement between the simulations and the asymptotic prediction (more comments on this in Section 3.1). The fourth plot represents the time history of the overall flame propagation speed. The three curves with the oscillations correspond to the finite ϵ direct simulations, the monotone curve corresponds to the asymptotic simulation. There are two phases in the computations: from time $t = -0.5$ to $t = 0$, only the intermediate scale velocity $\mathbf{v}'$ is active; at time $t = 0$, the large scale shear is turned on and the computations are carried on until the flame speed converges to a steady mean. Again, good convergence to the asymptotic prediction is observed as $\epsilon, \epsilon^{\alpha} \to 0$.

2.2. Asymptotic predictions

The rigorous asymptotic theory to predict the behavior of the solution in the limit of $\epsilon \to 0$ was developed in [Majda and Souganidis, 1994]. We omit here the technical details and give only a brief sketch of the predictions for the present test-case:

- In the limit of $\epsilon \to 0$, the reaction zone tends to an infinitely thin front moving normal to itself with an enhanced burning speed $S_T(x, y, t) > S_L$ (with $S_L = 1$ the laminar burning speed) as well as due to advection by the large scale flow.

- The enhanced speed $S_T(x, y, t)$ is obtained via the minimization over a scalar of the eigenvalue of a nonlinear Hamilton-Jacobi partial differential equation, the "cell-problem". This is a challenging computational problem, see [Khouider and Bourlioux, 2001, Khouider et al., 2001] for details on the equations and their numerical solution.

- In the present case, $S_T(x, y, t)$ depends to leading order on the following local parameters: the local large scale flow intensity $|\overline{v}(x, y)|$ and the local inclination angle $\theta(x, y, t)$ of the front. (Given the present set-up, only those two parameters control the speed at which the front is advected across a small periodic box, the faster the front is swept along, the smaller the burning speed enhancement).

The asymptotic predictions for the front shown in Fig.17.2 were obtained based on those theoretical results by performing an "asymptotic" large eddy simulation according to the following procedure:
1. Preliminary step: generate an "asymptotic flamelet library" for the subgrid enhanced burning speed $S_T = S_T(|\overline{v}|, \theta)$ by solving the ap-

propriate cell-problem over the expected range of the two independent variables;

2. Front evolution: use the level-set technique to evolve the front interface due to advection by the large scale shear and to burning normal to itself with local speed S_T. S_T is obtained at each grid point, each time step, by interpolation from the library given the local alignment of the front at the previous time-step and the local shear intensity.

3. Discussion: revisiting two practical issues related to the LES of premixed flames

In this section, two examples are given to illustrate the type of physical intuition regarding the dynamics of the fronts that can be extracted from the idealized mathematical framework. It is hoped that this intuition could be very useful, along with experiments and direct numerical simulations, in improving LES strategies for realistic flames.

3.1. A priori identification of a LES flame front in DNS

Practical issue: in a direct numerical simulation of a premixed flame, all the scales, down to the very small reaction zone thickness, are explicitly resolved. How should one use the DNS data a priori to predict the shape of a LES flame front, i.e. how do you extrapolate the resolved results to a zero-thickness front? This is an important question because many attempts at devising ad-hoc subgrid models for the front speed will take this geometric information into account explicitly, for example to propose a model for the subgrid "wrinkling" or to quantify curvature effects to control the smoothness of the front.

Theoretical consideration: it is well known that, in the laminar case, the burning speed of KPP flames is governed by the behavior of the reaction rate on the cold side (near $T = 0$). The rigorous asymptotic theory of Majda and Souganidis generalizes this argument: the cell-problem that leads to the prediction of the turbulent burning speed describes the interaction of mixing, diffusion, and reaction near the cold side of the flame [Majda and Souganidis, 1994, Khouider et al., 2001].

Numerical observation: analyzing the sequence of pictures in Fig.17.2, it appears that the reaction contour snapshots converge to the asymptotic front prediction from the right: at small but finite size wrinkles and reaction zone thickness, the asymptotic front appears to coincide with the cold side envelop of the wrinkled flame front.

Practical conclusion: for KPP flames, the LES front corresponds to the cold side of the flame and the proper way of generalizing the concept of cold side of the flame when the flame is wrinkled by turbulence is to consider the cold side envelop of the flame (hence the front might not be smooth). In particular, the LES front does not coincide with any specific iso-contour of the reaction rate or the temperature, nor does it coincide with any iso-contour of those variables after filtering using any usual LES filtering technique. For other types of reaction rates, this indicates that extrapolating a zero-thickness front from resolved data requires a sound understanding of the asymptotic behavior of the laminar flame.

3.2. Parameterizing the subgrid enhancement of the flame burning speed

Practical issue: The LES results in Section 2 required the generation of an extensive database for the subgrid enhancement. For more realistic cases, this would be too costly. Instead, one might want to use an inexpensive subgrid model of the type: $S_T = S_L + K\lambda^\beta$, with λ the subgrid turbulence (units of velocity) while the pre-constant K and the exponent β are parameters to be adjusted statically or dynamically. The value of the scaling exponent β in particular has been the object of many studies, as experimental results have indicated that it can vary depending on the intensity of turbulence etc.

Asymptotic and numerical considerations: We summarize here unambiguous results obtained with the model problem using a combination of careful numerical simulations and asymptotics (see [Khouider et al., 2001] for details). The results can be stated for fairly general velocity fields of the type used with the model, but for simplicity, we will restrict the presentation here only to the case when the perturbation velocity field $\mathbf{v}'$ is a time modulated shear normal to the front: $v'(y,t) = \lambda \sin(2\pi y)\cos(\omega t)$. The results are as follows:

- in the limit $\omega \to 0$, the burning speed enhancement corresponds exactly to $S_{E,0} = 2\lambda/\pi$ so that in that case, the scaling exponent is $\beta = 1$.

- when $\omega > 0$, the normalized speed enhancement $S_{E,N} = S_E/S_{E,0}$ is a function only of the nondimensional number $\tilde{t} \sim \sqrt{\lambda}/\omega$. This nondimensional number is the ratio of the modulation period $T \sim 1/\omega$ to the flame deformation time $t_{ref} \sim 1/\sqrt{\lambda}$ defined as the time it takes a steady shear to deform to steady state an initially planar flame.

- when $\tilde{t}$ is large (hence the flow frequency is very slow, so that the steady solution is a good approximation), $S_{E,N} = 1 - C/\tilde{t}$ so that, to leading order, S_E is proportional to λ and $\beta = 1$.

- when $\tilde{t}$ is small (hence the shear oscillates very fast, so that the flame has no time to respond and the enhancement is very small), $S_{E,N} = \tilde{t}^2/4$ so that S_E is to leading order proportional to λ^2 and $\beta = 2$

Those predictions of two distinct regimes for $S_{E,N}(\tilde{t})$ are verified in Fig.17.3 where the flame propagation speed has been processed from numerical simulations with a wide range of values of ω and λ. This "bending" of the burning speed enhancement has been observed in experimental data. It would be interesting to verify to what extent the scaling argument from the idealized case holds for more complex perturbation flows and chemical reaction rates.

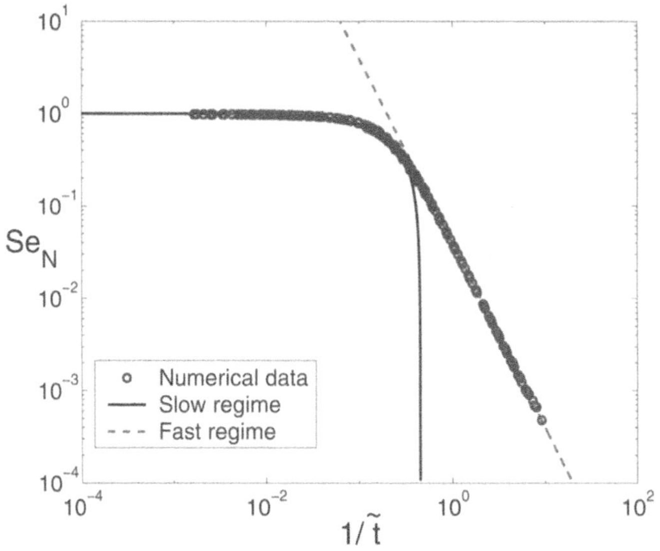

Figure 17.3. Normalized burning speed enhancement $S_{E,n} = S_E/\lambda$ as a function of the non-dimensional flame residence time $\tilde{t}$ = for the case of a time-modulated horizontal shear $v(y) = \lambda \sin(2\pi y)\cos(\omega t)$.

References

Bourlioux A. & Majda A. J., 2000. *Combustion Theory and Modelling*, 4, pp.173–188.

Khouider B., 2001. Ph.D. thesis, Université de Montréal.

Khouider B., Bourlioux A., & Majda A. J., 2001. *Combustion Theory and Modelling*, 5, 295–318.

Khouider B. & Bourlioux A., 2001. Submitted to *Siam J.Num.Anal.*.

Majda A. J. & Kramer P. R., 1999. *Physics Reports* **314(4-5)** 238–574.
Majda A. J., Souganidis P., 1994. *Nonlinearity*, **7**, 1.

Discussion

Charles Meneveau, Johns Hopkins "How difficult is it to extend your approach to analytic treatment of a multi scaled system if you had superposition of many eddies not just two."

Reply: That is something that we haven't tried. From the scaling I have the dominate contribution would be from the largest scale so I would assume that there would be a vamping the separate scale into one representative scale that is probably very close to the large scale but it something that we need to try. The results are fairly independent of the details of the small scale flow.

Emmanuel Villermaux, Marseilles "What would happen in your model if you let alpha to be larger than one?"

Reply: You cannot use homogenization. It makes no sense to look at the effective behavior of something that occurs on a pair that is larger than what you care about and then you resolve that. So it is a large eddy simulation. I am only talking about things on the time scale and the length scale you are looking at are sort of generic otherwise there is no point.

Or sorry that is the other way around. I am talking about alpha equals zero. You are talking about alpha larger than one that is the opposite to above. Then my turbulent velocity is smaller than my flame thickness and then your not in the flamelet regime and this doesn't apply.

Emmanuel Villermaux, Marseilles "Yes but this would be interesting because..."

Reply: Yes but theory applies in the flamelet regime

Emmanuel Villermaux, Marseilles "Most of the time this is the reality..."

Reply: I think that in some cases where this is a good approximation that turbulence does not penetrate the reaction zone. Here turbulence penetrates the preheat zone it does not penetrate the reaction zone.

Emmanuel Villermaux, Marseilles "Of course."

Santanu Sankar, San Diego "This result that the flame speed collapse with omega over square-root lambda is quite interesting. Is this because you have only two scales or do you think this result could generalize to a multi-scale model?"

Reply: It similar to the question Charles has asked. We have not looked at multi scaled problems what we looked at is applying this idea to a more complex flows, such as the one I showed you the snapshots. I've defined some non-dimensional flame residence time and the challenge for us so far is that it is very easy for this example, the one I discussed with you, once you start adding eddies and two dimensional flows and we've focused so far on can we extend this idea and use it to find a cutoff and a scaling law for this type of flow. I have not done the multi scale yet.

Chapter 18

STUDY OF MIXING IN SWIRLING TURBULENT JETS

Robert Zoltan Szasz, Laszlo Fuchs, and Doru Adrian Caraeni
Lund Institute of Technology, Sweden
szrobi@mail.vok.lth.se

Abstract Large Eddy Simulation of swirling flows in a typical gas turbine burner has been carried out. The goal has been to study the effects of swirl number, inlet velocity profile, confining wall distance, Reynolds- and Schmidt-number effects on the flow and turbulent mixing. The results show that the inlet velocity profile, the confinement and the swirl have a substantial effect on the resulting flow-field, while the values of the Reynolds- and Schmidt-numbers are of less importance in our case.

Keywords: Swirl, jet, mixing, gas turbine

1. Introduction

The turbulent, non-reacting flow field within a typical burner is considered for different geometrical and operating conditions. The current trend in gas turbine burner technology is to achieve high efficiency while keeping low levels of pollutant emissions (both CO and NO_x). One method to satisfy these highly non-compatible requirements is to use lean premixed, or partially premixed combustion [Gutmark and Grintsien, 1999]. However, under such conditions special care must be taken so as to maintain flame holding and stability. Commonly, one uses either a bluff body or swirling jets which act as flame holders. In contrast to flame holding using a bluff body, flame holding by swirl does not require contact between the flame and solid wall and hence causes less difficulties in terms of cooling the surface of the body.

Even at low swirl numbers the flow differs from the corresponding non-swirling jet case. In the swirling jet case the turbulence level is higher as well as the spreading rate of the jet, which will cause increased

A. Pollard and S. Candel (eds.), IUTAM Symposium on Turbulent Mixing and Combustion, 221-233.

entrainment of surrounding fluid. A more pronounced effect is observed for higher swirl numbers with the appearance of the so-called vortex breakdown phenomenon. At vortex breakdown the centrifugal forces are large enough so that the jet is pulled away from the center-line. Due to continuity, this leads to the formation of back-flow along some parts of the center line and the formation of an internal recirculation zone (IRZ). It should be emphasised that the extent of the IRZ may be time dependent and also undergo spanwise motion ("precession") at relatively low frequency. The presence of the IRZ implies higher levels of the turbulence due to the additional shear layers and the increased unsteadiness of the vortex breakdown profile. On the other hand this produces a better mixing of the fuel and air, improving the combustion efficiency and reducing the flame extent. The flame stabilization is improved also by the internal and external recirculation of the hot gases which help the re-ignition of the mixture in the case of extinction. Another difference between the swirling and non-swirling jets is the different mechanisms for the jet spread. While in the non-swirling case the jet spread is dominated by the turbulent mixing at the interface between the jet and the ambient fluid, in the case of the swirling jets the spread is driven by the static pressure gradients in both axial and radial directions [Farokhi and Taghavi, 1989].

The behaviour of swirling jets has been the subject of several studies (cf. [Grinstein et al., 1996] and [Billant et al., 1998]). The formation, steadiness and sensitivity of the recirculation zone has been found to be rather non-trivial. The experiments of Farokhi show that the principal contributor to the onset of the vortex breakdown is the strong adverse pressure gradient along the jet axis [Farokhi and Taghavi, 1989]. Champagne and Kromat [Champagne and Kromat, 2000] explained the development of appearance of the stagnation point and the internal recirculation zone in the following way: a divergence of the jet column produces negative azimuthal vorticity, which will induce negative axial velocity contribution on the jet axis, decelerating the flow. By continuity this will lead to a further increase of the jet diameter and subsequently of the negative azimuthal vorticity. In the theoretical study of Shtern and Hussain [Shtern and Hussain, 1999] the vortex breakdown development is attributed also to the pressure gradient: the swirl induces a pressure drop towards the axis, thus attracting other streamlines to the axis, this focuses further the swirl, thereby further decreasing the pressure core.

A major difficulty in studying swirling flows is related to the presence of *several* important parameters which have strong influence on the flow. One of these parameters is the swirl number, S, defined as the ratio of the axial flux of tangential momentum and the axial flux of axial momentum nondimensionalized with the jet radius [Panda and McLaughlin, 1994]:

$$S = \frac{\int_0^\infty \rho U W r^2 dr}{R \int_0^\infty \rho U^2 r dr} \tag{18.1}$$

where U is the axial component of the velocity, W the azimuthal component and ρ the fluid density.

Generally, for $S < 0.3$ no IRZ is formed, while for $S > 1.0$ the development of the IRZ is sure. In the intermediate range there are other parameters which can determine the appearance of IRZ. These parameters include the radial variation of the circumferential velocity, expansion rate, nozzle shape, the presence of combustion. The experiments of Billant et al. [Billant et al., 1998] show that the selection of the breakdown states can be influenced by temperature differences as small as 0.1^oC . Panda and McLaughlin [Panda and McLaughlin, 1994] found that the size, shape and location of the breakdown zone change significantly with the Reynolds number. The effects of confinement are also important: the introduction of walls confining the flame increases the recirculation zone size, while the use of interacting burners reduces it.

Buckley et al. [Buckley et al., 1983] have studied experimentally the effects of swirl profiles on combustor performance in the case of integral rockets/ramjets. They have considered four different cases with the same axial velocity profile and swirl number, but with different tangential velocity profiles, corresponding to the constant angle, forced vortex, free vortex and Rankine vortex cases. Their results indicate that there are significant influences of the tangential velocity profile on the combustion efficiency over a wide range of operating conditions.

Farokhi and Taghavi [Farokhi and Taghavi, 1989] carried out an experiment in order to study the influence of the tangential velocity profile at the inlet on the resulting flow-field. They studied two extreme cases, in which the tangential velocity profile followed the solid body rotation and the free swirl distribution, respectively, while maintaining the same swirl number as evaluated by equation (18.1). Their experimental results show completely different behaviour in the two cases, indicating that the use of the swirl number as a parameter characterizing swirling flows is not enough, other parameters are also required. This finding can be explained by the integral nature of the swirl number. In order to characterize the flow, Kim et al. [Kim et al., 1999] introduce another parameter

$$Rad = \frac{\int_0^R \rho U V r dr}{\int_0^R \rho U^2 r dr} \tag{18.2}$$

where U and V are the axial and radial components of the velocity, respectively. They observed that small increments in radial velocity profile significantly increase the strength of IRZ, even though the peak radial inlet velocity profile was much smaller than the other components. One may also add that the local radial variation of the azimuthal velocity is also important for the IRZ. Thus, it hardly can be expected that the integral parameters such as S and Rad are enough to uniquely characterize swirling flows.

The vortex breakdown profile can have different shapes, depending on the flow conditions. Billant et al. [Buckley et al., 1983] carried out a series of experiments in order to characterize the various breakdown states taking place in a swirling jet as the swirl ratio and the Reynolds number are varied. They found that, for the given set-up, the breakdown occurs when the swirl parameter reaches a well defined threshold S=1.3-1.4[1]. Four distinct forms of vortex breakdown have been identified: the bubble state, a cone configuration, an asymmetric bubble and an asymmetric cone state. The later two configurations were observed at high Reynolds numbers. They observed that below the onset of breakdown the swirling jet is highly asymmetric and takes the shape of a steady helix. The recirculating motions in the bubble case were strong, while the oscillations of the cone were slow and affected its whole structure. These aspects of swirling flows, namely the hysteresis of the appearance of the vortex breakdown, have been discussed in some detail and explained theoretically by [Shtern and Hussain, 1999].

Another characteristic of the swirling flows is that the swirl makes the organized structures weaker, suppressing vortex pairing and favouring the growth of random turbulence. This was demonstrated by Panda and McLaughlin [Panda and McLaughlin, 1994] who conducted a series of experiments in order to study the instabilities of the swirling jets. In their spectral measurement results neither a distinct peak nor any significant growth, which is characteristic of periodic, coherent structures, was observed. This implies that either the coherent structures are absent, or they are weak and random to leave a footprint in the velocity spectra. In the work of Shtern and Hussain [Shtern and Hussain, 1999] a theoretical explanation of the different features of a swirling flow (collapse, swirl generation, vortex breakdown, axi-symmetry, hysteresis) are described.

---

[1]The swirl parameter is defined in this reference as the ratio of the azimuthal and axial velocities, $S = \frac{2V_\theta}{V_x}$, where V_θ is the azimuthal velocity at the half radius and V_x is the axial velocity component

Here, we carry out some numerical investigations to determine the effects of some of the above-mentioned parameters. The geometry under consideration is a typical gas-turbine burner configuration with multiple swirling jets. We consider the turbulent flow regime and also the mixing effects of passive scalars. The turbulence is accounted for by using Large Eddy Simulations (LES). In the following we discuss the simulation method, the problem description and the main results.

2. Large Eddy Simulations

Modeling of turbulent flows has been and is still a topic for active research. This is so, and probably will remain so, due to the complexity of the physics and the lack of a universal turbulence model. There are many examples in the literature in which experimental results have been compared to different models ("validation" exercises). For swirling flows an example is the TECFLAM burner [TECFLAM webpage]. Besides the experimental study, Meier et al. [Meier et al., 2000] have investigated the ability of a commercial CFD software (RANS based models) to reproduce the experimental results . They observed considerable discrepancies between experimental and numerical results, the errors being attributed by the authors to the incorrect prediction of the cold turbulent flow field. In contrast to the Reynolds Averaged Navier-Stokes (RANS) framework, we believe that LES have the potential of remedying these shortcomings. LES can become, as the resolution is enhanced, an *approximation* to the governing equations, rather than being a *model*. We also believe that over time, LES will become accessible for many more complex industrial applications.

Some examples of recent LES application for gas turbine burners can be found in the literature. Caraeni et al. [Caraeni et al. 2001] consider the LES of liquid fuel injection into a gas turbine burner, break-up, evaporation of the liquid droplets as well as turbulent mixing. A more fundamental study of the LES as such for swirling flows (using a similar methodology) is carried out by Gullbrand et al. [Gullbrabd et al., 2001]. The last two cases use full 3-D configuration showing that LES is not only applicable but has clear advantages as compared to the RANS formalism.

The flow is considered to be incompressible and isothermal, so that the system of equations is:

$$u_{i,i} = 0 \tag{18.3}$$

$$\rho u_{i,t} + \rho u_i u_{j,j} = -p_{,i} + \frac{1}{Re}(u_{i,j})_{,j} \tag{18.4}$$

$$c_{i,t} + u_j c_{i,j} = \frac{1}{Sc}(c_{i,j})_{,j} \tag{18.5}$$

Where Re and Sc are the Reynolds- and the Schmidt-numbers, respectively. With LES these equations are space averaged leading to the presence of additional terms (due to the non-linear terms). These terms express the "sub-grid-scale" (SGS) effects on the resolved flow field. With increasing resolution the magnitude of the SGS terms reduces and for full spatial resolution they vanish identically. With the ambition of having high spatial resolution, one may neglect the SGS terms provided that the subgrid scale dissipation is accounted for. This is done here by a high (fourth) order dissipative term. Thus, in the calculations presented here, no explicit SGS terms are used. The reader is referred to [Gullbrabd et al., 2001] and [Olsson and Fuchs, 1998] for further discussion of the SGS modeling.

The flow solver is based on the finite differences method and uses an implicit discretization in time. The grid is Cartesian block-structured with the possibility of using local refinements. The discretization in space is 3rd and 4th order accurate. Further details are given in [Gullbrabd et al., 2001] and [Olsson and Fuchs, 1998].

3. Problem description

The computations have been carried out for the burner geometry as sketched in Figure 18.1. The burner has three co-rotating swirlers. The innermost swirler has an angle of $45^0, 38^0$ and 52^0, respectively. The amount of mass flow through each swirler is 8%, 17% and 75%, respectively. In the computations the axial components of the velocities at the three inlets were computed to provide the given values of the mass fluxes, while the azimuthal components were obtained with the help of the swirl angles. In the cross section in Figure 18.2 the grid is represented on the finest level in the burner zone. In the figure, Z represents the streamwise flow direction parallel to the axis of symmetry. All lengths have been made dimensionless by the outer diameter of the outer swirler. The zone left to the thick line in the figure is representing the wall and indicates that the inlets are not co-planar.

In order to study the Schmidt-number effects two scalars are injected simultaneously at the middle inlet, having Schmidt numbers 1 and 100, respectively.

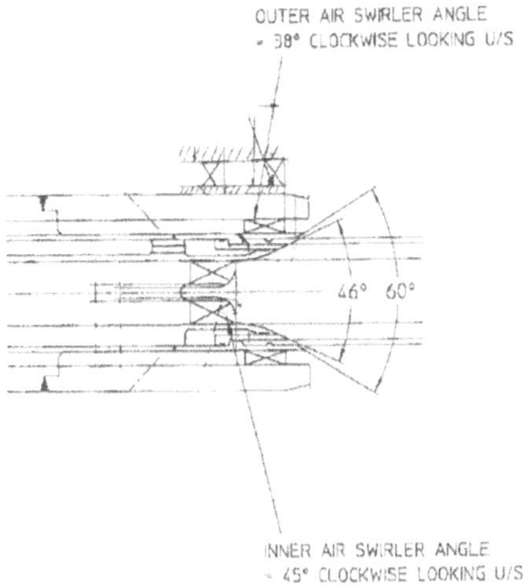

Figure 18.1. The geometry of the burner

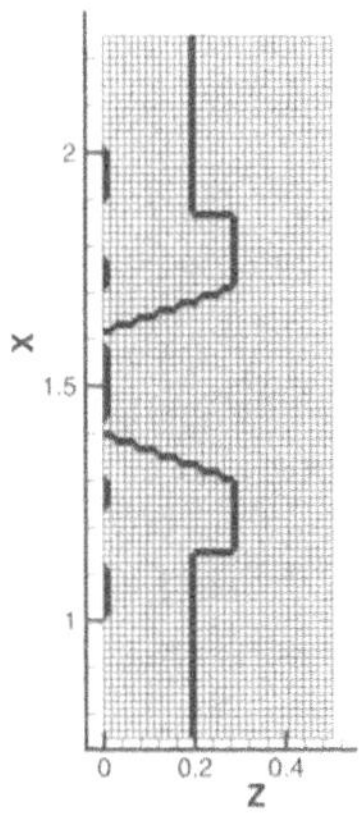

Figure 18.2. Part of the computational grid in the burner zone

4. Results

The objective is to study the influences of the inlet velocity profile, swirl, Reynolds-number and confinement on the vortex breakdown and mixing in the combustion chamber of a gas turbine. To achieve these goals different cases were studied and compared. A summary of the computed cases is given in Table 18.1.

Table 18.1. List of the studied cases

Case	*Geometry*	*Vel.Prof.*	*Re*	*Swirl*
1	3x6x7	flat	20000	.
2	3x6x7	flat	10000	.
3	3x6x7	polynomial	20000	.
4	3x6x7	polynomial	10000	.
5	3x6x7	flat	20000	+5%
6	2x2x8	polynomial	20000	.

The base case (*Case* 1) corresponds to the standard size of the combustion chamber (3x6x7 diameters), a flat velocity profile and a Reynolds number equal to 20000. *Case* 2 differs from the base case by a lower value of the Reynolds number. In *Cases* 3 and 4 the conditions are the same as in 1 and 2, except that the velocity profile is a polynomial. In *Case* 5 the swirl was increased with 5% in order to evaluate the effect of the swirl number, while in *Case* 6 the cross-section of the geometry was reduced to 2x2 units in order to study the confinement effects.

The spatial resolution may be assessed by considering the turbulent kinetic energy spectrum. It is found that the spectrum follows a -5/3 line for almost an order of magnitude. This shows that the spatial resolution is adequate and the need for an explicit SGS model is less acute. No presence of low-frequencies is observed in this case.

4.1. Velocity profile effect

Farokhi et al. [Farokhi and Taghavi, 1989] found experimentally that the initial profile of the tangential velocity component can have a crucial effect on the development of vortex breakdown profile. This effect is evaluated by considering two different axial velocity profiles: a flat one and one of third order polynomial, such that:

$$w(r) = c(r - r_i)^2(r_o - r) \tag{18.6}$$

where r_i and r_o are the swirler inner and outer radii, respectively, and c is adjusted to have the same flux as in the flat profile case.

Figure 18.3 depicts the mean radial velocity profiles for some axial distances. One observes a much higher spreading rate of the jets in the case of the non-uniform inlet velocity profile. This is due to the larger values of the centrifugal force at points close to the outer radius of the swirlers than at the points situated on the outer radius. This effect creates also an additional source of instability. Due to the higher spreading rate, in the case of the polynomial profile, the jets from the two main swirlers are already merged at one diameter from the inlet, while in the case of the flat velocity profile this phenomenon occurs further downstream.

The figure also shows that because of the wider opening of the jets in the second case, a relatively strong recirculation zone is developed at the symmetry axis between approximately 0.5 and 3.0 diameter distances downstream from the first inlet, typical for the bubble-type vortex breakdown profile. In the case of the flat velocity profile at the inlet, the jets are slowed down but no recirculation zone can be observed, and the shape of the vortex breakdown profile is closer to the conical type.

The corresponding results for the fluctuations show that for the polynomial velocity profile the fluctuations reach a higher level which is a result of the higher instability and the presence of the internal recirculation zone. This is in accordance with the experimental results of Billant et al. [Billant et al., 1998] who observed that strong recirculating motions occur within the bubble, while the oscillations of the cone are slow. We note also that at 3 diameters from the inlet the shear lay-

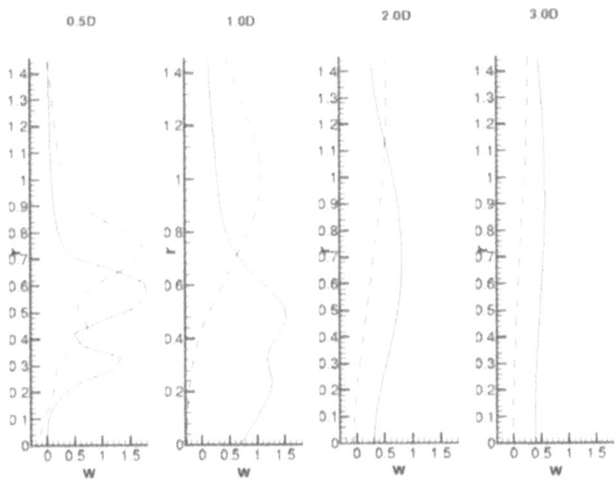

Figure 18.3. Radial profiles of the mean axial velocity component at 0.5, 1.0, 2.0 and 3.0 diameter distances from the inlet for the flat velocity profile (solid line) and the polynomial profile (dashed line)

ers have vanished. This shows that the potential core ends already at this distance, much earlier than in the case of a non-swirling jet. The same phenomenon was observed experimentally by Champagne et al. [Champagne and Kromat, 2000].

The evolution of the axial component of the velocities along the symmetry axis is presented in Figure 18.4. In both cases a recirculation zone is observed right after the first inlet in the small conical section. Further downstream both jets are accelerated by the streams coming from the second and the third swirlers. However, downstream at about 0.6 diameters from the first inlet, the two jets are behaving differently. In the case of the flat velocity profile (solid line) the jet is accelerated further until around 2 diameters downstream. Beyond that point due to the decrease of the pressure on the axis due to the swirling motion, the jet is slowed down. There is no negative values for the axial component of the velocity but there is a relative recirculation. Further downstream the jet is accelerated and close to the exit it will slow down again, approaching asymptotically the free-stream value.

In the case of the polynomial velocity profile the initial acceleration of the jets are stopped much earlier (at around 0.6 diameters downstream) due to the wider spreading of the jet which causes a more intense recirculation zone.

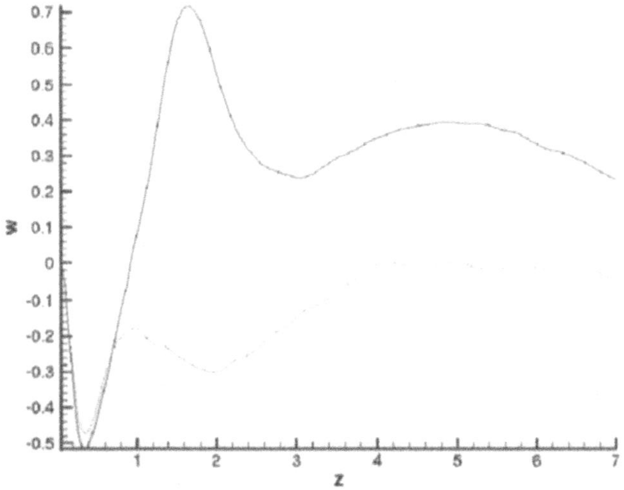

Figure 18.4. Mean axial velocity along the centreline for the flat velocity profile (solid line) and the polynomial velocity profile (dotted line)

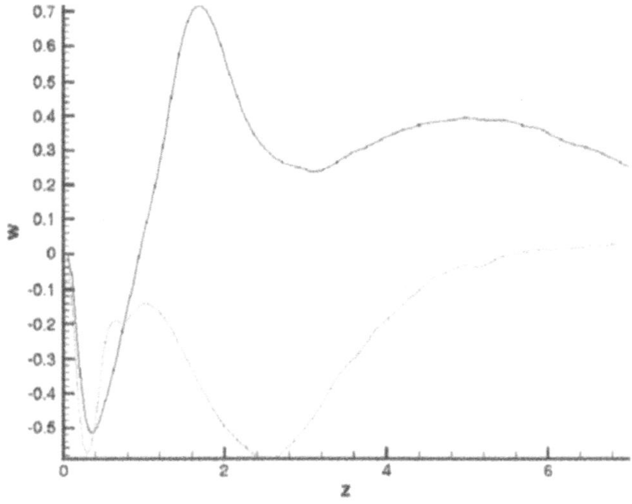

Figure 18.5. Comparison between the axial velocity component along the centreline for the base case (solid line) and the more confined case (dotted line)

4.2. Other effects

To study the effects of confinement, the results from the base case were compared with results obtained in a geometry which had a reduced cross section of 2x2 diameters. In Figure 18.5 the axial velocity component of the velocity is compared along the centreline for the base case (solid line) and for the smaller geometry (dotted line). As can be seen, the geometry has also a crucial effect on the flow field. In contrast with the base case, when the cross-section is reduced, a strong recirculation zone develops along the centreline, as can be deduced from the negative values of the velocity. The vortex breakdown profile in this case is of bubble-type. Due to the Coanda-effect, the flow is attached to the combustion chamber walls relatively close to the inlet.

Comparing the results of *Case* 5 with the base case, one finds that modifying the swirl angle by 5% has a substantial effect on the resulting velocity field: A strong recirculation zone is formed around the symmetry axis. Modifying the Reynolds or the Schmidt numbers had relatively small influence in comparison with the effects of the other parameters (not shown).

5. Conclusion

This article describes Large Eddy Simulations of swirling flows in a gas turbine burner geometry, consisting of three co-annular jets. It is shown that the radial distribution of the inlet (axial) velocity profile is an essential parameter for the characterization of the burner flow. The same conclusion can be drawn regarding the influence of confinement (i.e. distance to the side walls). When the dimensions of the combustion chamber are reduced, the recirculation zone becomes stronger and the jet is attached to the side-walls due to the Coanda effect. The swirl number has also an important effect, increasing it 5%, has a effect similar to dividing the cross-sectional area of the chamber by one half. The influence of the Reynolds- and Schmidt-numbers on the flow and mixing are less pronounced in these cases.

Acknowledgments

The authors would like to acknowledge the support of the Centre for Combustion Science and Technology (CECOST), the Swedish foundation for Scientific and Technical Research (TFR). Some of the computations have been carried out at the Supercomputer Centre at Lund Institute of Technology (LUNARC).

References

Billant,P., Chomaz,J.M., and Huerre,P. Experimental study of vortex breakdown in swirling jets. *J. Fluid Mech.*, 1998, vol. 376, pp. 183-219

Buckley,P.L., Craig,R.R., Davis,D.L., and Schwartzkopf,K.G. The design and combustion performance of practical swirlers for integral rocket/ramjets. *AIAA Journal*, 1983, vol. 21, pp. 733-740

Caraeni, D., Bergstrom, C. and Fuchs, L. Modeling of liquid fuel injection, evaporation and mixing in a gas turbine burner using LES. *J. Flow, Turbulence and Combustion*, 2000, vol. 65, pp. 223-244

Champagne,F.H., and Kromat,S. Experiments on the formation of a recirculation zone in swirling coaxial jets. *Exp. in Fluids*, 2000, vol. 29, pp. 494-504

Farokhi,S., and Taghavi,R. Effect of initial swirl distribution on the evolution of a turbulent jet. *AIAA Journal*, 1989, vol. 27, pp. 700-706

Grinstein, F.F., Gutmark, E.J., Parr, T.P. and Hanson-Parr, D.M. Streamwise and spanwise vortex interaction in an axisymmetric jet. A computational and experimental study. *Phys. Fluids* 1996, vol. 8, pp. 1515-1524

Gullbrand, J., Bai,n X-S. and Fuchs, L. High-order cartesian grid method for calculation of incompressible turbulent flows. *Int. J. Numer. Meth. Fluids*, 2001, vol. 36, pp. 687-709

Hardalupas,Y., Pantelides,K., and Whitelaw,J.H. Particle tracking velocimetry in swirl-stabilised burners *Exp. in Fluids*, 2000, suppl., pp. 220-226

Gutmark E.J. and Grinstein, F.F. Flow control with noncircular jets. *Annu. Rev. Fluid Mech.* 1999, vol. 31, pp. 239-272

Kim, W. W., Menon,S., and Mongia,H. C. (1999) Large-Eddy Simulation of a Gas Turbine Combustor Flow. *Combust. Sci. and Techn.*, 1999, vol. 143, pp.25-62

Meier,W., Keck,O., Noll,B., Kunz,O., and Stricker,W. Investigations in the TECFLAM swirling diffusion flame: Laser Raman measurements and CFD calculations. *Appl. Phys. B*, 2000, vol. 71, pp. 725-731

Olsson, M. and Fuchs, L. LES of a forced semi-confined circular impinging jet. *Phys. Fluids*, 1998, vol. 10, pp. 476-486

Panda,J. and McLaughlin, D.K. Experiments on the instabilities of a swirling jet. *Phys. Fluids*, 1994, vol. 6(1), pp. 263-276

Shtern,V. and Hussain,F. Collapse, symmetry breaking, and hysteresis in swirling flows. *Annu. Rev. Fluid Mech.*, 1999, vol. 31, pp. 537-566

TECFLAM webpage www.tu-darmstadt.de/fb/mb/ekt/tecflam

Discussion

Johannes Janicka, Darmstadt "Did I understand right, that you don't use a SGS model?"

Reply: I am not using a SGS, but I can not say it is a DNS.

Johannes Janicka, Darmstadt "So what was the motivation for doing this? You rely on the dissipation of the numerics for the SGS model?"

Reply: The SGS are assumed small.

Johannes Janicka, Darmstadt "What was the number of points in your simulation?"

Reply: 1.2 million cells.

Dan Ewing, McMaster University "Just a comment first there was a fair bit of work done on this problem by Penn State and what they found that to solve through the swirl was the only way they could get the swirl profile right so you may find that even if you do it experimentally you have to solve through the swirler to get that right because it is very sensitive. You didn't really mention what swirl number you were at in your simulation, whereabouts were you?"

Reply: Just a moment - 45 degrees, 38 degrees and 52 degrees, if I remember but...

Dan Ewing, McMaster University "In terms of did you work out a swirl number angle of momentum to the axial momentum?"

Reply: We didn't work out the swirl number just the ratio of the azimuthal component and the axial component.

Chapter 19

LES OF PARTIALLY PREMIXED COMBUSTION

Luc Vervisch and Pascale Domingo
Institut National des Sciences Appliquées de Rouen
UMR-CNRS-6614 / CORIA
Campus du Madrillet - Avenue de l'Université - BP 8
76801 Saint Etienne du Rouvray Cedex, France
vervisch@coria.fr - domingo@coria.fr

Abstract Numerical studies devoted to partially premixed combustion are first reviewed. Then a Large Eddy Simulation procedure is proposed for these flames when they develop in turbulent flows. The corresponding subgrid combustion modeling is tested for the case of a turbulent lifted-flame.

Keywords: Combustion, partial premixing, triple flames, large eddy simulation.

1. Introduction

Combustion is usually considered from two idealized regimes: before entering the reaction zone, fuel and oxidizer are expected either fully premixed or fully nonpremixed. This formal distinction between premixed and nonpremixed flames is relevant for characterizing the injection of reactants into the combustion system, but the flame developing further downstream may not exactly feature premixed or nonpremixed combustion properties. This is for instance the case in a nonpremixed combustion chamber when fuel and oxidizer have mixed upstream of the flame zone; sometimes mixing is also organized with products to insure flame stabilization (fig. 19.1). The intermediate combustion regime between fully premixed and fully nonpremixed is called *partially premixed* and motivates questions: Where is observed partially premixed combustion? What are the major properties of partially premixed flames? What are the tools for the numerical modeling of these flames? Is Large

A. Pollard and S. Candel (eds.), IUTAM Symposium on Turbulent Mixing and Combustion, 235-249.

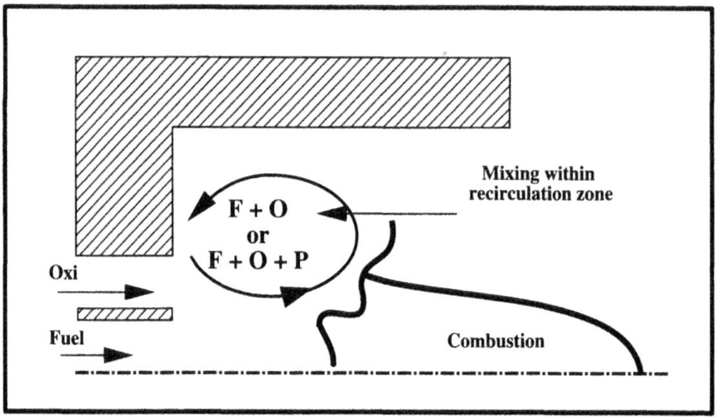

Figure 19.1. Schematic of a generic nonpremixed flame stabilization mechanism. The dilution of the reaction zone by products may be useful to control NOx production.

Eddy Simulation (LES) an interesting candidate for simulating partially premixed combustion? An attempt is made to start addressing these questions.

2. Major properties of partially premixed flames

Partial premixing is observed in nonpremixed combustion when fuel and oxidizer have mixed without burning. Even in premixed systems, the premixing of reactants is not always perfect and some partial premixing may be observed.

Direct Numerical simulation of a spray flame is shown in fig. 19.2. Fuel droplets are injected in a two-dimensional double-wake configuration, and flame stabilization results from the diffusion of heat vaporizing the liquid ahead of a partially premixed front [Réveillon and Vervisch, 2000]. The burning stoichiometric point is denoted triple point in fig. 19.2. At this location, three flames meet: a small fuel lean premixed flame and two separated trailing reaction zones composed of a rich premixed flame and a diffusion flame. The structure of the flame is strongly affected by the distribution of droplet sizes and a whole range of flame topology may be observed in spray partially premixed combustion. In fig. 19.2, the liquid penetrates between the triple points and combustion mainly occurs in a partially rich premixed regime, the amount of fuel burning in the diffusion flame is then rather small. This spray DNS illustrates how partially premixed combustion may dominate in an initially nonpremixed system.

The flame propagation velocity $S_o^l(Z)$ is one of the control parameters of premixed combustion. It is a function of the mixture composition that

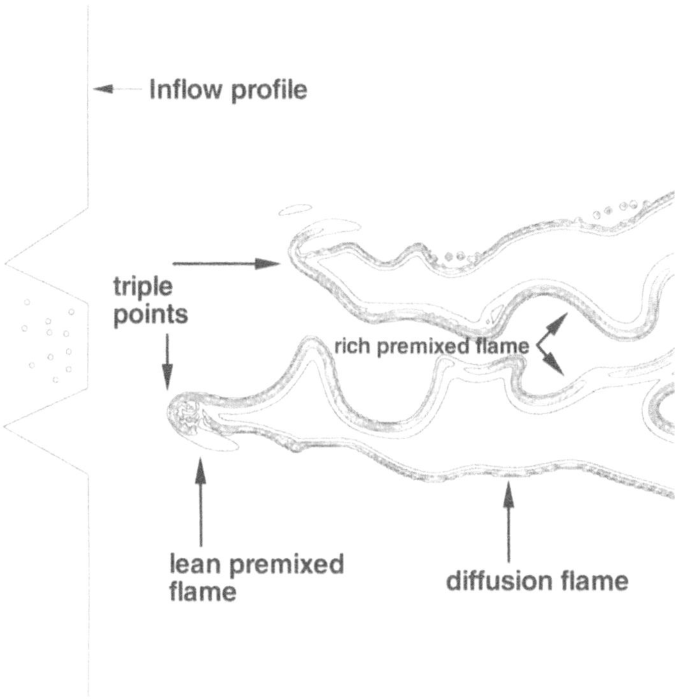

Figure 19.2. Snapshot of the reaction rate resulting from spray combustion as observed in DNS [Réveillon and Vervisch, 2000].

may be characterized by the mixture fraction Z, a passive scalar used to describe the amount of fuel mixed with air ($Z = 0$ in the air stream and $Z = 1$ in the fuel stream). Premixed flames propagate within flamability limits corresponding to a given range of mixture fraction [Poinsot and Veynante, 2001]. In nonpremixed combustion, the reactants must be mixed at the molecular level before reacting, and an approximation of the amplitude of mixing by molecular diffusion is obtained with the mixture fraction dissipation rate $\chi = \rho D |\nabla Z|^2$ [Liñán, 1974].

The detailed properties of partially premixed flames are thus expected to be related to three basic quantities Z, $S_o^l(Z)$ and χ. Various studies have illustrated how the structure of the flame depends on the amplitude of the variations of Z within the system. DNS of stratified propane-air mixture were recently reported [Haworth et al., 2000]. It was observed that there is no leakage of the primary fuel (propane) behind an initial thin premixed heat-release zone and that beyond three-to-four flame thicknesses behind the primary flame, most hydrocarbon fuel has broken down into CO and H_2. The rate of heat release in the secondary reaction zone was found to be governed by turbulent mixing and the ki-

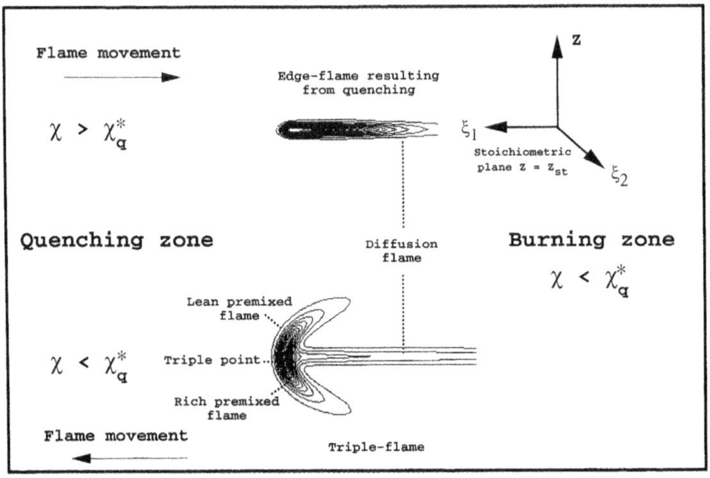

Figure 19.3. Two topologies for reaction zone extremities in diffusion flames. Top: the scalar dissipation rate exceeds the quenching limit χ_q^* and an edge flame appears. Bottom: a propagating triple flame separates a burning zone from a non-burning zone where $\chi < \chi_q^*$ [Vervisch, 2000].

netics of CO_2 production. These results suggest that more than a single approach may be needed to complete the modeling of partially premixed combustion. Specifically, these DNS motivate the introduction of a thin flamelet closure for the initial thin premixed heat-release zone, followed by a more mixing control closure for the secondary reaction zone.

When the mixture fraction ranges between zero and unity, the model problem for partially premixed combustion is the triple flame (fig. 19.3 - bottom). In a mixing layer configuration, the stoichiometric premixed kernel evolves to a rich partially premixed flame in the direction of the fuel stream, while a lean partially premixed flame develops on the air side (fig. 19.3 - bottom). These two premixed flames are curved because their respective propagation velocities decrease when moving away from the stoichiometric condition. The overall structure that is composed of the two premixed flames and of the diffusion flame is usually called a triple flame or tri-brachial flame [Dold, 1989, Hartley and Dold, 1991, Ruetsch et al., 1995, Kioni et al., 1993, Buckmaster, 1996, Ghosal and Vervisch, 2000, Ghosal and Vervisch, 2001]. The curvature of the partially premixed front increases with the scalar dissipation rate imposed in front of the flame. Curvature effects tend to diminish the flame speed by inducing heat fluxes in the direction that is perpendicular to the main direction of propagation. However, the effect of heat release is to deflect the flow upstream of the curved front, and via mass conservation, this makes the overall displacement triple flame speed greater than $S_o^l(Zs)$, the stoichiometric premixed flame speed [Ruetsch et al., 1995]. The effect of curvature on the triple flame velocity is known when the mixture fraction field is known. The ratio (U_{TF}/S_l^o), where U_{TF} is the triple flame speed, is a function of the scalar dissipation rate, χ, measured in front of the flame under stoichiometric conditions [Dold, 1989, Daou and Linan, 1998, Ghosal and Vervisch, 2000]. When the flame tip is approximated as a parabolic profile, results from expansions in parabolic-cylindrical coordinates provide an expression for the flame speed. The analysis includes small but finite heat release and gas expansion effects, and yields the following triple flame velocity [Ghosal and Vervisch, 2000]:

$$U_{TF} \approx S_l^o(Z_s)\,(1+\alpha) - \frac{\beta}{Z_{st}(1+\alpha)\sqrt{4\nu_F - 2}}\sqrt{\frac{\lambda}{\rho C_p}\chi}$$

where $\alpha = (T_{burnt} - T_{fresh})/T_{burnt}$ is defined from the temperatures on both sides of a stoichiometric premixed flame for the same mixture, β is the Zel'dovitch number [Williams, 1985], ν_F the stoichiometric coefficient of the fuel and χ is measured far upstream in the mixing layer where the triple flame propagates.

Diffusion flame quenching may also promote partially premixed combustion. Local quenching usually results from high level of scalar dissipation rate and reactants can be partially mixed within flame holes. For high value of the scalar dissipation rate, the partially premixed front cannot develop into triple flames. Burning edge flames are then observed at the extremity of the reaction zone that is quenched (fig. 19.3 - top). Edge flames were studied using asymptotics [Buckmaster and Weber, 1996, Buckmaster, 1996] and simulations suggest that the scalar dissipation rate controlling the growth of the flame hole is lower than the one that should be applied to first quench the flame [Favier and Vervisch, 2001]. This was also observed in experiments [Shay and Ronney, 1998] and is further discussed in this book by Boulanger and Vervisch.

Studies have analyzed Lewis number effects and concluded that fuel Lewis number larger than unity promotes the upstream pointing of the fuel-rich branch with possibility of extinction [Buckmaster and Matalon, 1988, Daou and Linan, 1998]. Because of the asymmetry in flamability limits, for most hydrocarbon combustion partially premixed fronts are non-symmetrical. A careful examination of diffusion flame extremities shows that the formation of intermediate species differs in a triple flame from the pure diffusion combustion regime. Near the triple point, only a small amount of fuel is left and essentially stable species like CO and H_2 are consumed in the trailing diffusion flame [Terhoeven and Peters, 1997, Echekki and Chen, 1998].

3. Large Eddy Simulation of partially premixed combustion

The lifted turbulent jet flame is a simplified laboratory problem for partially premixed combustion. At the turbulent flame base, combustion starts in a mixture composed of fuel and oxidizer that have been partially mixed. Downstream of the flame base, a turbulent diffusion flame develops. A variety of experimental and numerical studies have reported that triple flamelets are the basic ingredients of nonpremixed flame bases [Muñiz and Mungal, 1997, Schefer and Goix, 1998, Mahalingam et al., 1999, Plessing et al., 1998, Watson et al., 1999].

Numerical modeling of lifted flames requires the simultaneous description of the trailing diffusion flame and of locally thin propagating partially premixed flames defined for various Z. A procedure based on the topology of the species field was proposed to distinguish in simulations between premixed and diffusion combustion [Yamashita et al., 1996, Favier and Vervisch, 2001]. Combining this approach with Large Eddy Simulation (LES) for the premixed and non-premixed regimes, a subgrid combustion closure for partially premixed flame has been proposed [Domingo et al., 2001]. The statistical properties of the flame base are collected in the LES, they reproduce some of the experimental observations in term of flow velocity and reveal some properties of partially premixed turbulent combustion.

3.1. Partially premixed flamelets in LES

For LES of nonpremixed turbulent combustion including partially premixed flames, the mass weighted large eddy mass fractions fields $\widetilde{Y}_i$ are decomposed into $\widetilde{Y}_{i,p}$, a premixed part, and, $\widetilde{Y}_{i,d}$, a nonpremixed part:

$$\widetilde{Y}_i = \overline{\xi}_p \widetilde{Y}_{i,p} + (1 - \overline{\xi}_p)\widetilde{Y}_{i,d} \tag{19.1}$$

where $\widetilde{Y}_{i,p}$ and $\widetilde{Y}_{i,d}$ are calculated from a progress variable c for premixed combustion ($c = 0$ in fresh gases and $c = 1$ in fully burnt products), and a mixture fraction Z for nonpremixed flames. A possible definition for the progress variable reads:

$$c(Z; \underline{x}, t) = \frac{Y_{F,o}Z - Y_F(\underline{x}, t)}{Y_{F,o}Z - Y_F^{Eq}(Z)} = \frac{Y_{O,o}(1 - Z) - Y_O(\underline{x}, t)}{Y_{O,o}(1 - Z) - Y_O^{Eq}(Z)} \tag{19.2}$$

where ^{Eq} denotes chemical equilibrium conditions and $_o$ species concentrations in the feeding streams for either fuel or oxidizer. In the subgrid combustion model, c and Z are simultaneously resolved at the LES filtered level.

The starting point of the flame decomposition is the cross-scalar dissipation rate $\chi_{F,O}$ found in the transport equation for the product $Y_F Y_O$ [Favier and Vervisch, 2001]:

$$\chi_{F,O} = -D\nabla Y_F \cdot \nabla Y_O = -D|\nabla Y_F||\nabla Y_O| N_{F,O} \qquad (19.3)$$

where $N_{F,O} = \mathbf{n}_F \cdot \mathbf{n}_O$ is a cross-orientation factor with $\mathbf{n}_i = -\nabla Y_i / |\nabla Y_i|$ the normal vector to an iso-reactant surface. In diffusion flames, reactants are on both sides of the stoichiometric surface and $N_{F,O} < 0$ whereas in premixed flames, gradients of reactants are oriented towards the propagating front and $N_{F,O} > 0$ [Yamashita et al., 1996]. An indicator of partial premixing ξ_p, or flame index may be written using Eq. 19.3:

$$\xi_p = \frac{1}{2}\left(N_{F,O} + 1\right) = \frac{1}{2}\left(1 - \frac{\chi_{F,O}}{D|\nabla Y_F||\nabla Y_O|}\right) \qquad (19.4)$$

where $\xi_p = 1$ corresponds to fully premixed mixture and $\xi_p = 0$ to diffusion combustion.

After straightforward manipulations using Eq. 19.2 one may write:

$$\chi_{F,O} = -D|\nabla Y_F||\nabla Y_O| N_{F,O} = -Y_{F,o} Y_{O,o} \left(\mathcal{F}_c\, \chi_Z + \mathcal{F}_Z\, \chi_c + \mathcal{F}_{Z,c}\, \chi_{Z,c}\right) \qquad (19.5)$$

$\chi_Z = D|\nabla Z|^2$ and $\chi_c = D|\nabla c|^2$ are the so-called scalar dissipation rates of mixture fraction and progress variable respectively, while $\chi_{Z,c} = D\nabla Z \cdot \nabla c = D|\nabla Z||\nabla c| \mathbf{n}_Z \cdot \mathbf{n}_c$ is a cross-scalar dissipation rate. Other functions may be written:

$$\begin{aligned} \mathcal{F}_c &= -A_c\, B_c \\ \mathcal{F}_Z &= A_Z\, B_Z \\ \mathcal{F}_{Z,c} &= A_c B_Z - B_c A_z \end{aligned}$$

with A_c, B_c, A_Z and B_Z four positive functions:

$$\begin{aligned} A_c &= (1-c) - \frac{c}{Y_{O,o}} \frac{dY_O^{Eq}(Z)}{dZ} \\ B_c &= (1-c) + \frac{c}{Y_{F,o}} \frac{dY_F^{Eq}(Z)}{dZ} \\ A_Z &= (1-Z) - \frac{Y_O^{Eq}(Z)}{Y_{O,o}} \\ B_Z &= Z - \frac{Y_F^{Eq}(Z)}{Y_{F,o}} \end{aligned}$$

Therefore $\mathcal{F}_c < 0$, $\mathcal{F}_Z > 0$, while the sign of $\mathcal{F}_{Z,c}$ depends on local flow conditions.

The progress variable is used in the premixed flame part only, assuming thin flamelets propagate along iso-mixture fraction surface. Neglect interactions between these flamelets leads to $\chi_{Z,c} \approx 0$. A closure for $\overline{\chi}_{F,O}$ may then be proposed:

$$\overline{\chi}_{F,O} \approx -D\nabla\overline{Y}_F \cdot \nabla\overline{Y}_O - Y_{F,o}Y_{O,o}\left(\overline{\mathcal{F}}_Z\overline{\chi_c}^S + \overline{\mathcal{F}}_c\overline{\chi_Z}^S\right) \tag{19.6}$$

where $\overline{\chi}_Z^s$ and $\overline{\chi}_c^s$ are subgrid scalar dissipation rates. $\overline{\mathcal{F}}_c$ and $\overline{\mathcal{F}}_Z$ are filtered functions of above expressions estimated via a presumed mixture fraction pdf [Pierce and Moin, 1999]. The subgrid scalar dissipation rates, $\overline{\chi}_Z^s$ and $\overline{\chi}_c^s$, which are required to close the relation 19.6 are expressed using eddy break-up type closures [Bray, 1996]:

$$\overline{\chi}_Z^S \approx \frac{\widetilde{Z}_v}{\tau_t} \quad ; \quad \overline{\chi}_c^S \approx \frac{\widetilde{c}(1-\widetilde{c})}{\tau_t} \tag{19.7}$$

where the turbulent time is approximated by $\tau_t = \Delta^2\overline{\rho}/\mu_t$ and $\widetilde{Z}_v$ is the variance of the scalar field estimated from an equilibrium condition as $\widetilde{Z}_v \approx 0.2\Delta^2|\nabla\widetilde{Z}|^2$, Δ is the LES filter size.

From Eq. 19.4 and above developments, a modeled form may be proposed for the filtered flame index:

$$\overline{\xi}_p \approx \frac{1}{2}\left(1 - \frac{\overline{\chi}_{F,O}}{F_{norm}}\right) \tag{19.8}$$

where F_{norm} is used to normalize $\overline{\chi}_{F,O} = \overline{\chi}_{F,O}^r + \overline{\chi}_{F,O}^s$ and is obtained from the following expressions that depend on the values of the resolved cross-scalar dissipation $\overline{\chi}_{F,O}^r$ and of the subgrid correction $\overline{\chi}_{F,O}^s$:

$$\begin{array}{lll} \text{if} & \overline{\chi}_{F,O}^r\,\overline{\chi}_{F,O}^S > 0 & F_{norm} = D|\nabla\overline{Y}_F||\nabla\overline{Y}_O| + |\overline{\chi}_{F,O}^S| \\ \text{if} & \overline{\chi}_{F,O}^r\,\overline{\chi}_{F,O}^S < 0 \text{ and } |\overline{\chi}_{F,O}^r| > |\overline{\chi}_{F,O}^S| & F_{norm} = D|\nabla\overline{Y}_F||\nabla\overline{Y}_O| - |\overline{\chi}_{F,O}^S| \\ \text{if} & \overline{\chi}_{F,O}^r\,\overline{\chi}_{F,O}^S < 0 \text{ and } |\overline{\chi}_{F,O}^r| < |\overline{\chi}_{F,O}^S| & F_{norm} = -|\overline{\chi}_{F,O}^r| + |\overline{\chi}_{F,O}^S| \end{array}$$

This normalization insures that $\overline{\xi}_p$ stays between zero and unity.

The premixed flame featuring a non-uniform equivalence ratio is captured extending the subgrid combustion closure of Veynante and coworkers [Boger and Veynante, 2000]. A transport equation for the mass filtered progress variable $\widetilde{c}$ is derived and unknown terms are closed leading to:

$$\frac{\partial \overline{\rho}\widetilde{c}}{\partial t} + \nabla \cdot (\overline{\rho}\widetilde{\mathbf{u}}\widetilde{c}) = \nabla \cdot \left[\left(\overline{\rho}_u \frac{\widetilde{S}_L \Delta_c}{16\sqrt{6/\pi}} + \frac{\mu_t}{\sigma_t} \right) \nabla \widetilde{c} \right] + 4\overline{\rho}_u \widetilde{S}_L \Xi \sqrt{\frac{6}{\pi}} \frac{\widetilde{c}(1-\widetilde{c})}{\Delta_c} \tag{19.9}$$

where μ_t is the subgrid turbulent viscosity, σ_t is a turbulent Schmidt number and Δ_c a LES filter size. The wrinkling factor Ξ is assumed constant ($\Xi = 1.1$). In nonpremixed flames, to account for partially premixed flamelets, we compute the mean density of fresh gases $\overline{\rho}_u$ and the burning velocity $\widetilde{S}_L$ from the knowledge of $\rho_u(Z)$ and $S_L^o(Z)$, the response of these quantities to a non-uniform composition of fresh gases determined using the GRI methane - air chemical kinetics. $S_L^o(Z)$ is averaged using the mixture fraction pdf $\widetilde{P}(Z^*)$:

$$\widetilde{S}_L = \int_0^1 S_L^o(Z^*) \widetilde{P}(Z^*) \, dZ^* \tag{19.10}$$

The closed transport equation for the filtered mixture fraction is written:

$$\frac{\partial \overline{\rho}\widetilde{Z}}{\partial t} + \nabla \cdot (\overline{\rho}\widetilde{\mathbf{u}}\widetilde{Z}) = \nabla \cdot \left[\left(D + \frac{\mu_t}{\sigma_t} \right) \nabla \widetilde{Z} \right] \tag{19.11}$$

$\widetilde{Z}$ and its subgrid fluctuations are retained to parameterize the beta mixture fraction pdf $\widetilde{P}(Z^*)$. A basic flamelet model is utilized for the diffusion flame part $\widetilde{Y}_{i,d}$ [Peters, 2000].

The links between the premixed and nonpremixed parts are evidenced assuming thin partially premixed flame fronts, in a BML context [Bray and Libby, 1994]:

$$\widetilde{Y}_{F,p} = (1-\overline{c})\, Y_{F,o}\widetilde{Z} + \overline{c}\, \widetilde{Y}_F^{Eq} \tag{19.12}$$

$$\widetilde{Y}_{O,p} = (1-\overline{c})\, Y_{O,o}(1-\widetilde{Z}) + \overline{c}\, \widetilde{Y}_O^{Eq} \tag{19.13}$$

$\overline{c}$ is estimated from $\widetilde{c}$ using the BML relation, $\overline{c\overline{\rho}_u\widetilde{c}/(\overline{\rho}_b + (\overline{\rho}_u - \overline{\rho}_b)\widetilde{c})}$, where the subscripts $_u$ and $_b$ denote fresh and burnt gases respectively.

The Lagrangian dynamic model of Meneveau et al. [Meneveau et al., 1996] is retained to calculate μ_T and σ_T. The fourth order finite volume skew-symmetric-like scheme of Ducros et al. [Ducros et al., 2000] is chosen for spatial discretization, combined with a second order Runge Kutta time stepping. The flow is first analyzed using two-dimensional simulations.

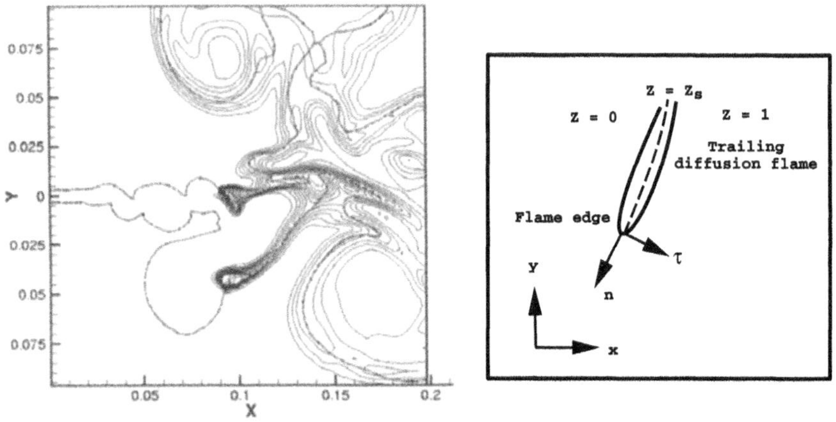

Figure 19.4. Left: Iso-contour of temperature at the flame base, Line: Stoichiometric mixture. Lengths are in meter. Right: Sketch of the local coordinates attached to a reaction zone extremity.

3.2. LES of methane - air turbulent lifted flames

The lifted flame configuration was studied experimentally [Muñiz and Mungal, 1997]. A round methane jet of diameter $d_o = 4.8\,mm$ is surrounded by an air coflow. The inlet velocities are $U_{CH_4} = 14\,m.s^{-1}$ and $U_{Air} = 0.27\,m.s^{-1}$, corresponding to case III of [Muñiz and Mungal, 1997]. The flame is lifted and combustion starts in a partially premixed regime at a mean streamwise position located at $14\,d_o$. In the simulations the grid is 128X128 clustered near the burner exit. Only the flame base is simulated, the length of the square computational domain is $41\,d_o$.

Figure. 19.4-left shows the iso-stoichiometric line and the temperature field. Unsteady mixing resulting from the roll-up of the shear layers is visible. When studying the topology of the flame base in their experiments, Muniz and Mungal found that velocity profiles taken across the flame base feature properties similar to those observed in triple flames. Suggesting that the reaction zone extremities have strong similarities with triple flames and edge flames.

To focus on the possible role of those reaction zone extremities found at the flame base in LES, quantities are plotted using a local coordinate system attached to one representative reaction zone extremity observed along the partially premixed front. Values are collected along the direction **n** as defined in Fig. 19.4-right, thus following the stoichiometric line.

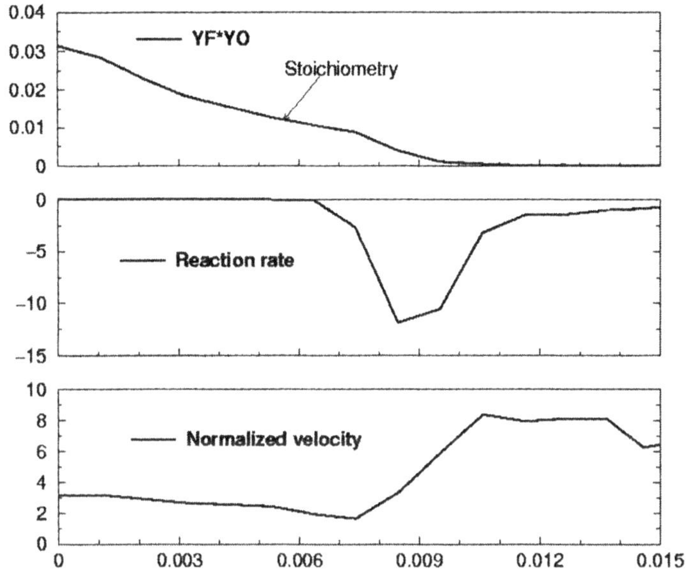

Figure 19.5. Along the stoichiometric line of a reaction zone extremity. Top: $\widetilde{\phi} = \widetilde{Y}_F\widetilde{Y}_O$. Middle: Filtered reaction rate. Bottom: Flow velocity.

The filtered partially premixed fraction $\widetilde{\phi} = \widetilde{Y}_F\,\widetilde{Y}_O$ indicates that premixing has developed upstream of the local edge flame where $\overline{\xi}_p \approx 1$ (Fig. 19.5-top). Then, the local premixed front consumes the reactants for the stoichiometric value $\widetilde{\phi}_s = Y_{F,o}Y_{O,o}Z_s(1 - Z_s)$, this is visible from $\widetilde{\phi}$ and the burning rate (Fig. 19.5-top and middle). The velocity profile following the stoichiometric line of a turbulent edge flame as captured by LES is shown in Fig. 19.5-bottom. The velocity is normalized by $S_L^o(Z_s)$, the stoichiometric burning velocity of a premixed flame. The complex tribrachial structure of the triple flame cannot be resolved by LES, however in both LES and measurements, the velocity profile features strong similarities with those observed in triple flames [Muñiz and Mungal, 1997], with an upstream velocity of the order of three times $S_L^o(Z_s)$. In other words, in the experiment and in the LES, the collection of reaction zone extremities (or edge flames) stabilizing the lifted flame, continuously follows within the turbulent mixture, the zone of low velocity that is of the order of a triple flame speed.

4. Conclusion

The numerical modeling of partially premixed combustion appears a real challenge because the reaction zones result from the combination of both thin propagating premixed flamelets and more mixing controlled diffusion flames. The modeling of these flames is at a very early stage, but is strongly needed since more and more industrial devices operate with partially premixed combustion. Large Eddy Simulation is a promising tool because the information gained at the resolved unsteady field may be used as a guide to determine the subgrid flame regime.

References

Boger, M. and Veynante, D. (2000). Large eddy simulation of a turbulent premixed v-shaped flame. In Dopazo, C., editor, *Advances in Turbulence*, pages 449 – 452. CIMNE, BARCELONA, isbn: 84-89925-65-8 edition.

Bray, K. and Libby, P. A. (1994). Recent developments in the bml model of premixed turbulent combustion. In Libby, P. and Williams, F., editors, *Turbulent Reacting Flows*, pages 116–151. Academic Press London.

Bray, K. N. C. (1996). The challenge of turbulent combustion. In *Proceedings of the 26th Symp. (Int.) on Combustion, Naples.* The Combustion Institute, Pittsburgh.

Buckmaster, J. (1996). Edge-flames and their stability. *Combust. Sci. Tech.*, 115:41–68.

Buckmaster, J. and Matalon, M. (1988). Anomalous lewis number effects in tribrachial flames. In *Proceedings of the 22nd Symp. (Int.) on Combustion*, pages 1527–1535. The Combustion Institute, Pittsburgh.

Buckmaster, J. and Weber, R. (1996). Edge-flame holding. In *Proceedings of the 26th Symp. (Int.) on Combustion.* The Combustion Institute, Pittsburgh.

Daou, J. and Linan, A. (1998). Triple flames in mixing layers with nonunity lewis numbers. pages 667–674.

Dold, J. W. (1989). Flame propagation in a nonuniform mixture: analysis of a slowly varying triple flame. *Combust. Flame*, 76:71–88.

Domingo, P., Vervisch, L., and Bray, K. (Seattle, July 29 - August 3, 2001). Large eddy simulation of lifted jet diffusion flames. In *18th International Colloquium on the Dynamics of Explosions and Reactive systems.*

Ducros, F., Laporte, F., Soulères, T., Guinot, V., Moinat, P., and Caruelle, B. (2000). High-order fluxes for conservative skew-symmetric-like schemes in structured meshes: application to compressible flows. *J. Comput. Phys.*, 161:114–139.

Echekki, T. and Chen, J. H. (1998). Structure and propagation of methanol-air triple flames. *Combust. Flame*, 114(1/2):231–245.

Favier, V. and Vervisch, L. (2001). Edge flames and partially premixed combustion in diffusion flame quenching. *Combust. Flame*, 125(1/2):788–803.

Ghosal, S. and Vervisch, L. (2000). Theoretical and numerical investigation of symmetrical triple flame using a parabolic flame tip approximation. *J. Fluid Mech.*, 415:227–260.

Ghosal, S. and Vervisch, L. (2001). Stability diagram for lift-off and blowout of a round jet laminar diffusion flame. *Combust. Flame*, 124(4):646-655.

Hartley, L. J. and Dold, J. W. (1991). Flame propagation in a nonuniform mixture: analysis of a propagating triple-flame. *Combust. Sci. and Tech.*, 80:23–46.

Haworth, D., Cuenot, B., Poinsot, T., and Blint, R. (2000). Numerical simulation of turbulent propane-air combustion with non-homogeneous reactants. *Combust. Flame*, 121(3):395–417.

Kioni, P. N., Rogg, B., Bray, K. N. C., and Liñán, A. (1993). Flame spread in laminar mixing layers: the triple flame. *Combust. Flame*, 95:276.

Liñán, A. (1974). The asymptotic structure of counterflow diffusion flames for large activation energies. *Acta Astronautica*, 1007(1).

Mahalingam, S., Thévenin, D., Candel, S., and Veynante, D. (1999). Analysis and numerical simulation of a nonpremixed flame in a corner. *Combust. Flame*, 118(1/2):221–232.

Meneveau, C., Lund, T. S., and Cabot, W. (1996). A lagrangian dynamic subgrid-scale model of turbulence. *J. Fluid Mech.*, pages 353–386.

Muñiz, L. and Mungal, M. G. (1997). Instantaneous flame-stabilization velocities in lifted-jet diffusion flames. *Combust. Flame*, 111(1/2):16–31.

Peters, N. (2000). *Turbulent Combustion.* Cambridge University Press.

Pierce, C. and Moin, P. (1999). A dynamic model for subgrid-scale variance and dissipation rate. *Physics of Fluids*, 10(12):3041–3044.

Plessing, T., Terhoeven, P., Peters, N., and Mansour, M. S. (1998). An experimental and numerical study on a laminar triple flame. *Combust. Flame*, 115(3):335–353.

Poinsot, T. and Veynante, D. (2001). *Theoretical and Numerical Combustion.* R. T. Edwards, Inc.

Réveillon, J. and Vervisch, L. (2000). Partially premixed combustion in turbulent spray. In *8th European Conference on turbulence, June 27-30, Barcelona.*

Ruetsch, G., Vervisch, L., and Liñán, A. (1995). Effects of heat release on triple flame. *Phys. Fluids*, 6(7):1447–1454.

Schefer, R. W. and Goix, P. (1998). Mechanism of flame stabilization in turbulent, lifted jet flames. *Combust. Flame*, 112(4):559–574.

Shay, M. L. and Ronney, P. D. (1998). Nonpremixed edge flames in spatially varying straining flows. *Combust. and Flame*, 112(1/2):112–171.

Terhoeven, P. and Peters, N. (1997). Basic flame structures of turbulent combustion for application in diesel engines. *Combust. and Flame*, page to appear.

Vervisch, L. (2000). Using numerics to help understand nonpremixed turbulent flames. *Proc. Combust. Inst.*, 28:11–24.

Watson, K. A., Lyons, K. M., Donbar, J. M., and Carter, C. D. (1999). Observations on the leading edge in lifted flame stabilization. *Combust. and Flame*, 119(1/2):199–202.

Williams, F. A. (1985). *Combustion theory.* The Benjamin/Cummings Publishing Company, Inc.

Yamashita, H., Shimada, M., and Takeno, T. (1996). A numerical study on flame stability at the transition point of jet diffusion flames. *Proc. Combust. Inst*, 26:27–34.

Discussion

N. Peters, Stanford University "Well I have one question when you have not a one set, but a multi-set chemistry and you convert the fuel into CO and H2 behind the premix flame front then your product of fuel plus oxygen should be zero as in the unburnt mixture."

Reply: Yeah you are right. The product in the unburnt mixture...?

N. Peters, Stanford University "In the unburnt you have - - If you are one side or the other side you either have fuel oxidizing equal to zero in the diffusion flame and here you are behind the premix flame but in front of the diffusion flame but you do not have fuel."

Reply: You mean just after the triple point when you burn through? Certainly the criterion should include the CO. Yes because after the diffusion fuel burns CO and H2 so you should change the fuel to CO.

Sebastien Candel, EM2C, Paris "I like very much your lecture. It was beautiful. In the triangle that you are drawing, where do you start in fact the triangle where should you or can you cut it somewhere?"

Reply: The earth is round so you start wherever you want you get back to the same point. I'd like to be honest I know that you would like me to start with the flame surface. I'd to start with the scale dissipation rate because combustion is somewhere mixing limited so the scalar dissipation rate is something you find everywhere and also when you integrate the flame surface but in fact you can start everywhere because everything is a function of everything and vice versa The triangle is not a - I should put a circle actually. Just to add something the difference between the scalar dissipation rate if now you see a description, the difference between the scalar dissipation on flame surface that within the scalar dissipation you keep only the small scales because the mean gradient disappears. So you do not have information on length scale wrinkling of the flame that you do have on the flame surface. So that the information on the length scale is included on the flame surface but not the scalar dissipation.

Johannes Janicka, Darmstadt "I have a question related to other models. You propose a very interesting model for partially premixed combustion. I can remember that there is an alternative model published by Norbert Peters and coworkers (Mueller, Breibach and Peters, 25th Symposium on Combustion, pp 1099, ed). I am interested in this problem because we are just integrating this model into LES and I would like you to comment on the special advantages of the work you have developed."

Reply: You refer to the model that uses the G equation. In fact you can replace the description of the premixed front, that is the C equation that we have, you can replace it with the G - no problem. I'm not sure how you do the transfer you know the transition between partially premix and diffusion in your approach. How do you do that? How do you switch clearly from the G equation to the diffusion flamelet part?

Johannes Janicka, Darmstadt "While the premix flame should be at the position after the G equation formula. If the G equation moves after this you know you have a diffusion flame."

Reply: So you have decided where the diffusion flame is. In that case I don't decide at all it depends on the flow and my resolution because when you decrease the resolution the subgrid orientation factor start to play a role and if you do some partial pre mixing of the sub grid you should see it. So I think that is that advantage. In your case you fix the transition. You say after premix its diffusion but if you quench somewhere and you get a lot of premixing because of obstacles or sediment in your system how would you do that? Because if you come back to premixing, this model apriori comes back to the premix model.

Cyrus Madnia, SUNY Buffalo "I want to know how good your model captures the local extinction and if you have ignition in the flow."

Reply: Well ignition I don't know how to do. I know how to do propagation. Now concerning the other aspect of your question. Local extinction in the premix part what we do see sometimes is local quenching of the premix part and then for the diffusion flame part it depends on the quality of your flamlet model. You have to realize that what I am talking about here is not a new turbulent combustion model it is a just a way of using the resolved field to combine things. You know for instance when you do PDF transport most of the mixing model have problems to propagate fronts so if people like Steve Pope and Dan Haworth develop things to incorporate the propagation with the mixing model. If you use a PDF approach within a LES context you may use this criterion to switch from one mixing model to the other. The fact is, we know part of the resolved scale.

Steve Pope, Cornell University "Just a quick comment on this triangle question. Of course in the triangle you had the conditional mean scalar dissipation but if we are interested in extinction for non premix combustion we would have to know more than just the mean. That is, we would have to know something about the distribution of scale dissipation. From my view point, the one fundamental quantity from which you can get everything else is the joint PDF of the scalar and its gradient because you can get all of those things plus the distribution of scale dissipation."

Reply: Sure in fact the relation between the flame surface and the PDF comes from the joint PDF of C and gradient in C and you are right.

Chapter 20

NONPREMIXED TURBULENT FLAMES INVESTIGATED WITH DIRECT NUMERICAL SIMULATION USING DETAILED CHEMISTRY

Renan Hilbert
Dominique Thévenin
Laboratoire EM2C, Ecole Centrale Paris
Grande Voie des Vignes, 92295 Châtenay-Malabry, FRANCE
hilbert@em2c.ecp.fr, thevenin@em2c.ecp.fr

Abstract Turbulent nonpremixed H_2/Air flames are simulated using direct numerical simulations coupled with detailed chemical schemes, complex transport models and accurate thermodynamic parameters. Typical results and post-processing procedures based on these results are presented in this paper. The choice and definition of a mixture fraction appropriate to describe turbulent mixing is discussed. The influence of differential diffusion is investigated comparing the present results with similar computations using a unity Lewis number hypothesis. Differences due to differential diffusion concerning global flame structure as well as conditionally-averaged quantities are emphasized.

Keywords: Direct Numerical Simulation, Detailed Chemistry, Differential Diffusion, Turbulent Combustion, Nonpremixed Flame

1. Introduction

Combustion is an important phenomenon in many industrial configurations. The flow in which reactions take place is most of the time turbulent, the chemical transformations involve a large number of elementary reactions and these two processes are closely coupled. A full understanding and modelling of turbulent combustion is still not completed yet and much work remains to be done. In particular the lack of generality of present models indicates that more information is needed on turbulent reactive flows. When using Direct Numerical Simulation of

A. Pollard and S. Candel (eds.), IUTAM Symposium on Turbulent Mixing and Combustion, 251-258.

the governing equations, all the flow and reactive space- and time-scales are solved without any turbulent closure model. Despite the inherent limitations of DNS in terms of computing time and avalaible computing ressources, it offers an interesting alternative to experiments to assess the importance of various physical mechanisms, get complementary informations and therefore help turbulent combustion modelling [Vervisch, 1998]. In the present study a H_2/N_2/Air turbulent nonpremixed flame is studied using DNS incorporating detailed models for chemistry and transport. Results of simulation will be presented and afterwards post-processed. The choice of the definition for the mixture fraction will be discussed and the influence of differential diffusion will be examined.

2. Numerical code and practical computations

The DNS code employed in this work, called *parcomb*, is described in detail in previous publications (see [Thevenin, 1996] and references therein). This code solves the full compressible Navier-Stokes equations. By coupling the flow solver with existing libraries [Maas, 1989] for chemistry and transport, we can take into account detailed reaction schemes, multicomponent diffusion velocities and an accurate description of thermodynamic coefficients (determined using fifth-order polynomial fits of experimental measurements). In this study, a chemical scheme involving 9 species (H_2, O_2, H_2O, H, O, OH, HO_2, H_2O_2 and N_2) and 37 elementary reactions [Maas, 1988] is used to describe combustion of H_2 in Air. The multicomponent diffusion velocities are computed in a standard way [Hirschfelder, 1954]. The Navier-Stokes Characteristic Boundary Conditions [Poinsot, 1992] taking into account detailed chemistry and thermodynamics are used. A pseudo-turbulent velocity field defined in Fourier space is used to initiate the simulation. The parameters chosen by the user are the r.m.s. velocity u' and the integral length-scale of turbulence Λ. The turbulent kinetic energy distribution is given by a Von-Kármán spectrum with Pao correction, in the form given by [Hinze, 1975]. The phases of the turbulent velocity components in Fourier space are considered to be uncorrelated and are fixed by a random-number generator.

A laminar nonpremixed flame of hydrogen diluted in nitrogen and air is first ignited in a one-dimensional configuration, using a source term in the energy equation. Resulting profiles are then uniformly extended in the $y-$direction and a pseudo-turbulent velocity field is imposed on top of it. In the $x-$direction, both boundaries are considered as non-reflecting outflow boundaries with pressure relaxation, while periodicity is assumed along the $y-$ direction. A uniform grid spacing of 50 μ is used in both directions, resulting in a 301 × 301 grid mesh. Global mixture ratio Φ is equal to 0.8. The initial turbulent field for the results presented afterwards corresponds to u' =2.46 m.s^{-1} and Λ =1.32 mm, resulting in a turbulent Reynolds number Re_t = 210, associated with an eddy turn-over time $\tau_t = \Lambda/u' = 0.5$ ms. A CRAY T3E parallel supercomputer using 12 processors was used for the simulation and the resulting computing time was about 17 hours (or 200 hours single-processor CPU time).

3. Results and Discussions

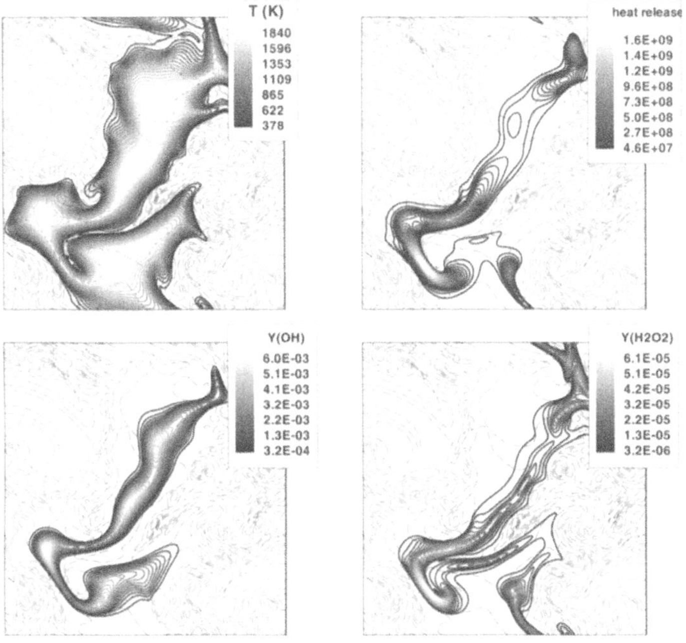

Figure 20.1. Typical result of DNS of a turbulent nonpremixed hydrogen flame. Results are shown after two eddy-turn-over times $t = 2\tau_t = 1$ ms.

As time evolves during the simulation, turbulence is slowly decaying and interacts with the initially-flat flame. Turbulence wrinkles the flame front and makes it thicker or thinner depending on the local conditions. We show in Fig.20.1 the instantaneous fields of temperature, heat release, mass fractions of the intermediate radicals OH and H_2O_2 after two eddy turn-over times. The flame appears to be highly contorted by the turbulent velocity field and it is clear from these pictures that the chemical species behave very differently, leading to serious difficulties in the interpretation. These simulations deliver a large quantity of data to post-process in order to examine the influence of different physical processes on the turbulent flame structure [Thevenin, 1999]. In this study, post-processing is focussed on the importance of transport models, and in particular of differential diffusion.

As stated in [Smith, 1995] for example, a notable consequence of taking into account differential diffusion is the impossibility of giving a unique definition of the mixture fraction. In the case of diluted flames, a classical way to define a conserved scalar is to base the definition on the mass fraction of an inert species (e.g. N_2). One can also use elemental mass fractions of H and O to get a passive scalar, such as in the definition proposed by [Bilger, 1993] :

$$Z = \frac{(1/2)Z_H/W_H + (Z_{O,0} - Z_O)/W_O}{(1/2)Z_{H,0}/W_H + Z_{O,0}/W_O} \tag{20.1}$$

where W denotes molar mass, Z_j mass fraction of element j in the mixture, and 0 initial compositions on respectively oxidizer and fuel sides.

Fig. 20.2 is a scatter plot representing all the grid points in the turbulent flame. We therefore check these two possible definitions of the mixture fraction using exactly the same data points. It is clear that the results do not follow a classical mixing line but are closer to a sigmoidal curve, as in [Smith, 1995]. This shows the importance of the definition of the mixture fraction used to get a simplified, global description of such complex flows. This point cannot be neglected, since it could lead to first-order errors if differential diffusion is not properly taken into account.

One can also calculate normalized values of the elemental mass fractions of H and O so that both are 0 on the air side and 1 on the fuel side.

$$F_H = \frac{Z_H}{Z_{H,0}} \text{ and } F_O = \frac{Z_{O,0} - Z_O}{Z_{O,0}} \tag{20.2}$$

The difference between these quantities can be taken as a differential diffusion parameter [Barlow, 2000]. In the absence of preferential diffusion, F should be 0. A scatter plot of F as a function of the previously-defined Z is presented in Fig.20.3. This plot clearly shows that F is far from 0 in this simulation. Differential diffusion thus plays a significant role on flame structure, as expected in particular for hydrogen flames where light species such as H or H_2 are especially important.

To get an exact idea of this role, the same calculation as previously described was carried out again with exactly the same initial turbulent

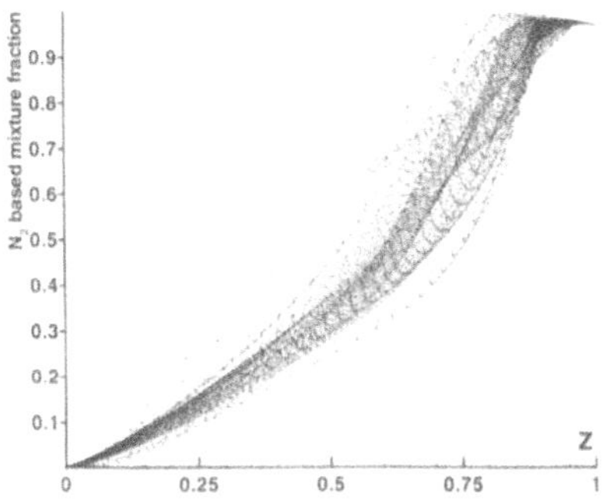

Figure 20.2. Comparison of two different definitions for the mixture fraction

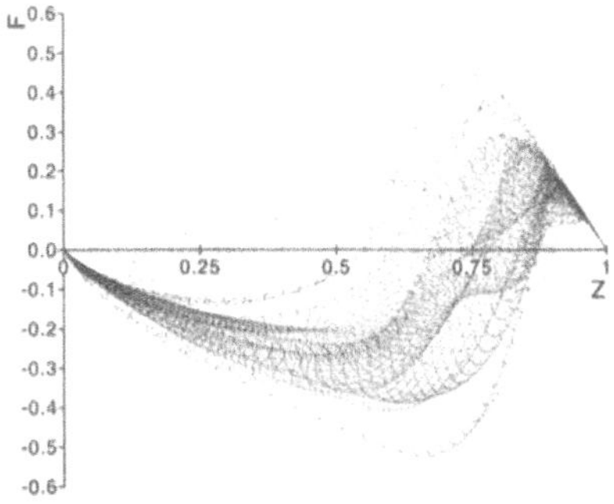

Figure 20.3. Scatter plot of the differential diffusion parameter $F = F_H - F_O$ given by eq.(20.2)

field, but with Lewis numbers constant and set equal to unity. Comparisons between these two simulations are shown for scatter plots of global quantities on Fig.20.4. An important difference is the scattering of the points, which is much stronger when using detailed transport models. This is true for temperature as well as for reduced scalar dissipation $|\nabla Z|^2$, which characterizes turbulent mixing.

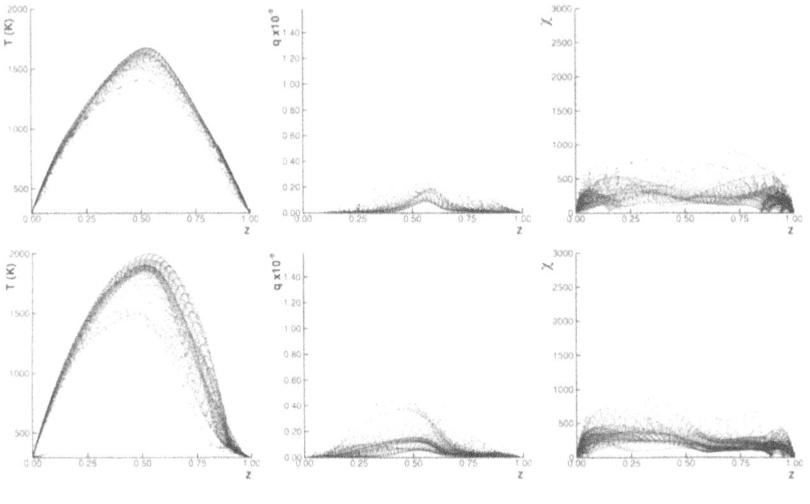

Figure 20.4. Scatter plots of temperature (left), heat release (center) and reduced scalar dissipation (right). Top: unity Lewis number. Bottom: multicomponent diffusion velocities.

The observed dispersion can be quantified by calculating averaged values of these quantities conditionned on the value of Z, and corresponding standard deviations. For temperature, these conditionally-averaged profiles only exhibit a small difference for the position of the peak value, but large differences are seen in the fluctuation profiles. This fact is consistent with [Nilsen, 1999]. The tendencies are the same when comparing averaged profiles of $|\nabla Z|^2$ and for mass fractions of reactants and products (not shown). For heat release, even the averaged profiles are very different. The fact that the lightest species are carried faster towards the reaction zone completely change the burning rate distribution in the mixing space. The main difference is that reaction is enhanced in the $Z < Z_{st}$ domain due to preferential diffusion.

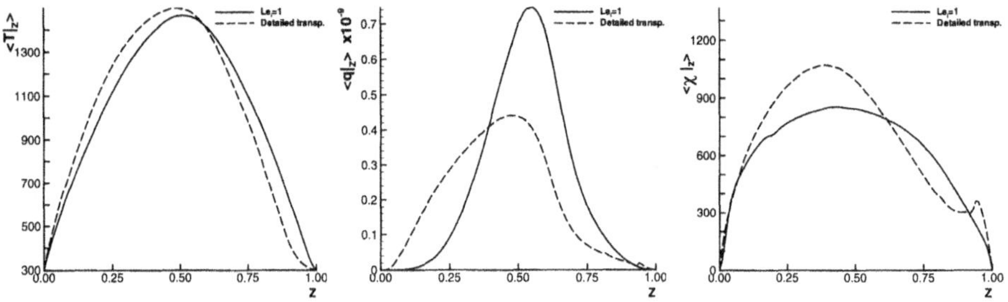

Figure 20.5. Averaged values conditioned on the value of Z (temperature, left. Heat release, center. Reduced scalar dissipation, right). Solid line: unity Lewis number. Dashed line: multicomponent diffusion velocities.

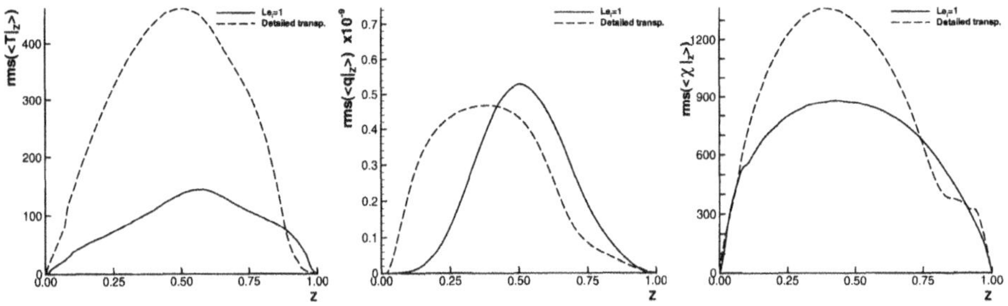

Figure 20.6. Standard deviation of averaged values conditioned on the value of Z (temperature, left. Heat release, center. Reduced scalar dissipation, right). Solid line: unity Lewis number. Dashed line: multicomponent diffusion velocities.

Conclusion

Direct Numerical Simulations using detailed models for chemistry and transport were carried out and the results were post-processed to analyze the influence of differential diffusion. This investigation indicates that great care must be taken concerning the most appropriate definition of the mixture fraction. The most visible influence of differential diffusion is that it leads to a larger dispersion for conditionally-averaged values of T and $|\nabla Z|^2$. This is particularly clear when comparing with similar results obtained for constant unity Lewis numbers for all species. The main consequences of this influence of differential diffusion on mixing and chemical reactions are investigated by [Gicquel, 2001], in particular regarding possible deviations from local chemical equilibrium. These results are based on the same DNS calculation as presented here.

Acknowledgments

The support of the french national supercomputer center IDRIS is gratefully acknowledged. We also thank N. Darabiha and O. Gicquel for fruitful discussions.

References

Bilger, R. W. (1993). The Structure of Turbulent Nonpremixed Flames. *Proc. Comb. Inst.* 22:475-488.

R. S. Barlow,G. J. Fiechtner, C. D. Carter and J.-Y. Chen (2000). Experiments on the scalar structure of turbulent $CO/H_2/N_2$ jet flames. *Combust. Flame* 120:549-569.

O. Gicquel, R. Hilbert, D. Thévenin, N. Darabiha (2001). Influence of Differential Diffusion on Local Equilibrium and Super-Adiabatic Combustion in Turbulent Non-Premixed Flames. *IUTAM Symposium on Turbulent Mixing and Combustion* Kingston, Canada.

J. O. Hinze (1975). Turbulence in Fluids. *McGraw Hill Book Company*

J. O. Hirschfelder, C. F. Curtis and R. B. Bird (1954). Molecular Theory of Gases and Liquids. *John Wiley & Sons*

U. Maas and J. Warnatz (1988). Ignition Processes in Carbon-Monoxide-Hydrogen-Oxygen Mixtures. *Proc. Comb. Inst.* 22:1695-1704

U. Maas and J. Warnatz (1989). Simulation of chemically reacting flows in two-dimensional geometries. *Impact of Computing in Science and Engineering.* 1:394-420

V. Nilsen and G. Kosàly (1999). Differential Diffusion in Turbulent Reacting Flows. *Combust. Flame.* 117:493-513

T. Poinsot and S. Lele (1992). Boundary conditions for direct simulations of compressible viscous flows. *J. Comput. Phys.* 101:104-129

N. S. A. Smith (1995). Modeling complex chemical effects in turbulent nonpremixed combustion. *Annual Research Briefs.* Center for Turbulent Research, p. 301-321.

D. Thvenin, Behrendt, F., Maas, U., Przywara, B., and Warnatz, J. (1996) Development of a Parallel Direct Simulation Code to Investigate Reactive Flows. *Computers and Fluids.* 5:485-496.

D. Thvenin and R. Baron (1999) Investigation of turbulent non-premixed flames using Direct Simulations with detailed chemistry. *Kluwer Academic.* p. 323-334.

Vervisch, L. and Poinsot, T. (1998) Direct Numerical Simulation of non premixed turbulent flames. *Ann. Rev. Fluid Mech.* 30:655-692.

Warnatz, J. (1989) Numerical Simulation of Ignition Processes. *Lecture Notes in Physics/Springer-Verlag.* p. 351

Discussion

Paul Dimotakis, Caltech. "It seems to be said often that differential diffusion effects go away as the Reynolds number goes up. I don't understand why that is. I mean, unless the implication is that at higher Reynolds numbers you have much better diffusion, much better homogenization, so that you won't have interfaces, so that is a different issue and it is not a Reynolds number effect. In fact as long as you are non premixed as the Reynolds number goes up the gradients get steeper so I would argue that it should get worse not better. In fact we have experiments in the hydrogen-fluorine facility where we kept the same stoichiometry and switched the hydrogen and the fluorine at very high Reynolds numbers, in the tens of thousands if not hundred thousands, and we saw measurable effects so I don't subscribe to do that and I was wondering what you think about it since you claimed it."

Reply: I must profess I don't really understand your question.

Paul Dimotakis, Caltech. "The question is why do you think the differential diffusion effects are larger at lower Reynolds numbers."

Reply: That is the thing I read in the literature on turbulent hydrogen jet flames by Barlow, that said that the effects of differential diffusion could be not so important for high values of Reynolds number. Of course I don't have any sort of personal opinion we this kind of calculation when the Reynolds number are what they are for DNS not having a conclusive answer to give you.

Paul Dimotakis, Caltech. "And my second question was and you show the scatter plot of the scalar dissipation rate versus mixture fraction, Z. My question is which Z?"

Reply: Here is the definition of Z, I have here taken into account elements...

Paul Dimotakis, Caltech. "Thank-you."

Chapter 21

COUNTER-GRADIENT SCALAR TRANSPORT IN LARGE EDDY SIMULATION OF TURBULENT PREMIXED FLAMES

S. Tullis and R.S. Cant
CFD Laboratory, Department of Engineering,
Cambridge University,
Cambridge, CB2 1PZ, U.K.

Abstract Counter-gradient transport is examined in *a priori* large eddy simulations of turbulent premixed combustion using direct numerical simulations (DNS) of 3D planar flames. It is found that the size of the LES filter, and consequent resolution of the flame, have an important effect on the scalar flux and that this can be considered explicitly by decomposing the flux into flame interior and non-reacting components. The LES form of a scalar flux transport equation was written and models developed for the two main counter-gradient terms, the pressure gradient and velocity-reaction rate correlation.

Keywords: Turbulent premixed combustion, scalar flux, large eddy simulation

1. Introduction

Large Eddy Simulations (LES) fully resolve large scale turbulent structures; however, structures smaller than the filter size must be modelled. For premixed combustion a transport equation can be written for a reaction progress variable c:

$$\partial_t \bar{\rho}\tilde{c} + \partial_j \bar{\rho}\tilde{u}_j\tilde{c} = \partial_j \bar{\rho} D \partial_j \tilde{c} - \partial_j \bar{\rho}\mathcal{T}(u_j, c) + \bar{\omega} \tag{21.1}$$

where an overbar signifies spatially filtered quantities and a tilde denotes Favre filtering. There are two subgrid scale (SGS) terms that must be modelled in this equation, the filtered reaction rate $\bar{\omega}$ and the (SGS) turbulent scalar flux term $\mathcal{T}(u_j, c) = \widetilde{u_j c} - \tilde{u}_j\tilde{c}$. This paper deals only with the subgrid scale turbulent scalar flux.

A. Pollard and S. Candel (eds.), IUTAM Symposium on Turbulent Mixing and Combustion, 259-267.

Gradient transport assumptions are typically used in both LES and Reynolds-averaged (RANS) models for the scalar flux. However, it has been recognized that counter-gradient transport exists due to the dilatation across the flame front. Counter-gradient transport is greater than turbulence-generated gradient transport when there is a high ratio of heat release to turbulence intensity, expressed in the dimensionless form:

$$N_B = \frac{\tau s_L}{2\alpha u'} \tag{21.2}$$

where τ is the heat release factor, s_L the laminar flame speed, u' the turbulence intensity (in the reactants) and α is an efficiency function that, in an LES context, measures how well the turbulence can produce SGS wrinkling [Veynante *et al.*, 1997].

2. A Priori LES

Here, two direct numerical simulations (DNS) of premixed turbulent flames have been spatially filtered to examine the LES terms and models involved in SGS scalar transport. The DNS flames have been chosen to provide different ratios of heat release to turbulence intensity. The details of the numerical methods used in the DNS results are provided in the appropriate references. Both of these flames are nominally planar with one step chemistry. The main differences between the DNS datasets are as follows:

Case R Planar flame with turbulent inflow (decaying as it is convected through the domain). Low Mach number approximation. Periodic sides [Zhang and Rutland, 1995].

Case S Back-to-back planar flames in decaying turbulence. Fully compressible. Periodic sides with 2 outlets [Jenkins & Cant, 1999].

DNS Flame	*R*	*S*
Grid	256x128x128	96^3
Heat release factor (τ)	2.3	5
Relative turbulence intensity (u'_0/s_L)	1	10
Relative turbulence length scale (l_T/δ)	~ 30	~ 3
$N_B = \frac{\tau s_L}{2\alpha u'}$ at flame location	~ 10	~ 1

Table 21.1. DNS database parameters : u'_0 is the initial (or inlet) turbulence intensity; l_T is the integral length scale; and δ is the DNS flame thickness.

The *a priori* LES is performed by imposing a range of filter sizes on the DNS data. Both Gaussian and top-hat filters (with $\Delta_{filter}/\Delta_{grid} = 2$) have been used with little difference observed between the filter types.

3. Turbulent Scalar Transport

The results discussed here are flame normal scalar fluxes where the normal is based on the gradient of the resolved progress variable.

The low turbulence case (R) exhibits counter-gradient transport for all filter sizes in almost all regions of the flame (see figure 21.1). For small filter sizes the scalar flux tends towards zero while for very large filter sizes, the magnitude of the scalar flux tends towards the difference between the mean product P and reactant R velocities $\bar{u}_P - \bar{u}_R$. The shape of these curves (particularly for large filter sizes) is similar to the parabolic behavior based on $\tilde{c}(1 - \tilde{c})$ as observed in the BML formulation. Between these extremes, the behavior is due mainly to the varying fraction of the filter volume that is occupied by the flame. As the filter size Δ decreases, this fraction increases and counter-gradient transport (which is mainly due to SGS dilatation) decreases as it becomes spread over the increasingly resolved flame interior. This effect is related to the variation of $(\bar{c} - \tilde{c})$ with filter size [Boger *et al.*, 1998, Hawkes *et al.*, 2001]. Turbulence driven gradient transport also tends to zero for small filter sizes as SGS flame wrinkling and turbulence levels decrease, although in the present case, this seems to be a smaller effect given the low turbulence levels.

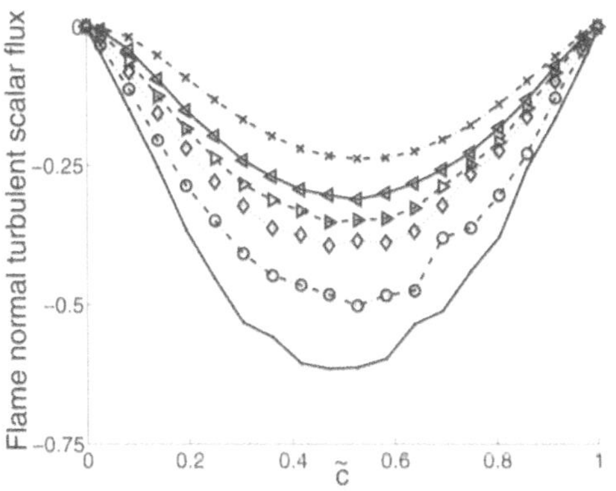

Figure 21.1. Turbulent scalar flux as a function of filter size Δ for the low turbulence Case R. Averaging to remove data scatter is done by values of the resolved progress variable $\tilde{c}$. Filter sizes Δ/δ shown are : $\times$ 1.2; $\triangleleft$ 2; $\triangleright$ 2.8; $\diamond$ 4; $\circ$ 8; $-$16

Case S, with a higher turbulence intensity relative to the heat release, still exhibits mainly counter-gradient transport. There are, however, significant regions of the flame surface which show gradient transport. These regions are mainly associated with areas where the SGS turbulent flow is counter to the dilatation. This emphasizes the local nature of the scalar transport.

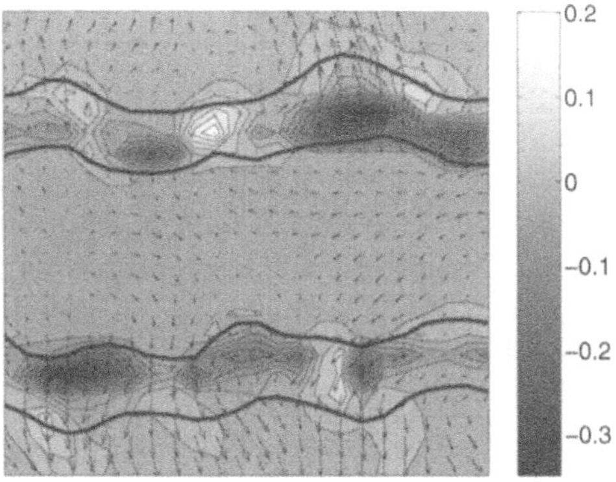

Figure 21.2. Contours of turbulent scalar flux on a slice through Case S ($\Delta/\delta = 0.6$). Also shown are velocity vectors on the slice plane and the bold contours show the flame location ($\tilde{c} = 0.1$ and 0.9)

As mentioned earlier, the variation of scalar flux appears to be mainly due to variations in the resolution of the flame interior. Consequently, the Bray-Moss-Libby, or BML [Bray *et al.*, 1985] assumption of an entirely bimodal probability density function (PDF) (where the volume of the filter consists of only reactants and products) is no longer valid (see Hawkes *et al.*, 2001 and Tullis, 2001, for details of this variation with filter size). However, the breakdown of the bimodal PDF can be considered directly by splitting the flux into two components:

$$\begin{aligned} \mathcal{T}(u_j, c) &= \mathcal{T}(u_j, c)|_{Non-Reacting} + \mathcal{T}(u_j, c)|_{Flame} \qquad (21.3) \\ &= (\alpha + \beta)[\widetilde{u_j c}|_{Non-Reacting} - \tilde{u}_j \tilde{c}] + \gamma[\widetilde{u_j c}|_{Flame} - \tilde{u}_j \tilde{c}] \end{aligned}$$

where α, β and γ are the volume fractions of the filter occupied by reactants, products and flame, respectively. The non-reacting term $\mathcal{T}(u_j, c)|_{Non-Reacting}$ has both dilatation and turbulent components. The dilatation can be expressed as the product of BML dilatation of a laminar flamelet: $\tilde{c}(1 - \tilde{c})\tau s_L$, and the quantity of flame surface over which this occurs: $\Sigma/|\nabla \bar{c}|$, where Σ the flame surface density. The turbulent component varies with the filter resolution of the flame wrinkles so one could use a term such as: $-2\alpha u'$, where the SGS turbulent intensity (in the reactants) u' is a function of Δ, and α is the efficiency function from equation 21.2. This efficiency function, measuring how well the turbulence can produce SGS wrinkling, is a strong function of Δ [Veynante and Poinsot, 1997].

The reacting term $\mathcal{T}(u_j, c)|_{Flame}$ is mainly due to dilatation and depends on the laminar flamelet structure (which occupies the volume $\gamma \approx \delta\Sigma$).

4. Transport equation for turbulent scalar flux

To help understand the processes involved the SGS scalar flux, a transport equation can examined. The LES form of the turbulent scalar flux transport equation can be written as:

$$\begin{aligned}\underbrace{\partial_t \bar{\rho}\mathcal{T}(u_j, c)}_{\text{time}} + \underbrace{\partial_j \bar{\rho}\tilde{u}_j \mathcal{T}(u_j, c)}_{\text{convection}} &= -\underbrace{\partial_j \bar{\rho}\mathcal{T}(u_j, u_i, c)}_{\text{turb transport}} - \underbrace{\bar{\rho}\mathcal{T}(u_j, u_i)\partial_j \tilde{c}}_{\text{grad } c \text{ production}} \\ &- \underbrace{\bar{\rho}\mathcal{T}(u_j, c)\partial_j \tilde{u}_i}_{\text{grad } u \text{ production}} - \underbrace{\mathcal{T}(c, \partial_i p)}_{\text{pressure gradient}} - \underbrace{\mathcal{T}(u_i, \partial_j J_j)}_{\text{molecular diffusion}} \\ &+ \underbrace{\mathcal{T}(c, \partial_j \tau_{ij})}_{\text{viscous dissipation}} + \underbrace{\mathcal{T}(u_i, \omega)}_{\text{velocity-reaction rate correlation}}\end{aligned} \tag{21.4}$$

where the triple term $\mathcal{T}(u_j, u_i, c) = \widetilde{u_j u_i c} - \tilde{u}_j \mathcal{T}(u_i, c) - \tilde{u}_i \mathcal{T}(u_j, c) - \tilde{c}\mathcal{T}(u_i, u_j) - \tilde{u}_j \tilde{u}_i \tilde{c}$ The relative importance of these terms can be seen in figure 21.3 for cases R and S. It can be seen that the important terms

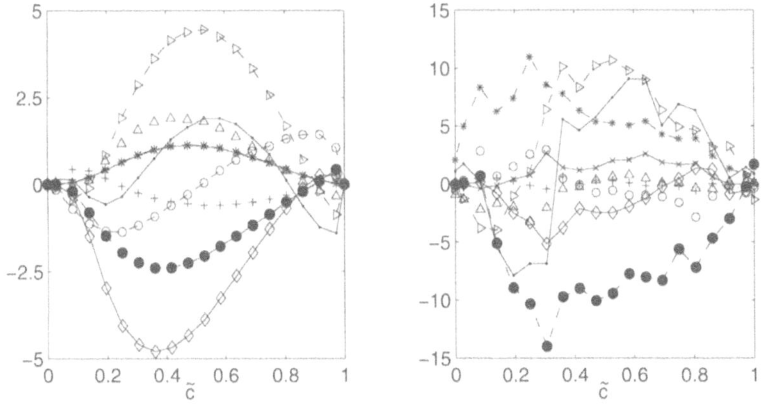

Figure 21.3. Relative importance of terms in the scalar flux transport equation for case R (weaker turbulence) with $\Delta/\delta \simeq 1.2$ on the left and case S (stronger turbulence) with $\Delta/\delta \simeq 1$ on the right. Terms are: $\cdot$ time; $\circ$ convection; $+$ turbulent transport; $*$ $\partial\tilde{c}$ production; $\times$ $\partial\tilde{u}$ production; $\bullet$ pressure gradient; $\triangle$ molecular diffusion; $\triangleright$ viscous dissipation; and, $\diamond$ velocity-reaction rate correlation.

contributing to gradient transport are production, viscous dissipation and molecular diffusion. The turbulent transport term appears small for both cases. The main contributions to counter-gradient transport come from the pressure term and the velocity-reaction rate correlation term. These two terms are discussed and modelled below.

4.1. Pressure Term

This term ($\mathcal{T}(c, \partial_j p) = \overline{c\partial_j p} - \tilde{c}\overline{\partial_j p}$) represents the effect of the pressure gradient acting on variable density packets of gas, preferentially accelerating light reaction products towards the products side of the flame (counter to the progress variable gradient). The term can be decomposed as

$$\mathcal{T}(c, \partial_j p) = \mathcal{T}(c, \partial_j p)|_{Flame} + \mathcal{T}(c, \partial_j p)|_{Non-Reacting} \tag{21.5}$$

with the non-reacting term further split into: $\mathcal{T}(c, \partial_j p)|_{external\ dp}$ and $\mathcal{T}(c, \partial_j p)|_{turbulence}$ in a manner similar to that used by Veynante & Poinsot (1997).

A model for the flame interior component can be based on the volume of the filter occupied by the flame $\gamma \approx \delta\Sigma$ and an approximation of the laminar flame shape (with c^* the average value of c in the flame) so that:

$$\begin{aligned} \mathcal{T}(c, \partial_j p)|_{Flame} &\approx \gamma[c^* - \frac{\tilde{c}}{2}]\partial_j p|_{Flame} \\ &\approx \delta\Sigma[c^* - \frac{\tilde{c}}{2}]C_1 \frac{\rho_0 \tau s_L^2}{\delta}\tilde{n}_i \end{aligned} \tag{21.6}$$

where $\tilde{n}_i$ is the resolved flame normal direction and C_1 a constant.

The external pressure gradient can be approximated as:

$$\mathcal{T}(c, \partial_j p)|_{external\ dp} \approx (\bar{c} - \tilde{c})\partial_j p|_{external} \tag{21.7}$$

and since,

$$\begin{aligned} (\bar{c} - \tilde{c}) &\rightarrow 0 && \text{as } \Delta \rightarrow 0 \\ &\rightarrow \tfrac{\tau\tilde{c}(1-\tilde{c})}{1+\tau\tilde{c}} && \text{as } \Delta \rightarrow \infty \end{aligned}$$

$$\mathcal{T}(c, \partial_j p)|_{external\ dp} \approx \frac{\Delta}{\Delta + 2\delta}[\frac{\tau\tilde{c}(1-\tilde{c})}{1+\tau\tilde{c}}]\partial\bar{p} \tag{21.9}$$

A similar process can be applied to the turbulence component:

$$\mathcal{T}(c, \partial_j p)|_{turbulence} \approx (\bar{c} - \tilde{c})\partial_j p|_{turb} \tag{21.10}$$

with

$$\partial_j p|_{turb} \approx C_2 \rho_0 \frac{\overline{u'^2}}{l_t} \approx C_2 \rho_0 \frac{k}{\Delta}\tilde{n}_i \tag{21.11}$$

4.2. Velocity - Reaction Rate Correlation

In case R, the close match between the behavior of this term ($\mathcal{T}(u_i, \omega) = \overline{u_i\omega} - \tilde{u}_i\bar{\omega}$) and the scalar flux itself suggests that BML type assumptions [Bray *et al.*, 1985] could be used in modelling it:

$$\mathcal{T}(u_i, \omega) = (\psi - \tilde{c})\frac{\mathcal{T}(u_i, c)}{\tilde{c}(1-\tilde{c})}\bar{\omega} \tag{21.12}$$

This is essentially a turbulent flame surface velocity [Veynante *et al.*, 1997]

$$\langle u'' \rangle = u_s - \tilde{u} \approx (\psi - \tilde{c})\frac{\mathcal{T}(u, c)}{\tilde{c}(1-\tilde{c})} \tag{21.13}$$

multiplied by the filtered reaction rate $\bar{\omega}$ (from, for instance, a flame surface density model where $\bar{\omega} = \rho_0 s_L \Sigma$). Figure 21.4 shows this relationship as a function of filter size Δ. The slopes of $\langle u'' \rangle$ agree fairly

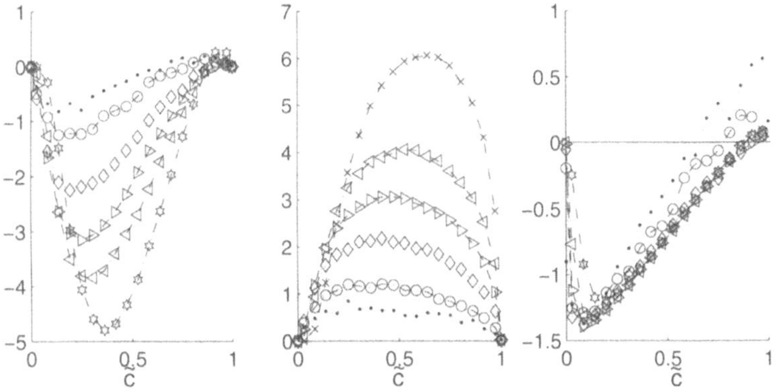

Figure 21.4. Case R velocity-reaction rate correlation term dependence on filter size, left: velocity-reaction rate correlation $\mathcal{T}(u_n, \omega)$, middle: filtered reaction rate $\bar{\omega}$, right: turbulent flame surface velocity $\langle u_n'' \rangle$ as the ratio of these. Averaging by resolved progress variable $\tilde{c}$ to remove data scatter . Filter sizes Δ/δ shown are : $\times$ 1.2; $\triangleleft$ 2; $\triangleright$ 2.8; $\diamond$ 4; $\circ$ 8; -16

well with the model slopes over the wide range of filter sizes while there is a small variation in the intercept ψ, particularly for larger filter sizes. Bray *et al.* (1985), in a RANS context, obtained a value of ψ (of 0.83) from the form of their laminar flamelet reaction term $\omega(c)$ which is close to the range seen in Figure 21.4.

When the turbulence levels are higher (case S), this BML type approach is unlikely to work so well. Even still, this term remains strongly correlated with the filtered reaction reaction rate and there is some evidence that back-calculated values of $(\psi - \tilde{c})$ collapse over a range of filter sizes [Tullis, 2001].

5. Summary

LES of the turbulent scalar flux requires consideration of counter-gradient transport and here a BML based approach was combined with a gradient turbulence-based model similar to that used by Veynante *et al.* (1997). Decomposition into flame and non-reacting components was used to account for the filter size dependence.

The terms in the developed LES transport equation for the scalar flux have been examined and the pressure gradient and velocity reaction rate terms identified as the main contributors to counter-gradient transport. A model for the pressure gradient term was proposed and it was shown that a BML based approach can be used to model the velocity reaction rate term.

Acknowledgments

The authors would like to thank Chris Rutland for access to his DNS dataset, NSERC and Alstom Power Ltd for financial support.

References

Boger, M., Veynante, D., Boughanem, H. and Trouvé, A. (1998). Direct numerical simulation of flame surface density concept for large eddy simulation of turbulent premixed combustion. *27th Symp. (Int.) on Combustion*, pages 917–927. The Combustion Institute, Pittsburgh.

Bray, K. N. C., Libby, P. & Moss, J. B. (1985). Unified modelling approach for premixed turbulent combustion - part I: general formulation. *Combust. Flame*, 61:87–102.

Hawkes, E. R., Tullis, S. W. and Cant, R. S. (2001). Direct numerical simulation analysis of the flame surface density balance equation for large eddy simulation. In preparation.

Jenkins, K. W. & Cant, R. S. (1999). Direct numerical simulation of turbulent flame kernels. In *Recent Advances in DNS and LES*, Knight, D. & Sakell, L., eds., pages 191–202. Kluwer Academic, New York.

Tullis, S. W. (2001). A Priori Large Eddy Simulation of Scalar Transport in Turbulent Premixed Flames. Ph.D Thesis, In preparation.

Veynante, D., Trouvé, A., Bray, K. N. C. and Mantel, T. (1997). Gradient and counter-gradient scalar transport in turbulent premixed flames. *J. Fluid Mech.*, 332:263–293.

Veynante, D. and Poinsot, T. (1997). Effects of pressure gradients on turbulent premixed flames. *J. Fluid Mech.* 353:83–114.

Zhang, S. and Rutland, C. J. (1995). Premixed Flame effects on turbulence and pressure-related terms. *Combust. Flame* 102:447–461.

Discussion

Darko Matovic, Queen's University "What do you mean by inclusion of models a posteriori in an LES code, as opposed to using it for actual predictions?"

Reply: Well, that would be using it for actual predictions. To be realistic, nobody here is going to develop turbulent scalar flux transport equation and throw it into their LES code. It is more that we will develop algebraic expressions for that. But for the transport equation it is useful to split out to see where the physics are coming from. That is the main way to do it.

Chapter 22

LOCAL FLAME STRUCTURE IN HYDROGEN-AIR TURBULENT PREMIXED FLAMES

Mamoru Tanahashi
Yoshikazu Ito
Masanobu Fujimura
and Toshio Miyauchi
Department of Mechanical and Aerospace Engineering,
Tokyo Institute of Technology
2-12-1 Ohokayama, Meguro-ku, Tokyo 152-8552, Japan

Abstract Local flame structures of the turbulent premixed flame are discussed. The study is based on the hydrogen transport near the flame front using DNS data of three-dimensional H_2-air turbulent premixed flame under the condition of $u'/S_L = 1.74$ and $l/\delta_L = 2.00$. In DNS, a detailed kinetic mechanism including 12 reactive species and 27 elementary reactions was used to represent the H_2-air reaction in turbulence. In a turbulent flow, the convection and diffusion effects on the hydrogen transport near the flame front depend on the curvature of the flame front. The convection effect is enhanced for the flame elements convex toward the burnt side, whereas the diffusion effect becomes dominant for the flame elements convex toward the unburned side.

Keywords: Turbulent premixed flames, direct numerical simulation, detailed kinetic mechanism, local flame structure

1. Introduction

Turbulent premixed flames are frequently observed in many engineering applications such as IC engines and gas turbines. However, complicated interactions between chemical reaction and turbulence lead to difficulties in understanding the detailed structure of the turbulent premixed flames. Recent developments of computer technology provided

A. Pollard and S. Candel (eds.), IUTAM Symposium on Turbulent Mixing and Combustion, 269-277.

a new tool for turbulence research: direct numerical simulation (DNS). DNS has allowed developments of turbulence models and provided further understanding of the physics of turbulence.

Similar approaches have been applied to turbulent combustion which may be viewed as an extension of fluid dynamics. In the recent DNS of turbulent premixed flames, researchers have tried to include realistic chemical reaction for hydrogen and hydrocarbons ([Baum *et al.* (1994)], [Echekki and Chen (1996)], [Miyauchi *et al.* (1997)], [Chen *et al.* (1998)], [Tanahashi *et al.* (1999a)]). These previous DNS of turbulent premixed flames gave a lot of information about the premixed flames in turbulence, while almost all of the previous studies were restricted to two-dimensional calculations due to the limitation of computer resources.

The resent studies on fine scale structure of turbulence ([Tanahashi *et al.* (1997)], [Tanahashi *et al.* (1999b)]) show that turbulence consists of many coherent tube-like eddies. The mean diameter of these eddies is about 10 times the Kolmogorov microscale η, and the maximum azimuthal velocity is about a half of the root mean square velocity fluctuation u'. As the coherent fine scale eddies of turbulence show a swirling motion that is quite strong and have a distinct three-dimensional structure([Tanahashi *et al.* (1999c)]), they influence the local flame structure in turbulent premixed flames. In the previous study ([Tanahashi *et al.* (2000)]), we have conducted three-dimensional DNS of a hydrogen-air turbulent premixed flame to investigate the interaction between the coherent fine-scale eddies and premixed flames. In this study, the previous DNS data are analyzed to clarify the local flame structures in turbulence.

2. DNS data of a hydrogen-air turbulent premixed flames

DNS of a hydrogen-air turbulent premixed flame propagating in three-dimensional isotropic turbulence have been carried out with a detailed kinetic mechanism and temperature dependence of transport and thermal properties ([Tanahashi *et al.* (2000)]). The detailed kinetic mechanism which includes 27 elementary reactions and 12 reactive species was used to represent H_2-air combustion in turbulence. Inflow boundary condition for chemical species is set equal to H_2-air mixture with $\phi = 1.0$ at 0.1MPa and 700K.

Inflow boundary condition for the velocity field is given as $u_{in}(y, z, t) = S_L + u'(y, z, t)$, where $u'(y, z, t)$ is a fully developed three-dimensional homogeneous isotropic turbulence and S_L is a laminar burning velocity. Initial Reynolds number based on the Taylor micro scale and u' is about 37.9. The calculations correspond to $u'/S_L = 1.74$ and $l/\delta_L =$

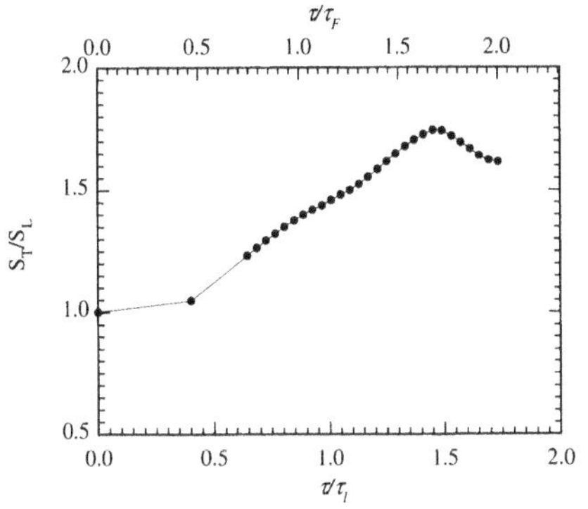

Figure 22.1. Development of the turbulent burning velocity

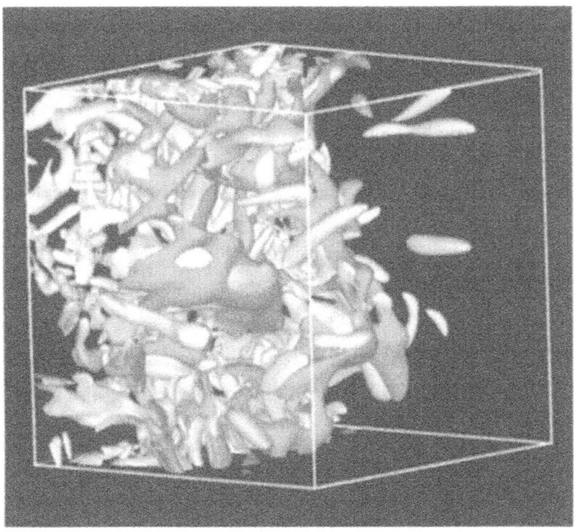

Figure 22.2. Contour surfaces of the second invariant of the velocity gradient tensor and heat release rate (white: $Q^+ = 0.02$, gray: $\Delta H^* = 1.0$)

2.00. Here, l is integral length scale of turbulence and δ_L is a laminar flame thickness. The computational domain was selected to be $0.5cm \times 0.5cm \times 0.5cm$ and $257 \times 128 \times 128$ grid points were used.

Figure 22.1 shows the development of turbulent burning velocity. From the initial condition, the turbulent burning velocity increases with time due to the initial turbulence-flame interaction. In this study, to avoid the initial transient effects, DNS data at $t = 1.5\tau_l$ ($\approx 1.7\tau_F$) is analyzed to investigate the local flame structure, where $\tau_l = l/u'$ and $\tau_F = \delta/S_L$ denote an eddy turnover time and a flame time, respectively.

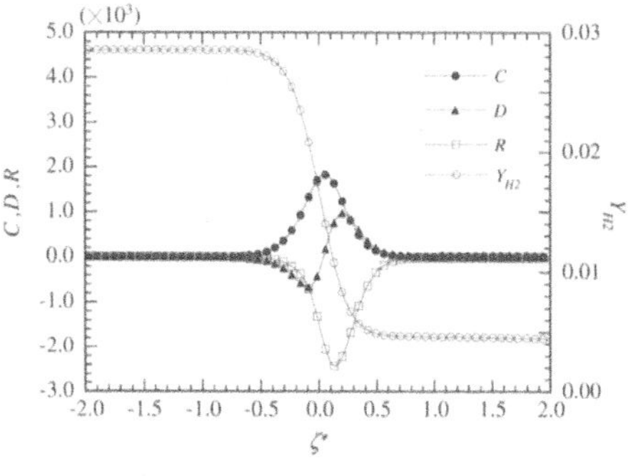

Figure 22.3. Contributions of the convection term (C), the diffusion term (D) and the reaction term (R) of the hydrogen transport in the laminar flame

Figure 22.2 shows the contour surfaces of the second invariant of the velocity gradient tensor and the heat release rate. The second invariant is defined by $Q = (W_{ij}W_{ij} - S_{ij}S_{ij})/2$, where S_{ij} and W_{ij} denote symmetric and anti-symmetric parts of the velocity gradient tensor($A_{ij} = \partial u_i/\partial x_j$). The second invariant is normalized by η and u' in the unburned side, and the local heat release rate is normalized by the maximum heat release rate of a laminar flame (ΔH_L). Hereafter, * denotes normalization by the corresponding laminar value. [Tanahashi *et al.* (1997)] have shown that the coherent fine scale eddies are well represented by the positive Q region in turbulence. As reported in the previous study [Tanahashi *et al.* (2000)], large modification of the heat release rate can be observed near the relatively strong coherent fine-scale eddies.

3. Local flame structure in turbulent premixed flame

In the present study, the flame fronts are defined as points at which the temperature gradient ($\partial T/\partial \boldsymbol{n}$) shows a maximum value, where $\boldsymbol{n}$ is a unit vector normal to the temperature gradient. Due to the turbulence-flame interaction, the local heat release rate of the flame element is fluctuating in turbulence as shown in Fig. 22.2. The maximum heat release rate reaches $1.2\Delta H_L$ and the minimum decreases to $0.5\Delta H_L$ [Tanahashi *et al.* (2000)]. In this study, local flame structure is discussed based on the hydrogen transport equation near the flame front.

Figure 22.3 shows contributions of the convection, diffusion and reaction terms for the hydrogen transport in the case of a laminar flame. ζ^* represents the coordinate normal to the flame front, which is non-dimensionalized by δ_L. The origin of ζ^* is set to be the location of the maximum temperature gradient. In front of the flame, the convection

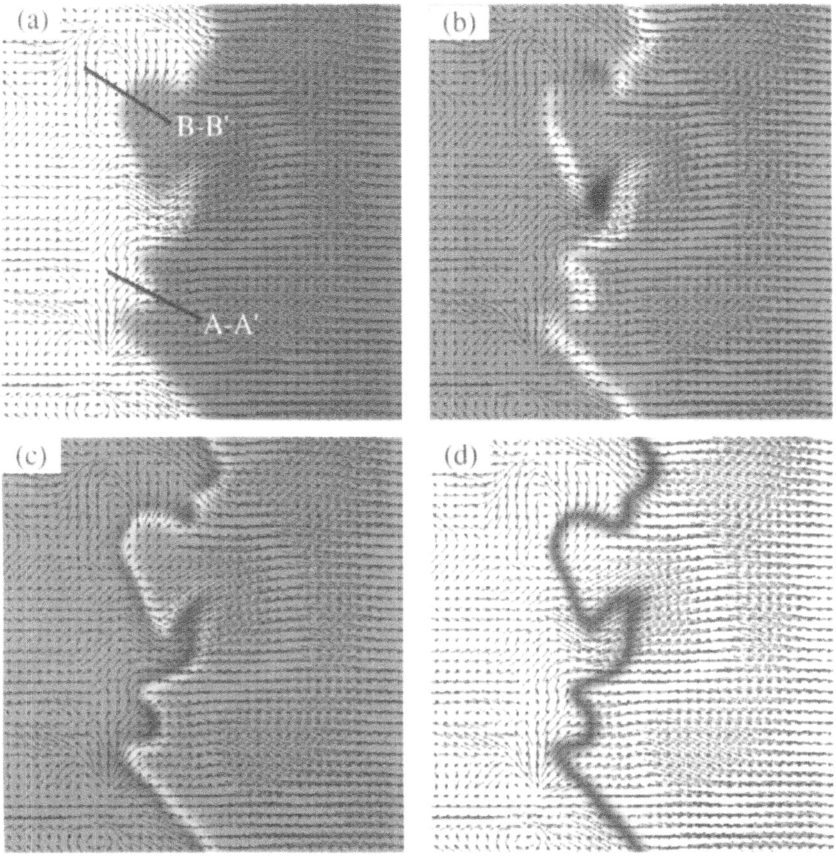

Figure 22.4. Distributions of H_2 mass fraction (a), the convection term (b), the diffusion term (c) and the reaction term (d) on a typical $x - y$ cross-section in the turbulent premixed flame

term makes positive contribution, whereas the diffusion and reaction terms make negative contributions. Behind the flame front, the diffusion term changes it's sign and starts to make a positive contribution.

Figure 22.4 shows the distributions of H_2 mass fraction and the convection, diffusion and reaction terms on a typical $x - y$ cross-section in the turbulent premixed flame. Velocity vectors are also plotted in Fig. 22.4. Due to the turbulent motion, the convection and diffusion terms show large fluctuations along the flame front. As shown in the previous study [Tanahashi *et al.* (2000)], the local heat release rate is strongly correlated with the curvature of the flame front. The flame elements convex toward the burnt side with large curvature tend to show a high

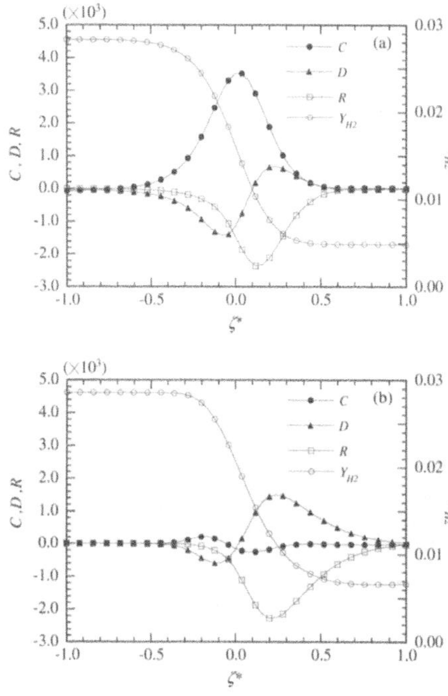

Figure 22.5. Contributions of the convection, diffusion and reaction terms of the hydrogen transport for two typical flame elements in turbulent premixed flame. (a): the flame element convex toward the burnt side and (b) the flame element convex toward the unburned side.

heat release rate. The convection and diffusion terms also depend on the flame geometry.

The contributions of three terms are shown in Fig. 22.5 for two typical flame elements which are denoted by A-A' and B-B' in Fig. 22.4. The A-A' flame element is convex toward the burnt side and the B-B' flame element is convex toward the unburned side. For the A-A' element, the convection term is much larger than that of the laminar flame. However, for the B-B' element, the convection term makes very weak contribution to the hydrogen transport and the diffusion term dominates the hydrogen transport process. The reaction term in the two flame elements do not differ markedly.

Fig. 22.6 (a) & (b) show the probability density functions (pdf) of the convection and diffusion terms. Here, the values of the convection and diffusion terms at the location of the minimum reaction term are selected as representative values. The pdf of the convection term shows a peak at 0.3 times the laminar value. Therefore, on the average, the convection term makes weaker contribution to the hydrogen transport in the case

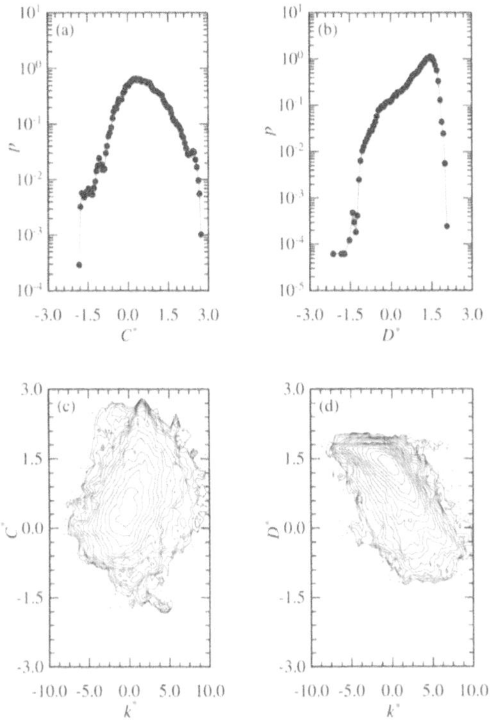

Figure 22.6. The pdf of the convection term (a), the pdf of the diffusion term (b), the joint pdf between the convection term and the curvature of the flame front(c) and the joint pdf between the diffusion term and the curvature (d).

of the turbulent premixed flame. However, the maximum value is about 3.0 times the laminar value and the minimum becomes negative. The pdf of the diffusion term shows a peak at about 1.5 times the laminar one. This result shows that turbulence-flame interactions enhance the diffusion of hydrogen in the local flame element in turbulence. The flame elements with the diffusion term larger than 1.5 times the laminar one is seldom observed.

The joint pdf between the convection term and the curvature of the flame front and that between the diffusion term and the curvature are presented in Fig. 22.6 (c) & (d). In these joint pdfs, the probability difference between neighboring contour lines is 1.2. The curvature is defined as $k = 1/R_1 + 1/R_2$, where R_1 and R_2 denote two principal radii of curvature on the flame surface. The flame elements convex toward the burnt side have positive values, and the curvature is normalized by δ_L.

From the joint pdf, it can be seen that the convection term has a weak correlation with the curvature of the flame front. The flame element convex toward the burnt side ($k > 0$) features a larger convection term. The diffusion term is also correlated with the curvature of the flame front. The diffusion term shows a tendency to become large for the flame element convex towards the unburned side ($k < 0$) contrary to the convection term. Even for the large fluctuations in the convection and diffusion terms, the fluctuation in the reaction term is not so large. The fluctuation level of the reaction term is nearly equal to that of the local heat release rate.

4. Summary

In this study, local flame structures of the turbulent premixed flame are discussed using DNS data of three-dimensional H_2-air turbulent premixed flame. Calculations correspond to $u'/S_L = 1.74$ and $l/\delta_L = 2.00$. The flame fronts are identified by using the local temperature gradient and the hydrogen transport mechanisms in individual flame elements are examined.

In the field of turbulence, the local convection and diffusion effects depend on the curvature of the flame front. The convection effect is enhanced for the flame elements convex toward the burnt side, whereas the diffusion effect becomes dominant for flame elements convex toward the unburned side. Fluctuation in the reaction term is relatively weak compared with the convection and diffusion terms. The fluctuation level of the reaction term coincides with that of the local heat release rate.

Acknowledgments

This work is partially supported by Grant-in-Aid for Scientific Research (A), Japan Society for the Promotion of Science and by a research project of the Center for Smart Control of Turbulence.

References

Baum, M., Poinsot, T. J., Haworth, D. C. and Darabiha, N. (1994) Direct Numerical Simulation of $H_2/O_2/N_2$ Flames with Complex Chemistry in Two-dimensional Turbulent Flows, *J. Fluid Mech.*, 281:1–32.

Echekki, T. and Chen, J. H. (1996) Unsteady Strain Rate and Curvature Effects in Turbulent Premixed Methane-Air Flames, *Combust. Flame*, 106:184–202.

Chen, J. H., Echekki, T. and Kollmann, W. (1998) The Mechanism of Two-Dimensional Pocket Formation in Lean Premixed Methane-Air

Flames with Implication to Turbulent Combustion, *Combust. Flame*, 116:15–48.

Miyauchi, T., Tanahashi, M., Nada, Y. and Imamura, Y., (1997) Structure and Fractal Characteristics of H_2-Air Turbulent Premixed Flames, *Proc. 1st Asia-Pacific Conference on Combustion*, pp.278–281.

Tanahashi, M., Saito, T., Shimamura, M. and Miyauchi, T. (1999) Local Extinction and NO_x Formation in Methane-air Turbulent Premixed Flames, *Proc. 2nd Asia-Pacific Conference on Combustion*, pp.500–503.

Tanahashi, M., Miyauchi, T. and Ikeda, J. (1997a) Scaling Law of Coherent Fine Scale Structure in Homogeneous Isotropic Turbulence, *Proc. 11th Symp. Turbulent Shear Flows*, Vol. 1, pp.4-17–4-22.

Tanahashi, M., Miyauchi, T. and Ikeda, J. (1999b) Simulation and Identification of Organized Structures in Flows, Proc. IUTAM Symposium (Eds. Sorensen, J. N., Hopfinger, E. J. and Auby, A.), p. 131, Kluwer Academic Publishers.

Tanahashi, M., Iwase, S., Uddin, Md. A. and Miyauchi, T. (1999c) Three-dimensional Features of Coherent Fine Sale Eddies in Turbulence, Turbulence and Shear Flow Phenomena - 1 (Proc. 1st Int. Symp. Turbulence and Shear Flow Phenomena), Eds. S. Banaerjee & J. K. Eaton, p. 79, Begell House Inc.

Tanahashi, M., Fujimura, M. and Miyauchi, T. (2000) Coherent Fine Scale Eddies in Turbulent Premixed Flames. *Proc. Combust. Inst.*, 28:529–535.

Discussion

Andrew Pollard, Queen's University "I noticed that you are using the Q criterion for the rate of rotation versus strain. Can you tell me if your conclusions would alter if you changed the threshold level for your..."

Reply: Yes, yes - visualization is always dependent on the threshold of the variables that is used for the visualization. But I skipped the details of the flow of eddy identification in this case in this study we identified the centre of the axis that does not depend on the threshold because of the eddy centre.

Paul Dimotakis, Caltech "You showed that the probability density function of curvature that if you normalize the curvature by the Kolmogorov scale that's between plus-minus 1. Is that the Komolgorov scale at cold conditions or hot conditions?"

Reply: Cold conditions because in this case we can define the Kolmogorov scale only in the ambient side.

Chapter 23

THREE-DIMENSIONAL DIRECT SIMULATIONS OF TURBULENT FLAMES USING REALISTIC CHEMISTRY MODELING

Dominique Thévenin*, Renan Hilbert and Olivier Gicquel
EM2C Laboratory
Ecole Centrale Paris (FRANCE)
thevenin,hilbert,gicquel@em2c.ecp.fr

Abstract In order to progress in our understanding of combustion in turbulent flows, Direct Numerical Simulations (DNS) are ideally suited. When considering such problems as pollutant emission or stability limits, detailed models must generally be employed to describe the combustion and diffusion processes with an acceptable accuracy. Due to the huge cost of such simulations, they have been generally restricted to two-dimensional configurations up to now, leading to unsolved questions concerning the applicability of the obtained results. Considering the lasting growth in computer power and the latest progress in algorithms and programming languages, we recently decided to develop a three-dimensional DNS code, with the aim of keeping reasonable computing times. This is made possible mainly by combining a low-Mach number approach with a new and accurate reduction technique for the chemistry. Some examples of the obtained results are shown in this paper.

Keywords: Direct numerical simulation, turbulence, chemistry reduction, low-Mach number approximation

1. Introduction

Most applications involving combustion (generation of heat or energy, car engines, aerospace propulsion...) take place in the turbulent regime. In order to investigate numerically combustion efficiency and pollutant

*corresponding author

A. Pollard and S. Candel (eds.), IUTAM Symposium on Turbulent Mixing and Combustion, 279-286.

formation, accurate models are therefore needed to describe turbulent combustion in a truly predictive way. Unfortunately such models are not really available at present time due to the complexity of this problem.

In order to progress in our understanding of the phenomena controlling combustion in a turbulent flow, detailed numerical simulations are essential. Since these simulations give access to all the values of all the quantities in the whole computational domain at every time-step, they can be used to compute all important quantities and in particular space- and time-correlations between different physical parameters. These correlations can then be used to try to identify much simpler turbulent combustion models.

Direct Numerical Simulations (DNS) are ideally suited for such studies since they do not rely on any particular hypothesis concerning the turbulence: all the physical spatial and temporal scales must be fully resolved during the simulation, and the numerical dissipation must remain below the real, physical dissipation associated with the mixture.

When considering such problems as pollutant emission, detailed models must of course be employed to describe accurately the combustion and diffusion processes. During the last ten years we have therefore developed a DNS code using full reaction schemes and taking into account multicomponent diffusion velocities as well as thermal diffusion. Individual reactions are described using Arrhenius parameters. Thermodynamic coefficients are computed using fifth-order polynomial fits of experimental measurements. Due to the huge cost of such simulations even on modern parallel supercomputers we were forced to restrict ourselves to two-dimensional simulations up to now. Even if such computations already lead to interesting conclusions [Thévenin and Baron, 1999], it is clear that they may miss many aspects that are important to describe the coupling between combustion and turbulence, since turbulence is intrinsically three-dimensional. In particular it is obvious that the development of the turbulent structures is completely different when the computations are restricted to two dimensions.

Considering the growth in computer power and the latest progress in algorithms and programming languages, we have therefore started in 2000 the development of a three-dimensional DNS code. This new-generation code, called π^3, is written in Fortran 95 and still employs realistic models to describe chemistry and transport. It is therefore appropriate to investigate the detailed structure of turbulent flames. A strong reduction of computing times has been obtained by combining a low-Mach number approach, well suited to most problems of interest, with a new and accurate reduction technique for the chemistry, called FPI for *Flame Prolongation of ILDM* [Gicquel *et al.*, 2000]. This method is an extension of the *Intrinsic Low-Dimensional Manifold* (ILDM) method of Maas and Pope [Maas and Pope, 1992, Gicquel *et al.*, 1999], improved in order to obtain a better description of the low-temperature zones. Thanks to these two methods we have been able to get down to acceptable computing times for three-dimensional direct simulations of turbulent flames using an accurate description of chemistry and transport. Preliminary results concerning a turbulent stoichiometric methane/air premixed flame are shown in a later section. We hope that π^3 will soon allow us to answer the questions concerning the main differences between turbulent flames computed in two- or three-dimensions.

2. Physical models and numerical methods

Many aspects of the new code, π^3, are similar to what had been used in the first-generation code, *parcomb*. This former code has been described in detail previously [Thévenin *et al.*, 1996]. The main differences are that π^3 relies on a low-Mach number approximation and not on a fully compressible formulation, and that chemical processes are described by a new chemistry reduction technique (FPI) and not any more using directly full reaction schemes.

2.1. Low-Mach number approximation

For most applications of interest, like for example furnaces and boilers, but also inside the combustion chambers for automotive or aeronautical/aerospace applications, the maximum Mach number is generally quite small, at least in the region where combustion takes place. This allows us to employ a low-Mach number formulation instead of working with the compressible equations. Thanks to this approximation an important speed-up can be obtained, since the stability restrictions associated in particular with the Courant-Friedrichs-Lewy (CFL) condition are practically released. Moreover the equations are considerably simplified since several viscous terms become negligible, in particular in the energy equation [Giovangigli, 1999].

In order to implement the low-Mach number approximation we employ a pressure-projection method following previous publications [Najm *et al.*, 1998]. The pressure is splitted in a mean thermodynamic pressure $\overline{p}$, depending only on time, and a fluctuating part $\tilde{p}$, varying in space and time. At low Mach numbers, $\tilde{p} \ll \overline{p}$. Starting from the (known) momentum value at step n, we begin with a prediction step to determine a first estimate of the new momentum terms :

$$(\rho \mathbf{v})^{\star} = (\rho \mathbf{v})^{n} + \Delta t \left[-\nabla \cdot (\rho \mathbf{v} \otimes \mathbf{v}) + \nabla \cdot \underline{\underline{\tau}} \right] \tag{23.1}$$

It is afterwards necessary to solve a Poisson equation for the pressure fluctuation $\tilde{p}$, using a finite-difference estimation of density changes:

$$\nabla^2 \tilde{p} = \frac{1}{\Delta t} \left[(\partial_t \rho)^{n+1} + \nabla \cdot (\rho \mathbf{v})^\star \right] \tag{23.2}$$

Finally the momentum values are corrected to take into account the actual pressure value by:

$$(\rho \mathbf{v})^{n+1} = (\rho \mathbf{v})^\star + \Delta t \left[-\nabla \tilde{p} \right] \tag{23.3}$$

The perfect gas relation used to couple density and temperature relies on the mean thermodynamic pressure $\overline{p}$ only: $\overline{p} = \rho r T$.

2.2. Description of chemical processes

Our previous code employed full reaction schemes to describe chemical processes. This is of course the most accurate technique but leads to unacceptable computing costs for fuels other than hydrogen, since tens or even hundreds (for higher hydrocarbons like n-decane) of supplementary transport equations must be solved during the computations. We therefore need an alternative solution to describe reaction mechanisms.

We choose to build on top of the *Intrinsic Low-Dimensional Manifold* technique [Maas and Pope, 1992]. This technique relies on the identification of low-dimensional subspaces in composition space, on which chemical processes are restricted. The identification of these inertial manifolds implies a decomposition of the system using Schur-eigenvectors. After computing this manifold all the information (all values of the mass fractions, mass production terms, possibly thermodynamic coefficients and binary diffusion coefficients,...) is stored in a data-base. This data-base is then called during the direct simulation in order to get the information needed to integrate the system in time. Instead of requiring a transport equation for each chemical species present in the reaction system, we only need a transport equation for the coordinates of the manifold. This means that, if the computed manifold is for example of dimension 2, only two coordinates and therefore two supplementary transport equations are needed to describe the full reaction system.

The classical ILDM method requires a much larger number of dimensions when low-temperature kinetics have to be included in the manifold. This leads of course to an increase in the needed CPU time and memory during the computation. Moreover the generation of a manifold with perhaps 10 or 12 coordinates becomes much more complex, and the quality and smoothness of the resulting manifolds are difficult to check. In order to avoid these problems, we have developed an alternative method, called FPI for *Flame Prolongation of ILDM* [Gicquel *et al.*, 2000]. In this case we still obtain the same attracting subspace as the standard ILDM technique in the high-temperature region, near the equilibrium point. But instead of increasing the dimension of this manifold when considering lower temperatures, we just prolongate the low-dimensional ILDM manifold by using the structure of the laminar premixed flame corresponding to the same fresh gas composition. This means that, when a point of the CFD computation does not fall any more on the standard ILDM manifold, we use for this missing information the correlation between species coming from the laminar premixed flame structure. At the difference of the ILDM method, there is no rigorous mathematical explanation to justify this approximation. On the other hand, this procedure takes into account exactly the correct boundary conditions both on the high- and low-temperature boundaries, and of course relies on a physically-exact flame structure to relate the species outside the ILDM manifold. Many tests have proved the interest of the FPI method to compute with excellent accuracy complex flame structures for methane [Gicquel *et al.*, 2000]. We are presently investigating other fuels with this same approach.

2.3. Numerics

In order to integrate the equations in time, a fourth-order Runge-Kutta procedure is used. Spatial derivatives are computed using a sixth-order centered approximation. The pressure perturbation is determined by solving a Poisson equation using a standard spectral method in the associated Fourier space. At present, the time-derivative of density needed in (23.2) is approximated by a third-order backward derivative, but this point certainly deserves further consideration to find an optimal solution. It is known that this term can lead to severe unstabilities and controls to a large extent the maximum speed-up that can be achieved using this form of the low-Mach number approximation.

3. Example of simulation

Quantitatively, we have reached an overall acceleration factor of about 100 compared to similar two-dimensional computations using our previous compressible code *parcomb* with full reaction schemes. Both the low-Mach number approximation (compared to the compressible formulation) and the FPI chemistry reduction technique (compared to full reaction schemes) lead to a speed-up of a factor roughly equal to 10. When using both techniques simultaneously, as has been done in π^3, we therefore obtain a global speed-up of about 100.

To illustrate this point, we have considered for a first computation a three-dimensional turbulent stoichiometric methane/air flame at atmospheric pressure. For this test we do not use the multicomponent diffusion velocities but we suppose instead a unity Lewis number for all species. At this point we have only implemented in π^3 periodic boundary conditions, allowing us to solve very efficiently the Poisson equation in the associated Fourier space. We therefore compute a double premixed flame structure, by initializing a slab of burnt gases in the middle of the computational domain, surrounded by a methane/air mixture (Fig.23.1).

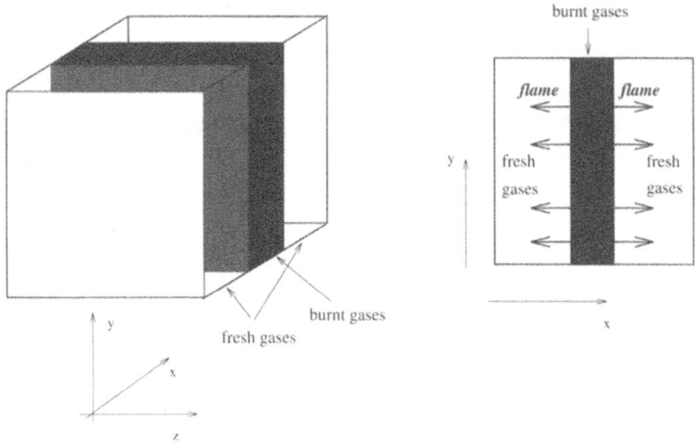

Figure 23.1. Initial conditions for the computation.

We then initialize a field of synthetic turbulence using the same technique as in the previous code [Thévenin and Baron, 1999]. For the results shown here, we generate a homogeneous isotropic 3D turbulence field corresponding to a Reynolds number of 176 based on the integral length of turbulence. We then iterate in time by solving the low-Mach number Navier-Stokes equations, including two transport equations for the chemical coordinates employed in the FPI method. The results shown on Fig.23.2 correspond to 0.8 turbulence time τ_t, defined as the ratio between integral length scale and turbulent fluctuation velocity $\tau_t = \Lambda/u'$. We show in this figure the field of CO_2 mass fraction, used to delineate the instantaneous flame front. The corrugation of the front by the turbulence as well as the propagation of the two flame fronts through the fresh gas are clearly visible in this figure.

This first computation has been carried out on a single standard PC with 1 GHz processor and 1 GB of memory. In fact, thanks to the FPI technique and to the dynamic memory allocation, the computation re-

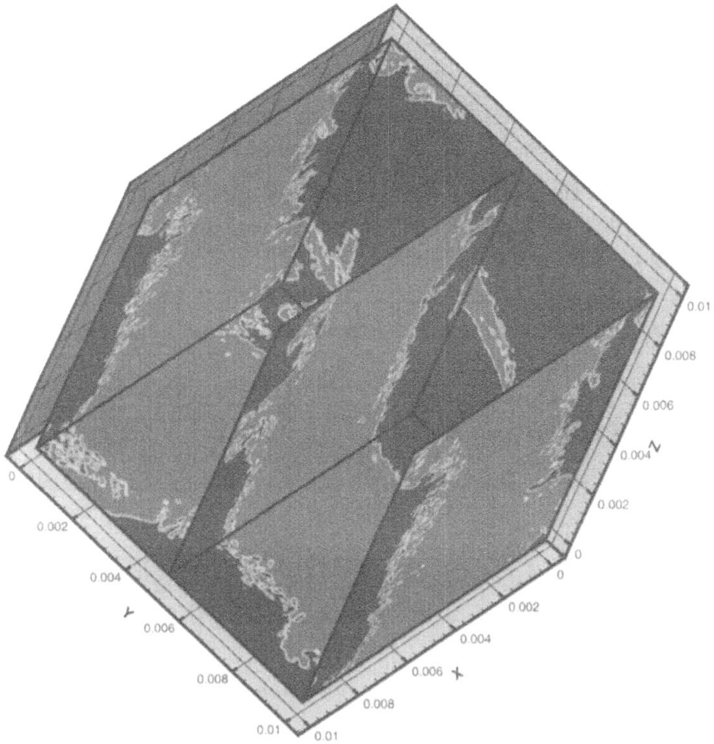

Figure 23.2. Field of CO_2 mass fraction (perspective view of specific planes through the solution) to show the position of the two turbulent flame fronts at $t = 0.8\tau_t$.

quires only about 650 MB of memory for one million of grid points (100^3). The needed computing time for this first three-dimensional computation is 60 hours, which is less than usual CPU times observed for previous two-dimensional computations of hydrogen flames using a compressible formulation and full reaction schemes, demonstrating the interest of our approach. It must nevertheless be kept in mind that further computations will use multicomponent diffusion velocities, and not as here a unity-Lewis number hypothesis, thereby increasing noticeably the requested computing time.

4. Conclusions and perspectives

We have successfully developed a three-dimensional direct simulation code that will be used to investigate accurately turbulent flames. Thanks to the low-Mach number approximation and to the FPI chemistry reduction technique, the needed computational times for the new code π^3 are similar to those of our previous, two-dimensional compressible code relying on detailed reaction schemes. We have in particular shown that a three-dimensional turbulent methane/air flame can for example be computed in less than 100 CPU-hours on a standard PC for a Reynolds number of about 200, using a unity Lewis number for all species. Further optimization steps should speed-up again the computations. Using supercomputers should allow us to reach much higher values of the Reynolds number. We therefore hope to be able in the near future to investigate in detail the structure of realistic turbulent flames, in order in particular to develop more accurate subgrid-scale models for Large-Eddy Simulations of turbulent combustion.

Acknowledgments

The support of the French national supercomputing center IDRIS is gratefully acknowledged. Many fruitful discussions with J. de Charentenay must also be mentioned here.

References

Gicquel, O., Thévenin, D., Hilka, M. and Darabiha, N., "Direct numerical simulation of turbulent premixed flames using intrinsic low-dimensional manifolds", *Combust. Theory Modelling*, **3**, pp.479-502, (1999).

Gicquel, O., Darabiha, N. & Thévenin, D., "Laminar premixed hydrogen/air counterflow flame simulations using Flame Prolongation of ILDM with differential diffusion", *Proc. Comb. Inst.*, **28**, pp.1901-1908, (2000).

Giovangigli, V., *Multi-Component Flow Modeling*, Birkhäuser Boston, (1999).

Maas, U. and Pope, S.B., "Simplifying chemical kinetics: intrinsic low-dimensional manifolds in composition space", *Combust. Flame*, **88**, pp.239-264, (1992).

Najm, H. N., Wyckoff, P. S. and Knio, O. M., "A semi-implicit numerical scheme for reacting flow", *J. Comput. Phys.*, **143**, pp.381-402, (1998).

Thévenin, D., Behrendt, F., Maas, U., Przywara, B. and Warnatz, J., "Development of a parallel direct simulation code to investigate reactive flows", *Comput. Fluids*, **5**, pp.485-496, (1996).

Thévenin, D. and Baron, R., "Investigation of turbulent non-premixed flames using DNS with detailed chemistry", in *Direct and Large Eddy Simulation III*, (Voke, P.R., Sandham, N.D. and Kleiser, L., Eds.), Kluwer Publishers, pp.323-334, (1999).

Chapter 24

SIMULATION OF A BUOYANCY-DRIVEN JET DIFFUSION FLAME

J. de Charentenay[1,2], D. Thévenin[2] and B. Zamuner[1]
(1) ONERA,
29, avenue de la Division Leclerc,
F-92322, Châtillon Cedex
(2) EM2C,
Ecole Centrale Paris and CNRS,
Grande Voie des Vignes,
F-92295, Châtenay-Malabry
j.mareschal@onera.fr, thevenin@em2c.ecp.fr, zamuner@onera.fr

Abstract In order to improve our knowledge in turbulent combustion, it is of great interest to thoroughly investigate the interaction between a flame front and a single vortex. The aim of the present study is to simulate a buoyant nonpremixed H_2/O_2 jet flame. The natural instability of the jet is excited by gravity so that coherent structures appear periodically at a frequency of about 15 Hz. The properties of the flame are highly dependent on diffusion and local mixing induced by these large-scale structures. To compute this unsteady flame, we develop a low-Mach number code derived from a DNS solver that relies on detailed chemistry and transport models. The numerical results are successfully compared to the experimental measurments of OH mole fraction and temperature. In particular, this comparison shows a very good agreement in both shape and level for the temperature field.

Keywords: Jet diffusion flame, low-Mach number, buoyancy instabilities

1. Introduction

The combustion process in diffusion flames is governed by the interaction of fluid dynamics and chemical reactions. In practical combustors these interactions can not be controlled easily. It is therefore necessary

A. Pollard and S. Candel (eds.), IUTAM Symposium on Turbulent Mixing and Combustion, 287-293.

to develop simplified cases. For example, the interactions between a counter-flow flame and vortices has been investigated both experimentally and numerically and has been used to establish a nonpremixed flame/vortex interaction diagram [Thévenin et al., 2000]. These studies have clearly demonstrated the need of numerical and experimental studies of unsteady diffusion flames.

Investigations of the stability of jet flames have shown several instability modes [Gutmark et al., 1988] that greatly depend on the flame. However, stability analysis yield a preferred frequency [Lingens et al., 1996], which also corresponds to the frequency of the buoyancy-driven instability [Yuan et al., 1994]. Therefore, the buoyancy-driven jet diffusion flame is a perfect test case to perform both numerical and experimental investigations [Katta et al., 1994, Grisch et al., 1996]. In this configuration, the flow is laminar and the coherent structures develop naturally with a highly reproducible frequency. Experimental and numerical investigations can be easily compared in this situation.

In the present paper, we numerically investigate a buoyancy-driven jet diffusion flame. The computational code is based on a DNS solver and is dedicated to the simulation of unsteady reactive flows using complex chemistry and detailed transport. This solver is used to predict the development of the instability and the interaction between the flame and the vortices. The results are compared to the experimental data.

2. Numerical Methods

2.1. Governing equations

In the present study, we use the low-Mach number approach [Majda and Sethian, 1985] to describe the reactive flow. In this approximation, the pressure p is split as follows: $p(x,t) = p_0(t) + \widetilde{p}(x,t)$, where p_0 and $\widetilde{p}$ respectively denote a spatially-uniform and a fluctuating pressure. Note that the use of the low-Mach number assumption, in which acoustic waves are neglected, is fully justified here since we focus on an open geometry. This assumption leads to the following equations [Giovangigli, 1999]:

$$\partial_t \rho + \partial_{\mathbf{x}} \cdot (\rho \mathbf{v}) = 0, \tag{24.1}$$

$$\rho \partial_t (Y_k) + \rho \mathbf{v} \cdot \partial_{\mathbf{x}} Y_k + \partial_{\mathbf{x}} \cdot \mathbf{F_k} = M_k \omega_k, \quad k = 1, N_s, \tag{24.2}$$

$$\partial_t (\rho \mathbf{v}) + \partial_{\mathbf{x}} \cdot (\rho \mathbf{v} \otimes \mathbf{v}) + \partial_{\mathbf{x}} \cdot \mathbf{S} = -\partial_{\mathbf{x}} \widetilde{p}, \tag{24.3}$$

$$\rho c_p \partial_t T + \rho c_p \mathbf{v} \cdot \partial_{\mathbf{x}} T + \partial_{\mathbf{x}} \cdot \left(\mathbf{Q} - \sum_k (h_k \mathbf{F}_k) \right) = - \sum_k \left(c_{p_k} \mathbf{F}_k \cdot \partial_{\mathbf{x}} T \right) - \sum_k (M_k \omega_k h_k), \tag{24.4}$$

where ρ denotes the density, $\mathbf{v}$ the velocity vector, $\partial_{\mathbf{x}}$ the spatial derivative operator, Y_k, $\mathbf{F_k}$, M_k and ω_k the mass fraction, flux diffusion vector, molar mass and molar production rate of the kth species, $\mathbf{S}$ the stress tensor, $\widetilde{p}$ the fluctuating pressure, T the temperature, $\mathbf{Q}$ the thermal diffusion flux. Chemistry and transport properties are computed using the dedicated libraries Chemkin II [Kee et al., 1980] and EGlib [Ern and Giovangigli, 1995].

2.2. Numerical scheme

Spatial discretization is done as follows: convective terms are discretized by a fourth-order accurate finite difference scheme on an unstaggered mesh, while second-order accurate schemes are used for other terms. The momentum equation (24.3) is split in predictor-corrector steps. Between the two steps, the fluctuating pressure $\widetilde{p}$ is determined using a Poisson equation. This equation is discretized using a five-points stencil of second-order accuracy and is solved using a LU decomposition method. The time-integration is performed using a fully implicit formulation and the Jacobian matrix are inverted using a GMres solver with a Jacobi preconditioning.

3. Configuration

The diffusion flame configuration is shown schematically in figure 24.1. The fuel jet is composed of a mixture of hydrogen ($Y_{H_2} = 0.2$) and nitrogen ($Y_{N_2} = 0.8$) that is injected in a 10 mm diameter nozzle at a velocity equal to 3.26 m/s. The flame is surrounded by a 150 mm diameter co-annular air jet at a velocity equal to 0.3 m/s.

The flame is anchored at the nozzle exit. Experimental results [Lynn et al., 1988] show that the flow is laminar in the region of interest and that large coherent axisymmetrical structures naturally appear at a 15 Hz frequency [Bressson, 1999]. The development process of these structures is well documented [Yuan et al., 1994, Durox et al., 1997] and is directly related to the gravity.

In the simulations, the computational domain is 300 mm high and 150 mm wide. The mesh size is 201 $\times$ 101 nodes and is refined inthe flame zone. The typical grid size, in the flame zone, is 250 μm in the x direction and 900 μm in the y direction. This grid is sufficient to

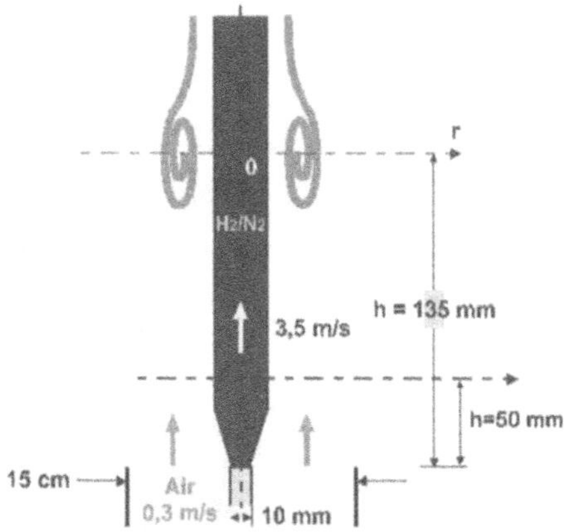

Figure 24.1. Configuration of the problem

accurately predict the development of the instabilities and their impact on the diffusion flame.

4. Results and Discussion

Temperature and species profiles are presented in figure 24.2 at the same physical time. This figure shows that the flame is anchored at the burner lip. In this region, the stabilization is driven by the diffusion process of fuel and oxidizer. Note that we did not use a hot point to account for the heating of the burner wall. Therefore, the stabilization process of the diffusion flame depends on the boundary conditions.

Figure 24.2 shows that the flow is laminar and that the instabilities appear at 5 cm from the burner exit. The vortical structures are fully developed between 10 and 13 cm. At this location, we observe a local minimum of the intermediate species OH, O and H mass fraction in the compressed regions of the flame, whereas these species are maximum in the stretched regions. These observations are in agreement with previous experimental results [Grisch et al., 1996, Bressson, 1999].

To compare simulations and experimental results, we use the following procedure to post-treat the results: we extract radial profiles at a height of 135 mm from the nozzle and we plot them as function of radius and time. Note that this procedure corresponds to a time description of the instability, which can be directly compared to the experimental data. The numerical and experimental fields of temperature and OH mole fraction are displayed in the left and right part of figures 24.3 and 24.4. These figures show that the frequency of the instability is well reproduce by the simulations. However, the wrinkling of the flame is less

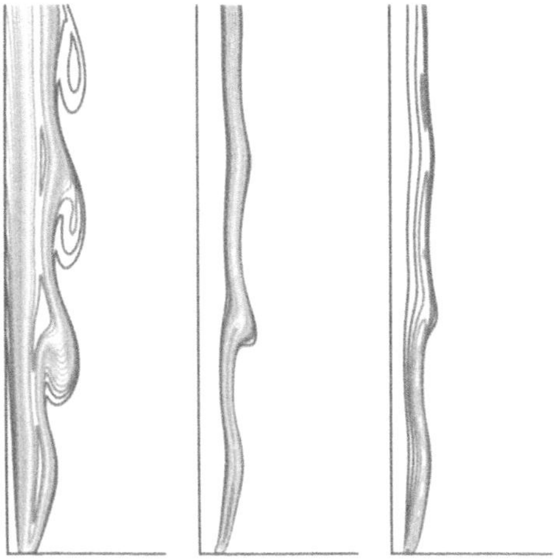

(a) T (b) Y_{OH} (c) Y_H

Figure 24.2. Computed temperature and OH, H mass fraction fields.

pronounced in the computations and, therefore, the spatial growth of the instability seems to be slightly underpredicted. The predicted maximum temperature is identical to the experimental one and the computed maximum level of OH mole fraction is of the same order as the experimental one. Moreover, both simulation and experimental measurement show that the interaction of the combustion process and the vorticies leads to a local increase of the temperature and of the consumption of radicals in the compressed region. In general the agreement between the simulation and the experimental results is good.

5. Conclusion

This paper describes a simulation code dedicated to laminar unsteady flame configurations. This code may be used to compare simulations and experimental measurements in order to validate chemical mechanisms and transport models. This solver has been applied to a buoyancy driven jet diffusion flame. The comparison of the experimental measurements and the numerical results indicates that the development of the instability is well reproduced in the simulation. In particular, the characteristic frequency of this buoyant flame (15 Hz) exactly matches that found in the simulations. Moreover, the interaction between the flame and the

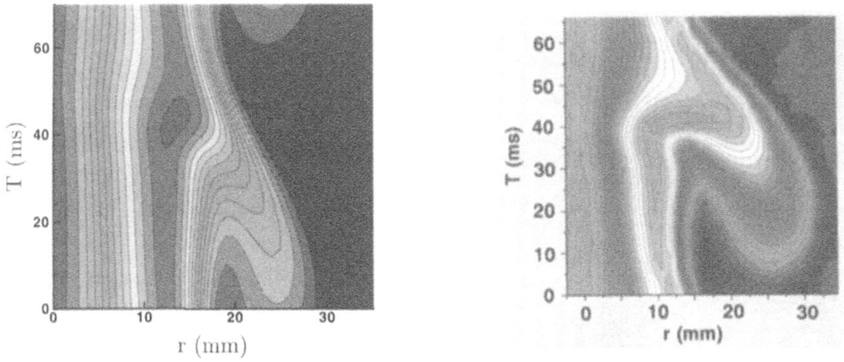

Figure 24.3. Temperature field (300K to 2350K). Numerical result (left) and experimental data (right).

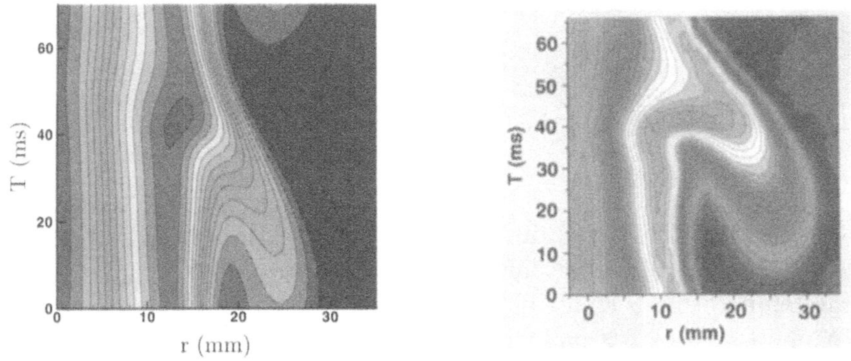

Figure 24.4. OH mole fraction field. Numerical result (left, 0 to 10300 ppm) and experimental data (right, 0 to 13500 ppm).

vorticities leads to a characteristic behavior of the temperature and OH mole fraction, which is clearly reproduced by the numerical solver. It may be concluded from this study that the low Mach number code is suitable to simulate unsteady reactive laminar configurations.

References

Bressson, A. (1999). PhD thesis.

Durox, D., Yuan, T., and Villermaux, E. (1997). The effect of buoyancy on flickering in diffusion flames. *Combust. Sci and Tech.*, 124:277–294.

Ern, A. and Giovangigli, V. (1995). Fast and accurate multicomponent transport property evaluation. *J. Comput. Phys.*, 120:105–116.

Giovangigli, V. (1999). *Multi-Component Flow Modeling*. Birkhäuser Boston.

Grisch, F., Attal-Tretout, B., Bouchardy, P., Katta, V., and Roquemore, W. (1996). A vortex-flame interaction study using four-wave mixing techniques. *J. of Nonlinear Optical Physical and Materials*, 5(3):505–526.

Gutmark, E., Parr, T. P., Parr, D. M., and Schadow, K. C. (1988). Evolution of vortical structures in flames. *Proc. Combust. Inst.*, 22:523–529.

Katta, V. R., Goss, L. P., and Roquemore, W. M. (1994). Numerical investigations of transitional H_2/N_2 jet diffusion flames. *AIAA J.*, 32(1):84–94.

Kee, R. J., Miller, J. A., and Jefferson, T. H. (1980). Chemkin : A general-purpose, problem-independant, transportable, fortran chemical-kinetics code package. Technical Report SAND80-8003, Sandia National Laboratories.

Lingens, A., Neemann, K., Meyer, J., and Schreiber, M. (1996). Instability of diffusion flames. *Proc. Combust. Inst.*, 26:1053–1061.

Lynn, W., Goss, L., Chen, T., and Trump, D. (1988). Two component velocity measurements on an axially symmetric H_2-N_2 jet diffusion flame. In *Proceedings of central states section meeting of the combustion inst.*

Majda, A. and Sethian, J. (1985). The derivation and numerical solution of the equations for zero Mach number combustion. *Combust. Sci. Tech.*, 42:185–205.

Thévenin, D., Renard, P., Fiechtner, G., Gord, J., and Rolon, J. (2000). Regimes of nonpremixed flame/vortex interactions. *Proc. Combust. Inst.*, 28

Yuan, T., Durox, D., and Villermaux, E. (1994). An analogue study for flame flickering. *Experiments in Fluids*, 17:337–349.

Chapter 25

VISUALIZATION OF THE FUEL STRIPPING MECHANISM FOR WAKE-STABILIZED DIFFUSION FLAMES IN A CROSSFLOW

M.R. Johnson and L.W. Kostiuk

Combustion and Environment Group

Dept. of Mech. Eng.,

University of Alberta Edmonton,

Alberta, Canada T6G 2G8

Abstract Recent publications have shown that diffusion flames stabilized in the wake of a burner tube exhibit combustion inefficiencies (i.e. incomplete combustion) that are sensitive to the crosswind velocity. Although data have shown that the inefficiencies appear to be a result of stripping of unburned fuel from the flame in coherent bursts, a clear picture of this mechanism is lacking. The goal of this paper is to help elucidate the mechanism by which these inefficiencies occur through flow visualization. Jet diffusion flames of natural gas were burned in a low turbulence wind tunnel. The gaseous fuel stream flowed from a 24.7 mm diameter open tube that protruded vertically into the 1.2 m high and 2.4 m wide test section of a wind tunnel. The fuel jet was seeded with fine silicone oil droplets that evaporated as they reached the flame and acted as a conditional marker for unburned fuel. Using pulsed laser sheet illumination, Mie scattering from the silicon droplets showed that the unburned fuel was contained in coherent structures related to primary vortex structures in the jet and are acted upon by the mean flow field surrounding the flame and burner tube.

Keywords: Visualization, Jet, Diffusion, Flames, Crossflow, Imaging, Experiments

1. Introduction

The diffusion flame burning in a crossflow of air is a basic combustion problem that has many applications including gas flaring in the energy and petrochemical industries and cross-stream fuel injection in some

A. Pollard and S. Candel (eds.), IUTAM Symposium on Turbulent Mixing and Combustion, 295-303.

types of furnaces and gas turbine combustors. At low momentum flux ratios ($R = \rho_j V_j^2 / \rho_\infty U_\infty^2$ where ρ is the density, V and U are velocity, and subscripts j and ∞ denote the fuel jet and crossflow respectively), the flame is "wake-stabilized" by a standing vortex that exists on the leeward side of the burner tube as shown in Figure 1. This paper focuses on flames in this low-momentum regime (R of the order of 1 or less) which have been the subject of a few recent publications (Huang and Chang, 1994; Gollahalli and Nanjundappa, 1995; Huang and Wang, 1999; Johnson and Kostiuk 2000; Kostiuk *et al*, 2000). These publications have proposed and provided data to support a three-zone structure to these flames as illustrated in Figure 1.

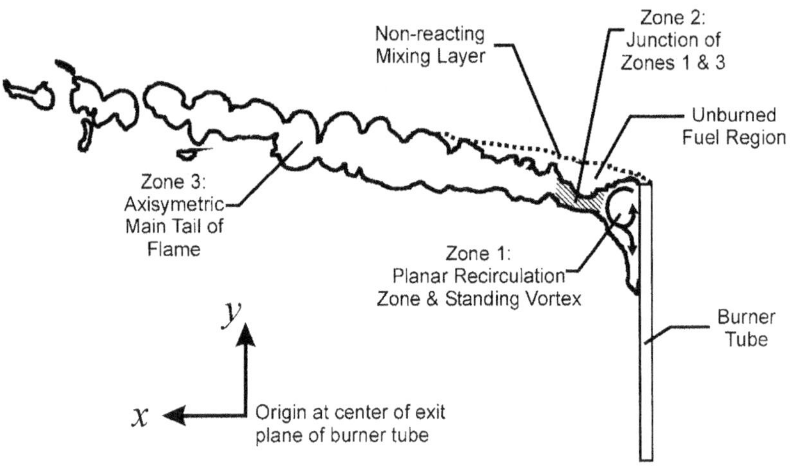

Figure 25.1. Three-zone structure of a wake-stabilized flame

Johnson and Kostiuk (2000) have shown that these wake-stabilized flames are susceptible to combustion inefficiencies that are a function of to the crosswind velocity. In this context, the combustion inefficiency (more accurately termed the carbon conversion inefficiency) is a measure of combustion completion and is defined as the fraction of carbon mass in the fuel stream that is not converted by the flame to carbon in the form of carbon dioxide. Gas chromatograph analysis of the gases emitted from these flames has shown that the majority of the inefficiency comes from raw fuel being emitted and dispersed into the crosswind. Recently, data presented by Johnson *et al* (2001) have been used to propose a fuel stripping mechanism that is believed to be responsible for the measured inefficiencies in these flames. It was proposed that the mean flow induced by the recirculating vortex on the leeward side of the stack draws unburned fuel from the upper mixing layer down through the flame and ejects it in bursts on the underside of the flame. The goal of this paper is to use flow visualization techniques to further elucidate and clarify this proposed mechanism.

2. Experimental methodology and facilities

Experiments were conducted in a large closed-loop wind tunnel with a 1.2 m high and 2.4 m wide test section where the winds could be varied from 2 to 35 m/s at low turbulence levels (~ 0.4 % turbulence intensity). The fuel used was natural gas (95.2% CH_4, 2.1% C_2H_6, 1.7% N_2, 0.8% CO_2, 0.2% other) and flow rates were set by a calibrated mass flow controller. The burner used was made of 24.7 mm in diameter (22.1 mm i.d.) quartz tubing and the jet exit velocity profile was laminar and was held at a mean velocity of 1 m/s. More complete details of the facility are available in Johnson and Kostiuk (2000).

Flow visualization was achieved by seeding the fuel stream with small droplets of silicone oil and using a pulsed laser sheet to illuminate the location of the fuel at an instant in time via Mie scattering. The light sheet used to image the fuel flow was created using a 400 mJ/pulse Nd-Yag laser. The laser beam was passed through two sets of cylindrical optics to expand the beam (originally 0.9 cm in diameter) in two directions. The light was subsequently reflected off a large (100 mm diameter) mirror and introduced into the wind tunnel through a glass window in the tunnel floor so that the flow downstream of the tube could be visualized as shown in Figure 2. The primary expansion of the sheet was in the streamwise (x) direction but the sheet was also expanded slowly in the cross-stream (z) direction so that the sheet was approximately 7 cm wide through the region of interest. This wide sheet permitted visualization across the full width of the fuel stream.

The droplets of silicone oil were created by a blast atomizer and seeded into to the fuel stream. The mean droplet size of this mist was estimated to be 5 micrometers (Kostiuk *et al*, 1989). When these droplets were located in the laser sheet they scattered light (Mie scattering) and could then be imaged. By using silicone oil, which evaporates at approximately 200^oC, the seed droplets in any fuel that participates in combustion evaporate and no longer scatter light. As a result, the Mie scattering provides a marker to show the structure of the fuel stream up to the point where the temperature reaches 200^oC.

To understand how the bursts of ejected fuel evolve and to distinguish between fuel that is on a path to be burnt and fuel that is escaping combustion, it was necessary to capture "instantaneous" and "simultaneous" images of both the fuel stream and the luminous flame. This raised several complexities. Since the fuel is illuminated by a pulse of 532 nm wavelength light and the natural luminosity of the flame is either blue or yellow, it was decided to use color to distinguish the various components of the flow. To provide high spatial resolution over the large flow area (up to approximately 1.4 m x 0.7 m) it was decided to use film. However, the exposure times necessary to prevent blurring in the images of the moving flow required the shutter to be open for only 1/500 s. Given the emitted light levels from the flame and the short exposure, film with the highest possible light sensitivity was required. After some experimentation, Kodak Ektachrome 1600 film was used which was subsequently pushed two additional stops during processing to an effective a.s.a rating of 6400.

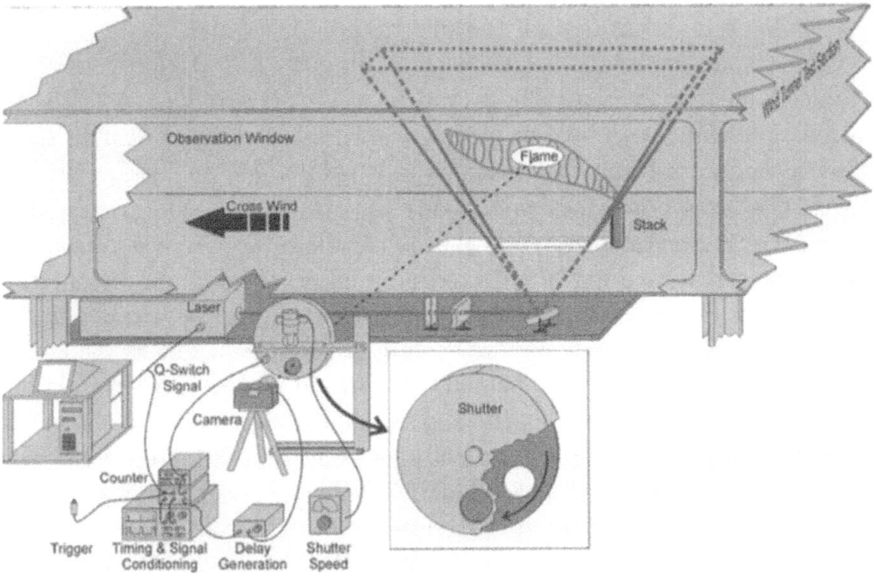

Figure 25.2. Schematic of the set-up used for flame visualization and photography.

Photographs were taken using a Nikon F1 camera and an f1.4 aperture, 55 mm lens that was mounted perpendicularly to the laser sheet. As with most commercial cameras, the minimum fully open shutter-time was limited to 1/125 s. This was a significant constraint since the shutter must be fully open when the 7 nanosecond duration laser sheet is firing but longer exposure times would blur the flame in the photographs. Thus, stroboscopic techniques, which would require the camera shutter to be left open while an external light source fired, were not an option due to excessive ambient light. To overcome this limitation an external mechanical shutter system was created which used a spinning disc

(30 cm in diameter and rotating at 3000 rpm) with an elliptical hole to create the required exposure time, as shown in Figure 2. Unfortunately, accurately synchronizing the firing of the laser, the opening of the camera shutter, and the opening of the spinning disc shutter was non-trivial. Because the high-power laser could not be easily pulsed on demand, timing was synchronized to the laser Q-switch signal, which was fixed at 10 Hz. An optical slot switch with a Schmidt trigger was mounted on the spinning disc and manually synchronized with the Q-switch signal of the laser using an oscilloscope such that the mechanical shutter was open for 1/500 s while the laser was firing. When the pulse of the laser and the opening of the disc shutter were synchronized, a signal generator and delay circuit were used to open the camera shutter and record a photograph. Thus, although the camera shutter was fully open for 1/125 s, light from the flame only passed through the disc shutter for 1/500 s resulting in 1/500 s flame exposures synchronized with the Mie scattering from the 7 ns pulse of laser sheet.

3. Discussion and Results

Figures 3 and 4 show photographic images of the natural gas flames at 1 m/s exit velocities (V_j) and various levels of crosswind. Unburned fuel in the images appears as light grey as the fine silicone oil droplets scatter the 532 nm light from the laser sheet. Close inspection of images 3b through 3e illustrates that the silicone oil is an excellent tracer for the fuel as dots of light grey fuel can be seen surrounded by the deep blue haze (contact authors, .eds) of the diffusion flame that is consuming them. At very low crosswind speeds ($\sim$ 1 m/s), the flame may be either rim stabilized or wake-stabilized as shown in images 3a and 3b. At higher crosswind speeds, the upwind side of the flame separates from the burner rim and the flame is stabilized in the wake of the burner tube. As the wind speed is further increased, the recirculating vortex on the leeward side of the burner tube grows stronger and the flame remains stabilized.

There is some hysterisis in the stabilization of the flame at low wind speeds, which has been noted by other researchers as well (Huang and Chang, 1994). In essence, if the crosswind on the rim stabilized flame of Figure 3a is increased, the flame will remain in this mode before undergoing a sudden transition to the partially lifted / wake stabilized case which is apparent in images 3b through 3j. However, once the flame has undergone this transition, the crosswind speed must be lowered to well below the initial transition velocity before the flame will again stabilize on the rim of the burner.

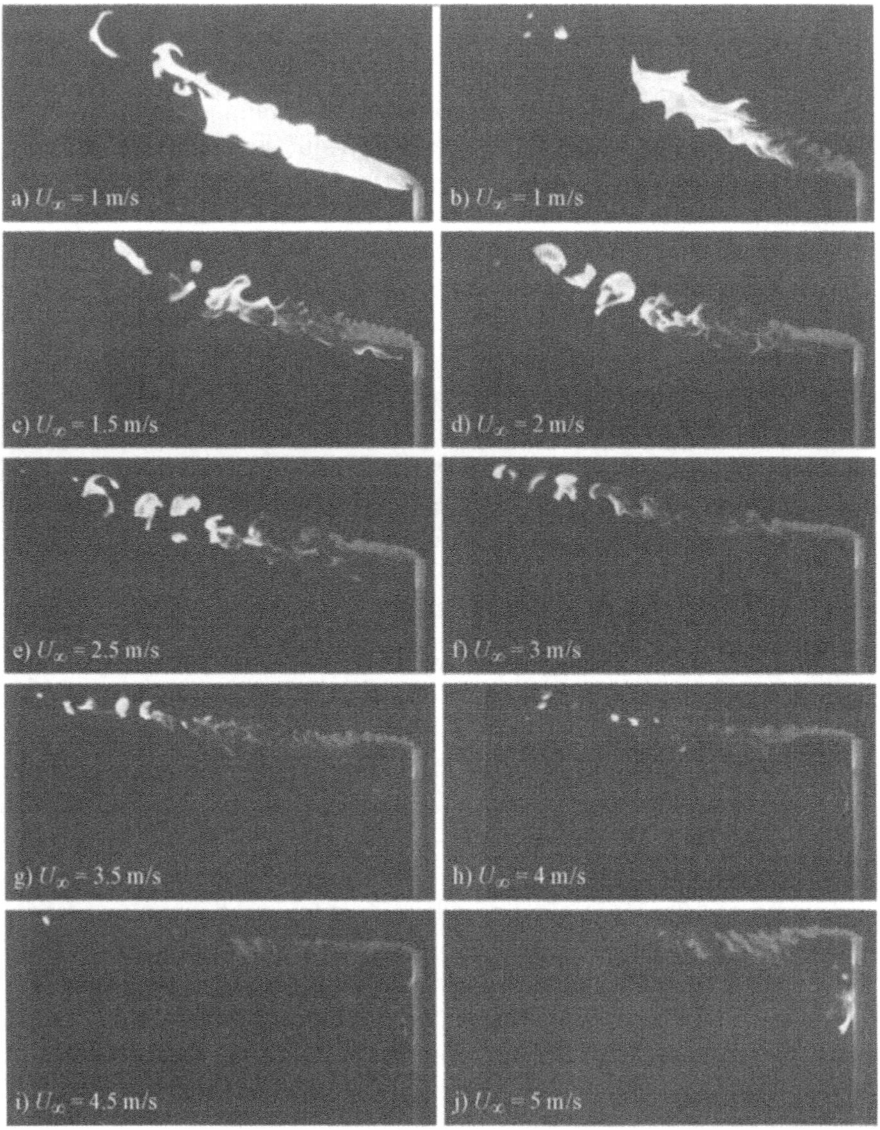

Figure 25.3. Colour images showing unburned fuel regions in wake-stabilized flames at various levels of crosswind speed. Light grey in images is Mie scattering of 532 nm laser light by fine silicone oil droplets seeded into the fuel stream.

In image 3c, vortex rings in the fuel the jet are readily apparent. These rings are created from the vorticity in the boundary layer on the inside of the stack and have been seen by several researchers studying cold-jets in crossflow (e.g. Kelso *et al* 1996 and Moussa *et al*, 1977). As the

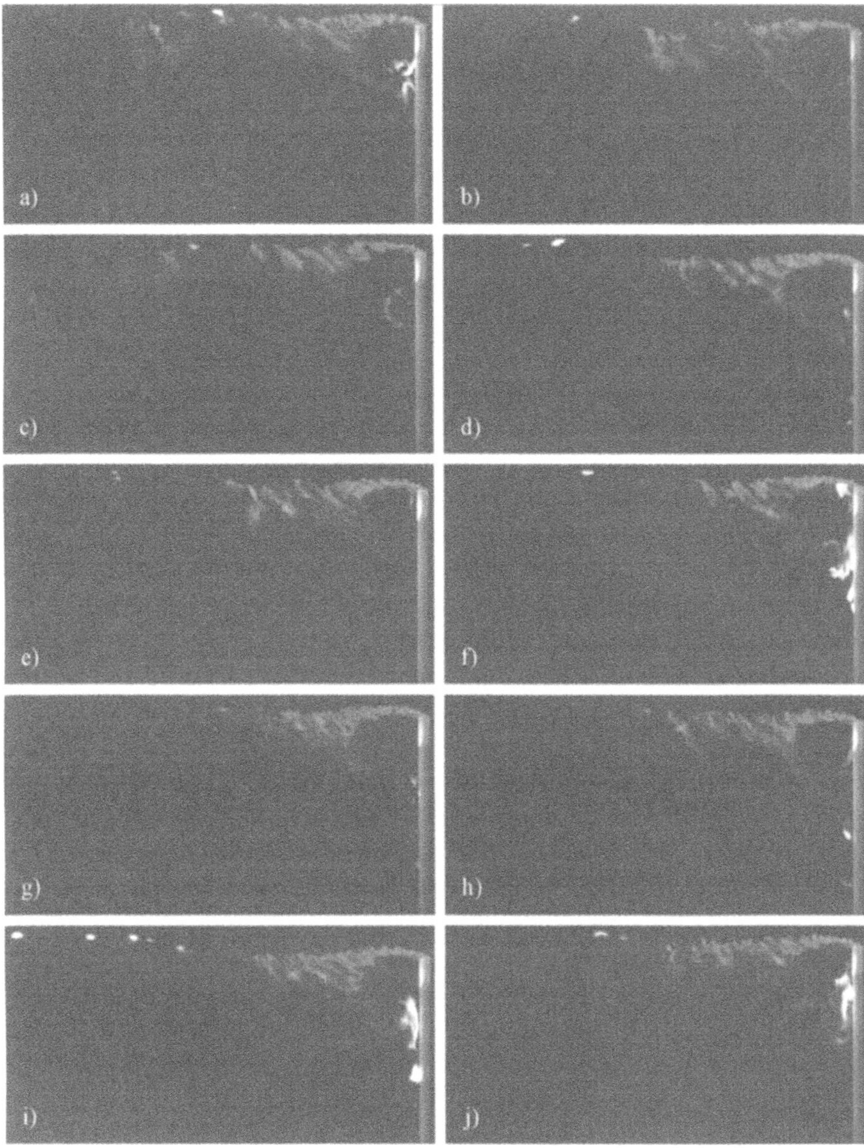

Figure 25.4. Multiple realizations of the wake-stabilized flame at $U_\infty = 5$m/s. Unburned fuel is apparent beneath the flame on a path to escape combustion. Images also show discrete pockets of combustion interspersed with coherent bits of unburned fuel.

crosswind speed is increased further, these rings begin to interact and distort. Close inspection of image 3g reveals evidence of vortex pairing and roll up in the upper mixing region. As the three zone structure of the flame becomes more clear, the images show that unburned fuel begins to penetrate into Zone 2 of the flame. Perhaps most interesting is the observation that at higher wind speeds, unburned fuel can be found to exist between the discrete pockets of combustion as shown in image 3j. This behavior was predicted in (Johnson *et al*, 2001) but the experimental techniques used could not conclusively demonstrate that these structures existed.

Figure 4 shows multiple realizations of the flame at a crosswind speed of 5.5 m/s where the overall inefficiency of the flame in approximately 3 % (Johnson and Kostiuk, 2000) and illustrates the stochastic nature of the flow. Images 4a, b, c, d, e, g, and h clearly show coherent bits of unburned fuel beneath the flame, which are on a path to escape combustion. These results substantiate data from previous measurements of instantaneous hydrocarbon concentration using a fast flame ionization detector (Johnson *et al*, 2001) that showed that the majority of the unburned fuel was being ejected on the underside of the flame. These images also confirm the observation that fuel is not stripped away from the non-reacting mixing layer on the upper surface of the flame.

Almost all of the images show discrete pockets of flame interspersed with coherent bits of unburned fuel. It appears that these coherent bits are periodically ejected through Zone 2 of the flame where they escape combustion. Previous research hypothesized the link between the appearance of discrete pockets of combustion and the measured inefficiencies (Johnson and Kostiuk, 2000; Johnson *et al*, 2001). These results seem to clarify that link.

Consideration of all of the images in Figure 3 and 4 would suggest that the structural nature of the ejected bursts of unburned fuel is directly attributable to the vortex rings in the upper shear layer, which were clearly apparent at lower wind speeds. Although these coherent bits are grossly distorted as they interact and stretch and are forced through Zone 2 of the flame, they are able to maintain some coherence and avoid being consumed via combustion. It is believed that this is the mechanism that leads to the measured inefficiencies in wake-stabilized jet diffusion flames in crossflow.

4. Conclusions

A conditional laser sheet imaging technique has been used to show that for a diffusion flame stabilized in the wake of a burner tube, as the crosswind speed is increased, the flame breaks into discrete pockets of combustion interspersed with coherent bits of unburned fuel. At higher wind speeds these bits of fuel are ejected through the underside of the flame in response to the mean flow induced by the standing vortex on the leeward side of the burner tube. The images presented help clarify the fuel stripping mechanism previously proposed and suggest that the mean flow induced by this standing vortex stack transports and stretches ring vortices of fuel from the upper shear layer and ejects them on the underside of the flame. These coherent bursts of fuel escape the flame without burning and ultimately cause the measured inefficiencies in the conversion of hydrocarbon fuel to carbon dioxide.

5. Acknowledgements

This work was supported by the Natural Sciences and Engineering Research Council of Canada through a Strategic Project Grant, Environment Canada through PERD funding, and the Canadian Association of Petroleum Producers through ERAC funding.

References

Gollahalli, S.R. and Nanjundappa, B. (1995). *Comb. Sci. Tech.*, 109:327–346.

Huang, R.F. and Chang, J.M. (1994). *Comb. and Flame*, 98:267–278.

Huang, R.F. and Wang, S.M. (1999). *Comb. and Flame*, 117:59–77.

Johnson, M.R. and Kostiuk, L.W. (2000). *Comb. and Flame* 123:189–200.

Johnson, M.R., Wilson, D.J., and Kostiuk, L.W., "A Fuel Stripping Mechanism for Wake-Stabilized Jet Diffusion Flames in Crossflow", *Comb. Sci. Tech.*, in press.

Kelso, R.M., Lim T.T., Perry, A.E. (1996). *J. Fluid Mech.*, 306:111–144.

Kostiuk, L.W., Majeski, A.J., Poudenx, P., Johnson, M.R., and Wilson, D.J. (2000). *Proc. Comb. Inst.* 28:553–559.

Kostiuk, L.W., Zhang, Y., Chew, T.C., Bray, K.N.C. and Britter, R.E. (1989). *Flow Visualisation V: Proc. Fifth Intnl. Symp. Visualisation*, pp. 751–757.

Moussa, Z.M., Trischka, J.W. and Eskinazi, S. (1977). *J. Fluid Mech.*, 80:49–80.

Chapter 26

PROGRESS IN CONTROL OF MIXING AND LARGE EDDY SIMULATION OF TURBULENT COMBUSTION

Parviz Moin
Center for Turbulence Research
Stanford University & NASA-Ames

Abstract Recent progress at CTR in the control of jet mixing using numerical simulations and experiments is summarized. These control experiments use a basic form of evolution strategies and have been successful in enhancing turbulent mixing. Progress in large eddy simulation of turbulent combustion is also discussed. Results from two new flamelet based techniques for simulation of diffusion flames are presented.

Mixing Control

Enhanced turbulent jet mixing has several technological applications. Both unsteady passive and active jet control strategies have been used to improve jet mixing. In direct numerical simulations jet Reynolds numbers have been typically low, $O(3000)$, but the experiments and large eddy simulations conducted to test the control strategies derived from DNS were run at high Reynolds numbers, $O(100,000)$. In passive control simulations (Freund & Moin, 2000), high amplitude, low mass flux pulsed jets blowing normal to the main jet's shear layer were placed near the nozzle. Forcing the jet at several frequencies causes the jet to expand rapidly in the plane parallel with the actuators and to contract in the plane normal to the actuators. It was found that forcing at Strouhal number 0.2 is more effective at spreading the jet farther downstream as compared to forcing at Sr=0.4 which affects the jet near the nozzle.

Helical and axial forcing (as in the experiments of W.C. Reynolds and co-workers, see Juvet & Reynolds, 1993) were also used to enhance jet mixing. Evolution optimization strategies were used to systematically

A. Pollard and S. Candel (eds.), IUTAM Symposium on Turbulent Mixing and Combustion, 305-314.

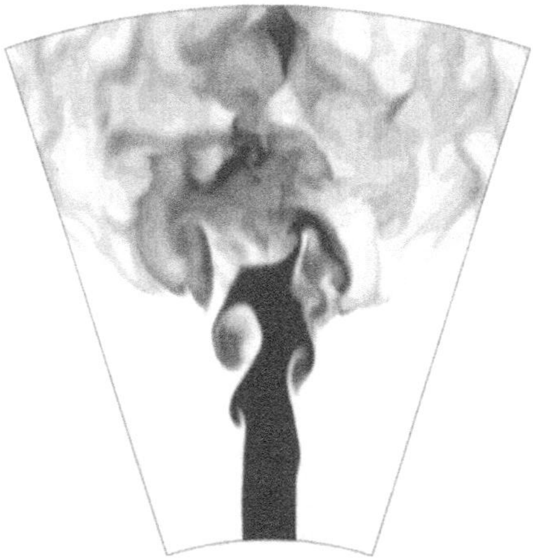

Figure 26.1. A snapshot of passive scalar concentration for a jet excited with a single-frequency actuation in the plane of the actuation; $St = 0.036$ and $A_h = 0.05$.

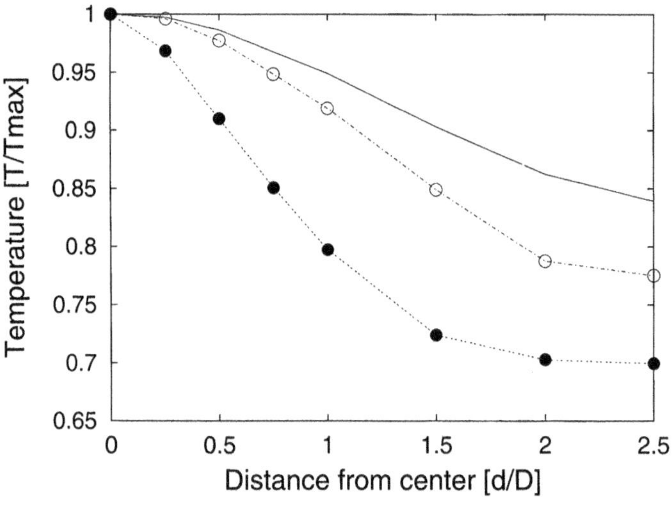

Figure 26.2. Normalized temperature profiles of natural jet (—•—), initial condition for ES (—○ —) and best result of ES (———).

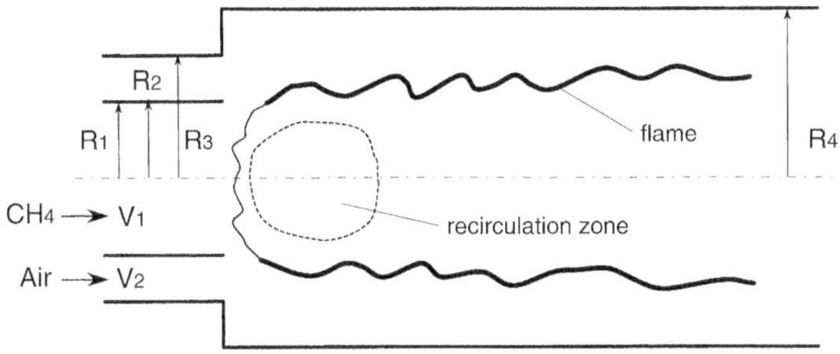

Figure 26.3. Schematic of the co-axial jet combustor showing a lifted flame and recirculation zone which stabilizes the flame

find the forcing amplitudes and frequencies. A very basic evolution strategy was used to reduce the cost or fitness function, which, in this case, was the variance of the temperature profile across the jet. As shown in Figure 26.1, this optimization strategy was very effective for mixing enhancement (Hilgers, 2000). Also indicated in the figure are the optimized axial and helical forcing parameters obtained by the optimization procedure. These parameters are near, but not the same as, those found previously in experiments.

There has been some skepticism on the effectiveness of the above control strategies in high Reynolds number jets. To extend the range of Reynolds numbers considered, some of the calculations were performed using large eddy simulation. In addition, the evolutionary optimization algorithm was implemented in the high Reynolds number jet facility of Juvet and Reynolds at Stanford. We believe this was the first time that evolutionary algorithms were implemented in a physical experiment. At jet Reynolds numbers of about 100,000, significant improvement in jet mixing characteristics were obtained. The starting point for the evolutionary optimization was the blooming jet of Juvet and Reynolds.

As shown in Figure 26.2, the evolution strategy (ES) flattens the jet temperature profile significantly and finds more effective control parameters than those used by Juvet and Reynolds (blooming jet).

LES of Turbulent Combustion

One of the major pacing items for accurate computation of turbulent diffusion flames occurring in gas turbine engines is the prediction of the lifted flame from the fuel source. A schematic of a lifted flame in a co-axial jet combustor is shown in Figure 26.3.

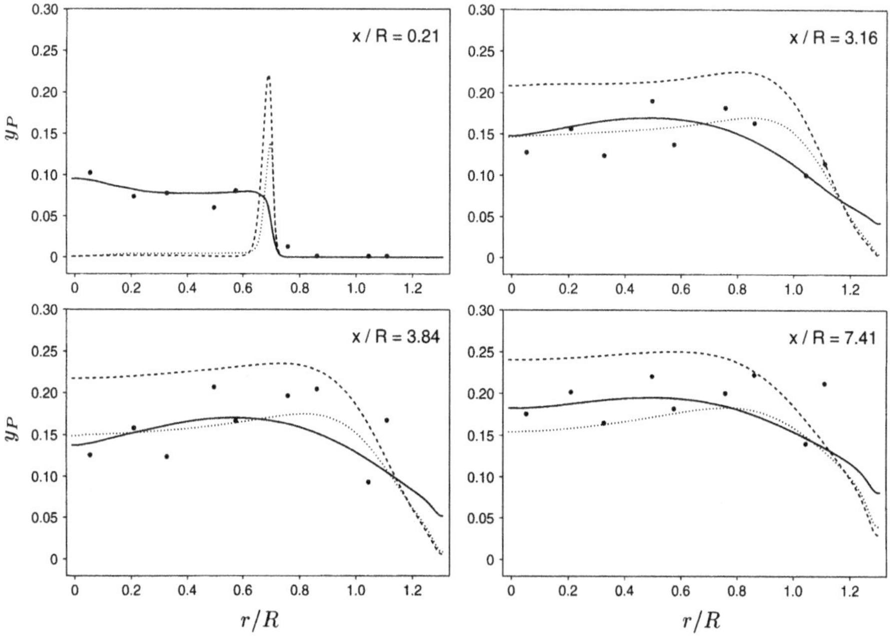

Figure 26.4. Radial profiles of time-averaged mass fraction; Fast chemistry(----), Steady flamelets (········), Progress variable approach (———), Experiment (•)

Experimental observations have shown that the flame is highly unsteady and chaotic with frequent extinctions and re-ignitions. To predict the flame lift-off phenomenon, Pierce & Moin (2001) have developed a new approach, which consists of using a scalar to track the progress of the reaction. Two scalar equations, one for the conserved mixture fraction and another for the non-conserved progress variable are solved in addition to the momentum and continuity equations. A modified flamelet (Peters, 1986) approach is used where the flamelet parameters are the mixture fraction and the progress variable instead of the scalar dissipation rate. The progress variable provides for a good representation of the unsteady dynamics of the flame. Very good agreement with the experimental data is obtained for quantities such as velocity, mixture fraction and product concentrations (Figure 26.4). The agreement of the so called intermediate species and pollutants such as concentration of carbon monoxide is only adequate.

The unsteady flamelet approach has enjoyed considerable success to predict the pollutants in jet flames. Pitsch and Steiner (2000) have predicted NO_x concentrations in a piloted methane-air diffusion flame using an unsteady flamelet model formulation. However, CO and H_2

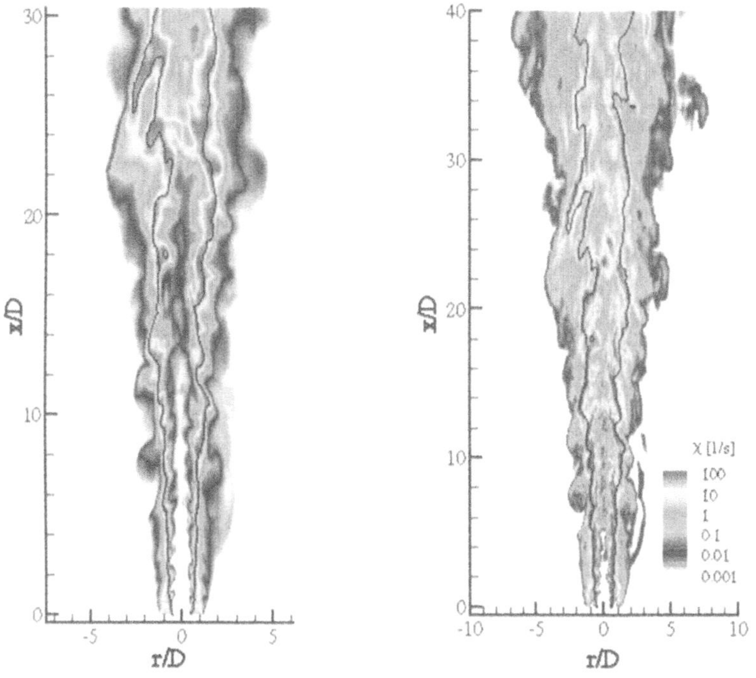

Figure 26.5. Instantaneous temperature (left) and scalar dissipation rate (right) distribution. Black line represents the contour of stoichiometric mixture fraction

were overpredicted in the fuel rich part of the flame. An example of the instantaneous temperature distribution from this simulation in the near field of the nozzle is shown in Fig. 26.5. The corresponding scalar dissipation rate, shown in Fig. 26.5, reveals strong fluctuations. Pitsch (2001) has shown that these fluctuations, which were neglected in the model by Pitsch and Steiner (2000), strongly influence the chemistry. Pitsch (2001) presents a new formulation of the unsteady flamelet model that incorporates the effect of scalar dissipation rate fluctuations on the chemistry. Apparently, accounting for scalar dissipation rate fluctuations (which is available in LES and not RANS) improves the predictions of CO and H_2. The conditional mean temperature, NO_x, CO, and H_2 mass fractions from Pitsch's simulation are compared to experimental data at $x/D = 30$ in Fig. 26.6. In order to extend this model to partially premixed combustion, and lifted flames, Pitsch and Duchamp (2001) have developed a combined conserved scalar/level-set flamelet model, which is currently being validated against experimental data for a series of lifted jet diffusion flames.

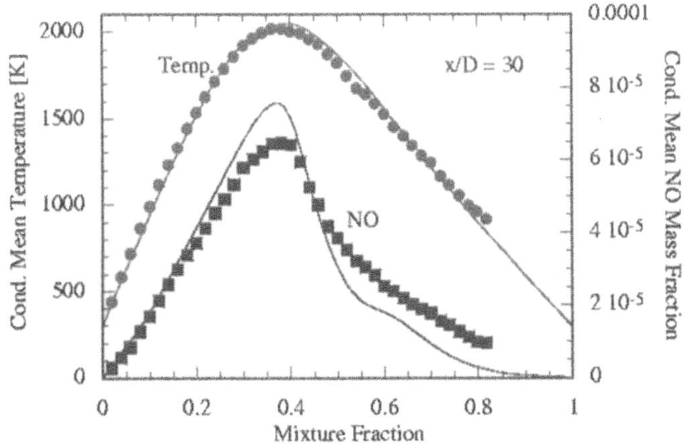

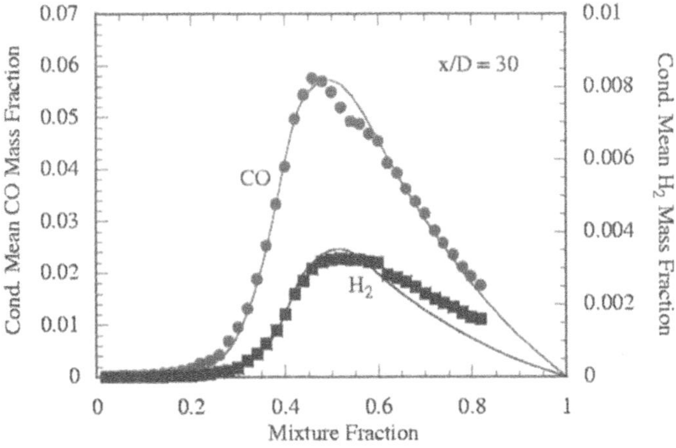

Figure 26.6. Conditional mean averages of temperature, NO, CO, and H_2 mass fractions at $x/D = 30$. Lines: numerical simulation; symbols: experimental data.

References

Fruend, J. B. and Moin, P. ,(2000). "Jet Mixing Enhancement by High-Amplitude Fluidic Actuation", AIAA J. vol 38, No. 10, pp. 1863-1870.

Hilgers, A. ,(2000). "Control and Optimization of Turbulent Jet Mixing", Annual Research Briefs, Center for Turbulence Research, pp. 45-54.

Juvet, P. J. D. and Reynolds, W. C., (1993). "Control of High Reynolds Number Round Jets", Report number TF-59, Thermosciences Division, Department of Mechanical Engineering, Stanford University.

Peters, N., (1986). "Laminar Flamelet Concepts in Turbulent Combustion", *Proc. of Combust. Inst.*,**21**, pp. 1231-1250.

Pierce, C., D. and Moin, P., (2001). "Progress Variable Approach for Large Eddy Simulation of Turbulent Combustion",Report number TF-80, Flow Physics and Computation Division, Stanford University.

Pitsch, H., (2002). "Eulerian Flamelet Model for LES of Non-Premixed Combustion", prepared for submission to Comb. Flame.

Pitsch, H., Duchamp de Lageneste, L. (2001). "Large-Eddy Simulation of Lifted Turbulent Diffusion Flames", 2nd Joint meeting of the US sections of the Combustion Institute, Oakland, CA, paper 263.

Pitsch, H. and Steiner H., (2000). "Large Eddy Simulation of a Turbulent Piloted Methane/Air Diffusion Flame (Sandia Flame D)", *Physics of fluids*, **12**,pp-2541–2554.

Sbalzarini, I., F., Su, L. and Koumoutsakos, P., (2000). "Evolutionary Optimization for Flow Experiments", Annual Research Briefs, Center for Turbulence Research, pp. 31-43.

Discussion

Josette Bellan, JPL "You mentioned in one of your initial slides that it was important to have the spray combustion coupled into this. Is it fair to say that at this stage that you don't have that in the model? "

Reply: Yes that is right, that is very fair to say.

Steve Pope, Cornell University "The question of differential diffusion has come up a few times before your lecture. I would suggest that those old experiments are not the best experiments in which to address the question of differential diffusion. In fact the more recent experiments from Sandia and elsewhere would lay the diagnostics are able to look at that question in some detail and the bulk of those experiments that I can remember seem to show that the high Reynolds number turbulent flames the flamelet calculations without detailed transport properties often do a better job of representing experimental data then qualifications with the detailed transport properties. So the sum effect that Paul (Dimotakis, ed) asked us to explain I have no explanation at all. It seems that a high Reynolds number these differential diffusion effects are less significant."

Reply: That may be -I remember- well by the way how high is high. These are high Reynolds numbers as high as jet aircraft engine manufacturers would like to have them, between 28, 000 - 30,000, somewhere between 30,000 to less than 100,000. So that is where this calculation was done. But that is a good point where you mention - I will show you that maybe the statement that you made though that the sum of the steady flamelet results can do a better job you saw that I think at the far downstream stations, that is true. Upstream where you have this lifted flame and rich mixture that is back there was clearly not true. Were the Sandia experiments piloted or did they have a fuel rich section that you have with the radicals there with a huge variation in the diffusivities?.

Steve Pope, Cornell University "This has been observed in both piloted flames and other flames like bluff body stabilized flame. With all that said I think that conclusions about differential diffusion would be better based on detailed laser measurements rather than those probe measurements. "

Reply: What you are looking at - we did not look at - we looked at the products of the reactions like the product, the temperature and if you can't predict those with or without, you see the effect of adding the differential diffusion and you see that you can predict them - what else would you like? You see that is the bottom line type of a test.

Steve Pope, Cornell University "It's just a question of the directness of evidence."

Paul Dimotakis, Caltech "I was wondering what in your model you would do different if Reynolds number had to go up by a factor of 2 or a factor of 4. I think it is not a correct statement that gas turbine engine manufacturers would not like to increase the Reynolds number, the measurements you showed were at 3.6 atmospheres there now up to 20 and 30 so that is not a correct statement. So I was wondering what or how your model would be different if the Reynolds number had to go up. That is one question. The second question is I couldn't quite understand why the progress variable temperature field ended up with higher temperatures at the wall than the fast chemistry model. The third is a comment, I was intrigued to note that cost optimization lead you to experiments."

Reply: Always costs leads us to try to use the best tools the cheapest tool. In this case clearly experiments have the edge. It takes a lot longer, you have run an ensemble of several hundred LES calculations to be able to reproduce and put it into evolution. So lets go backwards, to answer your question about why the fast chemistry and the flamelet approach seems to have this jet , cold air jet close to the wall versus flamelet, I really don't know rigorously versus a progress variable. It was based on those two snapshots that we showed that it appears that in that case at least the progress variable approach the flame brushes against the wall. You probably can track it back to what took place early on in terms of having the fast chemistry reactions taking place early on and tracking them. Now going back to the Reynolds numbers, I did not say that they don't like to have it. I say that these are the Reynolds numbers for this experiment and the Reynolds numbers that people at

Pratt and Whitney and others tell us are typically of the same order. We are not talking about two orders of magnitude difference in the Reynolds numbers we are talking typically of the order of 100000 or less now that brings us to the range factor of 2 or 4 that you mention. As far as what would be different in our calculation we would probably need to have a lot more points probably a factor of 4 to 8 more points to just capture the dynamics of the broadband spectrum. What else would I do to in terms of changing the model - probably nothing. Nothing in terms of changing the model after all these are supposed to be high Reynolds number models and you would hope that increasing the resolution the calculations would give you a signal that you don't have the resolution that is required at those high enough numbers and you would just add to it.

Paul Dimotakis, Caltech "The difficulty with that is it invalidates the premise of LES if to capture the higher Reynolds number I need higher resolution. I mean, after all the whole point of LES and subgrid scale modeling is that I am modeling presumably, hopefully, correctly what happens below the scales I don't resolve."

Reply: That premise has been dictated for the non-wall bounded flows. Walls set the Reynolds numbers. Now in this case, because you don't have attached boundary layers in the main part of the flow, I don't think the Reynolds number effect are all the important. But certainly the shear layers coming off the splitter plates are going to be dominant the Reynolds number affect the thickness of those layers; the scales that come out of the shear layer are going to part of the.. ,you have to capture them and that has been the pacing item in early LES. You cannot, you cannot leave the large scale feature locally large scale features, although they might be small compared to the overall dimension of the flow, but those thin shear layers coming from the splitter plates, you cannot ask the Smaogorinsky-type models to capture the effect of those. LES, the right LES, for those cases has to capture those. You have to put zonal grids, unstructured grids whatever numerical tools you have by brut force. In fact the resolution requirement for wall bonded LES is something like Reynolds number squared. Which is not that much better than Reynolds number to the 9-4th that you get for DNS. So essentially when walls come and attached boundary layers especially like what you have coming off of the splitter plate you have to capture that.

Murray Thompson, University of Toronto "I just wondered if you included Radiation transfer in the model?"

Reply: No, that's another thing that we have to do. We have some ideas how to do it. In the first step bring it into the flamelets and its library. That could also have some effect with regard to the temperature discrepancies.

Jeff Lienau, Pratt and Whitney "Could you put up to CO plot again I was just curious. Is it just fortuitous that the steady flamelet model is actually hitting the CO pretty well or is that like compensating errors type thing."

Reply: It could be. That is what I was talking to Steve about. I don't know the answer to that.

Jeff Lienau, Pratt and Whitney "We really don't have any basis to believe that, do we?"

Reply: No. I think I noticed it, especially downstream, right...

Jeff Lienau, Pratt and Whitney "Yeah, CO is our main pollutant we are concerned with in the industrial area."

Reply: Yes the downstream here.

Jeff Lienau, Pratt and Whitney "Yeah downstream. Even at the 3.84 location and 3.16, its not bad."

Reply: No, I agree.

Jeff Lienau, Pratt and Whitney "But really we have no basis to believe that. I mean..."

Reply: Just happen - fortuitous.

Jeff Lienau, Pratt and Whitney "Yeah fortuitous ."

Reply: Maybe - yes at this point yes, that is what I would say.

Terry Poinsot, CERFACS, Toulouse, France "I am still surprised by the power utilisation, which to me is the key way of doing this computation. You match a flamelet that is actually igniting, which is an unsteady flamelet on a steady diffusion flamelet. I am even thinking that there are regions for example where the flame is just igniting cannot be matched anywhere on a steady flamelet because they are all quenched there, so certainly at the beginning of the ignition you have to do something where your power- or you have to do something else?"

Reply: Well you know that in this case, we are using the unstable branch as well. That is the controversial point. We are using it but I can argue that with the progress variable approach in its unsteady context because the of the unsteadiness that C brings in here I think is appropriate. But that accounts - so if you do get a point on the unstable branch you will have a choice of going either immediately to the extinction or you go back to the ignition. Now whether that is doable but whether it is physically realistic in terms of modeling the ignition or extinction I don't know, but the calculations certainly can account for that because we take of that unstable branch.

Lorenz Sigurdson, University of Alberta "I have three questions. The first is just to clarify that using the control algorithm, that was experimentally done? That is the first thing."

Reply: Yes.

Lorenz Sigurdson, University of Alberta "And the second thing is could it find the blooming jet all by itself. I mean, I noticed that you started with the blooming jet and that start to optimize above that if you just started - you were kind a quick."

Reply: Blooming jet is not optimal that picture, what it showed, when even we started with it, it moved onto something else there are other calculations that were done completely random of initial conditions and I think, you went to a new optimal but I don't think it was blooming jet optimal case. We did not look at the instantaneous pictures to see if the actual instantaneous structures look like.

Lorenz Sigurdson, University of Alberta "So you could find something starting from an ad hoc random case hat was actually better than the blooming jet. So the question is, how much better?"

Reply: Absolutely. As much as this. As much as what you saw over here, which was 10-20% better. In fact what was interesting, the numbers that it converged to, was only slightly different of the order of 10% or so different of the blooming jet parameters but that slight difference made a big difference in the case of the mixing.

Lorenz Sigurdson, University of Alberta "10 or 20% in what factor?"

Reply: We looked at the variance, the deviation from the flat profile for temperature. The idea being that the flat temperature profile that everything being mixed.

Lorenz Sigurdson, University of Alberta "So the maximum amplitude?"

Reply: We normalized it with maximum and then looked at the profile. The shape of the profile deviation from the flat value.

Shows video, of lifted flame from Chuck Pierce of CTR.

Chapter 27

LARGE EDDY SIMULATION OF A NON-PREMIXED TURBULENT BURNER USING A DYNAMICALLY THICKENED FLAME MODEL

J. P. Légier, T. Poinsot
CERFACS
42, Av. G. Coriolis, 31057 Toulouse Cedex
France

B. Varoquié, F. Lacas and D. Veynante*
Laboratoire E.M2.C.
C.N.R.S. - Ecole Centrale Paris
92295 Châtenay-Malabry Cedex
France

Abstract A new LES subgrid scale turbulent combustion model, adapted to combustion regimes which are nor perfectly premixed or non-premixed, is tested in a simplified configuration. This model does not require any a priori assumption on the flame structure and is able to compute flows where both premixed and non-premixed flamelets coexist. Three combustion regimes are successfully recovered in numerical simulations.

Keywords: Large eddy simulation, turbulent combustion, modeling

Introduction

In most gas turbines, turbulent flames are neither perfectly premixed, nor perfectly non-premixed and require the development of large eddy simulation (LES) models adapted to this situation. Unfortunately, very

*Corresponding author, email: denis@em2c.ecp.fr

A. Pollard and S. Candel (eds.), IUTAM Symposium on Turbulent Mixing and Combustion, 315-326.

few studies have tried to address these problems because they gather the complexities of pure mixing (without combustion), of ignition, of partially or perfectly premixed combustion and of non-premixed combustion. All these regimes may be encountered simultaneously in a gas turbine and a proper model should be able to handle them all. This is especially true for recent technologies like LPP (Lean Premixed Prevaporized) combustors which are designed to operate in a lean premixed mode but are prone to flashback (i.e. a flame propagation upstream its designed location), a regime dominated by diffusion flames. Being able to predict flashback requires models which are not yet available.

Many LES studies have been published for mixing [Pierce and Moin, 1998], premixed flames [Veynante and Poinsot, 1997, Im et al., 1997, Piana et al., 1997, Boger et al., 1998, Colin et al., 2000] or diffusion flames [De Bruyn Kops et al., 1998, DesJardin and Frankel, 1999, Moin et al., 2000]. But all these models take explicitely into account, the flame topology, premixed or not, thereby limiting the predictive character of simulations when the exact regime of combustion is not known a priori.

A new model called DTF (Dynamic Thickened Flame) is proposed here to compute mixing, diffusion and premixed flames simultaneously. This is achieved by modifying the thickened flame model derived for premixed flames. Instead of using a constant thickening factor [Colin et al., 2000], a local thickening factor F is active only in the vicinity of the flame front. Outside flame zones, thickening is suppressed and mixing can be predicted correctly.

The DTF model presented in Section 1, is tested to simulate a two-dimensional burner described in Section 2. Section 3 presents LES runs and discusses numerical results.

1. Dynamically Thickened Flame (DTF) model

The propane/air chemistry is described using a single one-step scheme:

$$C_3H_8 + 5(O_2 + 3.76N_2) \rightarrow 3CO_2 + 4H_2O + 18.8N_2 \qquad (27.1)$$

where the fuel consumption rate is given by:

$$\dot{\omega}_F = A\nu_F W_F \left(\frac{\rho Y_F}{W_F}\right)^a \left(\frac{\rho Y_O}{W_O}\right)^b \exp\left(-\frac{T_a}{T}\right) \qquad (27.2)$$

T_a is the activation temperature, W_F and W_O are respectively the molar weights of propane ($W_F = 44$) and oxygen ($W_O = 32$). The prexponential constant A is fitted to provide correct flame speeds for lean premixed flames when compared to full chemistry results. Chemical parameters are: $A = 1.65.10^{11}$ cgs, $T_a = 15080\,\mathrm{K}$, $\nu_F = 1$, $a = 0.5$ and $b = 1$

The fuel mass balance equation is, with usual notations:

$$\frac{\partial \rho Y_F}{\partial t} + \nabla \cdot (\rho \mathbf{u}) = \nabla \cdot (\rho \mathcal{D} \nabla Y_F) - \dot{\omega}_F \tag{27.3}$$

The thicknened flame model [Colin et al., 2000] extends the pioneering work by [Butler and O'Rourke, 1977] who showed that, for a premixed laminar flame, multiplying species and heat diffusion coefficients by a factor F and decreasing the exponential constant by the same factor F provides a flame propagating at the same laminar flame speed than the non-thickened flame but is thickened by a factor F. Adjusting F to sufficiently large values allows the flame to be resolved on an LES grid. Unfortunately [Angelberger et al., 1998], the thickening procedure reduces the flame response to the smallest turbulent motions. An efficiency function, depending on the thickening factor, length and velocity ratios comparing flame and turbulence, is introduced to account for the unresolved flame wrinkling [Angelberger et al., 1998, Colin et al., 2000].

This initial model is easily extended to dynamic thickening, depending on time and spatial location, by recognizing that a premixed flame where the thickening factor F changes spatially still propagates at the same laminar flame speed (Cuenot, 2000, private communication). In the new DTF formalism, a local thickening factor F is active only in the vicinity of the flame front ($F > 1$) and relaxes to $F = 1$ (no effect) outside flame zones to suppress thickening and correctly predict mixing [Légier et al., 2000].

The thickening factor is determined as:

$$F = 1 + (F_m - 1)\tanh\left(\beta \frac{\Omega}{\Omega_m}\right) \text{ with } \Omega = (Y_F)^a (Y_O)^b \exp\left(-\Gamma \frac{T_a}{T}\right) \tag{27.4}$$

where F_m and Ω_m are respectively the maximum values of the thickening factor F and the sensor Ω detecting the reaction zone. Γ ($\Gamma < 1$) and β control respectively the thicknesses of the zone where the thickening is active and of the transition layer between thickened and non-thickened zones.

2. Experimental configuration and regime map

Fig. 27.1 displays the experimental configuration used to test the DTF model. Two propane streams are injected through small slots (5 mm height) into an air coflow. Two backward facing steps (25 mm height each) promote the flame stabilization. The combustion chamber, dowstream the fuel injector lips, is 300 mm long, 100 mm height and 80 mm depth. The experiment is designed to produce two-dimensional flows

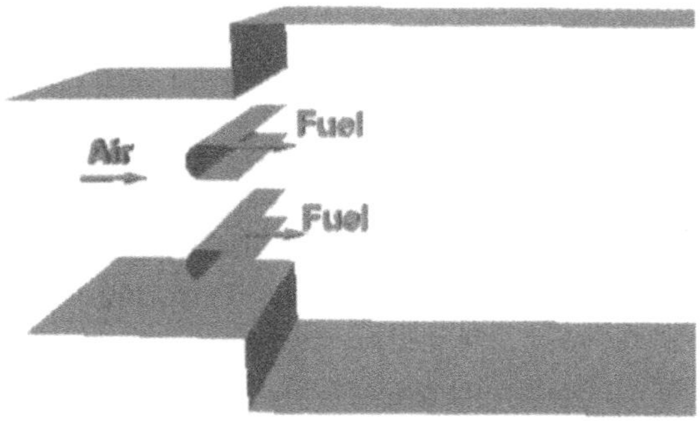

Figure 27.1. Dump-stabilized burner configuration.

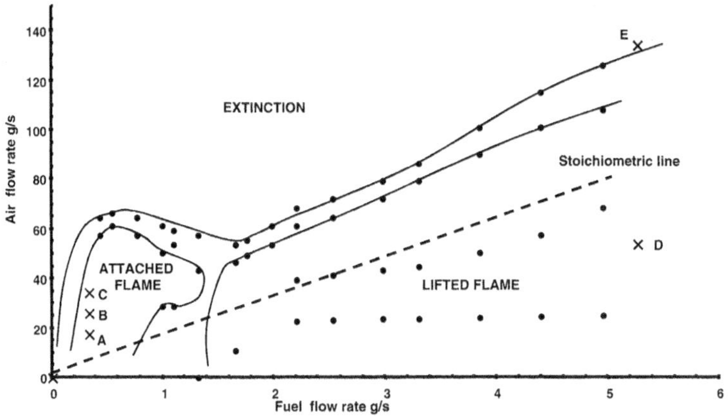

Figure 27.2. Combustion regimes observed in the burner displayed in Fig. 27.1 and plotted as function of the fuel and air mass flow rates. The global stoichiometric line, where fuel and air are injected in stoichiometric proportions, is also indicated.

to simplify optical diagnostics (CH and C_2 radical emission, laser induced fluorescence on OH radical,...) and model developments and validations. The maximum burner power is 300 kW.

The burner exhibits various operating regimes, briefly summarized on Fig. 27.2. For low fuel and air flow rates, flames are stabilized a few millimeters downstream of the fuel injectors ("anchored" flames, Fig. 27.3a). For higher reactant flow rates (higher powers) and rich overall equivalence ratio (excess of fuel compared to the among of air injected), the flames lift from the injectors and are stabilized by the hot gases recirculating behind the backward facing steps ("lifted" flames, Fig. 27.3b). For high reactant flow rates but excessively lean overall equivalence ratios, the flame gets quenched.

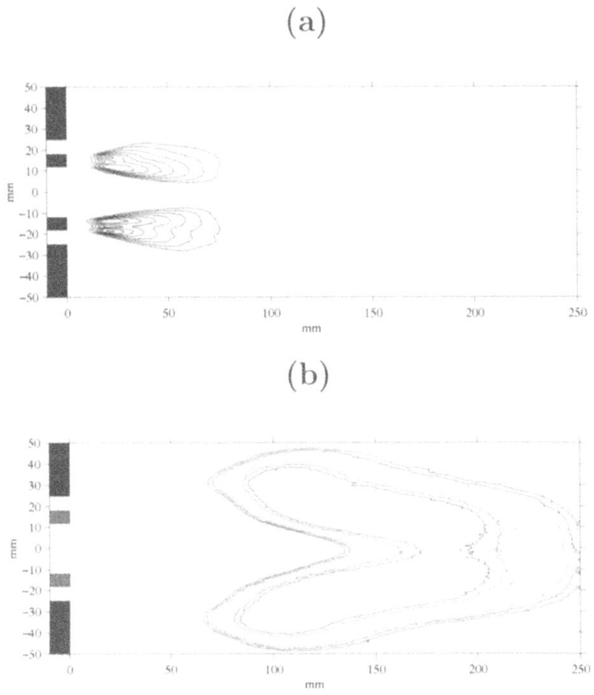

Figure 27.3. Combustion regimes visualized using mean CH radical spontaneous emission, corresponding to the mean reaction rate. (a): "anchored flame" regime. The flame is stabilized in the vicinity on the fuel injectors but is not anchored on the lips. (b) "lifted flame" regime where the flame is stabilized by recirculation zones induced by the two backward facing steps.

3. Large eddy simulations

The DTF model is implemented in the LES code AVBP, developed at CERFACS. The numerical scheme is third order both in space and time [Colin and Rudgyard, 2000]. LES were conducted for three combustion regimes, referred to as C, D and E in Fig. 27.3. The first case (C) corresponds to an anchored flame, in the second one (D), the flame is lifted whereas a flame blow-off is expected in regime (E). Unresolved fluxes are modeled using a filtered Smagorinsky model [Ducros et al., 1997]. For these first tests, two-dimensional simulations are performed and only the half upper part of the burner is computed (the flow field is assumed to be symmetrical along the burner axis). All model parameters were kept constant in all simulations.

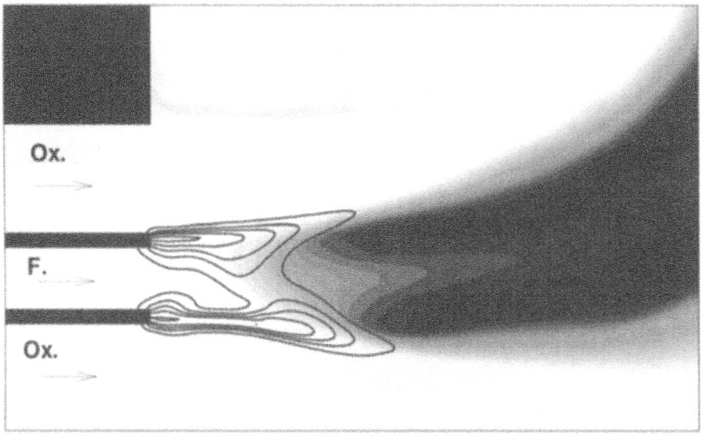

Figure 27.4. Anchored flame (case C): mixing index $Y_F Y_O$ (lines) and temperature (gray-scale) fields. Zoom in the vicinity of the upper fuel injector.

3.1. Anchored flame regime (case C)

For low reactant flow rates, and low powers, flames are stabilized in the vicinity of the fuel injector. A small lift-off distance is observed experimentally as well as in the LES results. Fig. 27.4 presents a view of the flow close to the fuel injector for case C: the mixing index $Y_F Y_O$ (lines) and the temperature (gray scale) fields are displayed.

The flame is stabilized by a couple of "triple flames": one for the upper air jet and some part of the fuel stream and another one for the central air jet and the rest of the fuel stream. Even though the attachment region flaps slightly, the structure of the zone close to the injectors appears rather steady. The recirculation zone does not contain burnt gases and is not involved in the stabilization process.

3.2. Lifted flame regime (case D)

For larger reactants flow rates, and higher burner powers, the flame becomes stabilized by the recirculation zones. A typical snapshot of the flowfield is presented in Fig. 27.5: the fuel mass fraction Y_F (gray scale) is superimposed to the reaction rate field and the two stoichiometric lines (bold lines). Even though fuel and oxidizer are injected separately into the burner, only a few of the flame zones exhibit a diffusion-like structure and lie around the stoichiometric iso-surface: in fact, strong mixing occurs before any combustion starts. When combustion begins, a strong rich premixed flame, leaving fuel in products, is observed (location B, Fig. 27.6b). This excess fuel then burns with air downstream

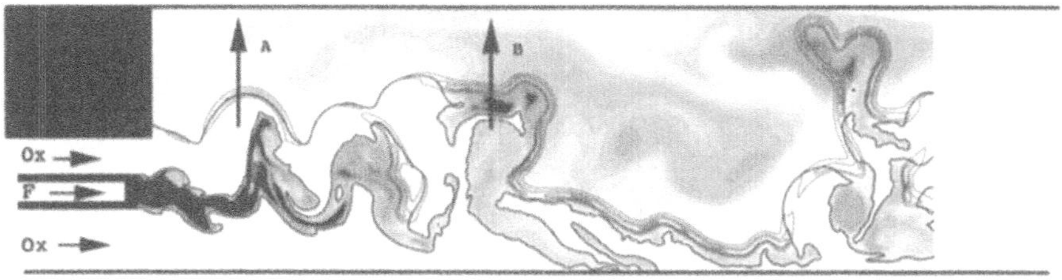

Figure 27.5. Lifted flame (case D): fuel mass fraction Y_F (gray scale) and reaction rate (contour lines) fields are superimposed to the stoichiometric iso-surface (bold lines). Arrows indicate location and direction of the cuts displayed on Fig. 27.6.

or in the recirculation zone. In this latter case, a non-premixed flame is observed between hot fuel diluted in burnt products and fresh air flow (location A, Fig. 27.6a). The diffusion flame structure observed for this point is very different from the usual fuel (cold) / oxidizer configuration used in flamelet models. First, the maximum fuel mass fraction is about $Y_F \approx 0.05$, far from the maximum value $Y_F = 1$ found in a pure propane / air diffusion flame (the fuel is diluted within burnt gases). The fuel temperature also corresponds to the burnt gases temperature of the previous rich premixed flames.

Starting from the stabilized lifted flame corresponding to point D, the air flow rate is increased to reach the regime E (Fig. 27.3). Very rapidly, the fresh air entering the combustion chamber dilutes the mixture involved in rich premixed flames and starts filling and cooling the recirculation zone. As soon as this zone is too cold, the whole stabilization process is compromised and the flame quenches. This test confirms that blow-off can be predicted using the DTF model (usual simple flamelet models cannot predict blow-off because the flame is generally assumed to burn in a steady state regime). Moreover, this blow-off is found for flow rate values corresponding to experimental observations. Of course, more test points are required to determine the exact quenching limits and to validate the DTF model but this finding is quite promising.

3.3. Mean flow fields (lifted case)

The LES data can also be averaged to be compared to measurements. Fig. 27.7 shows mean fuel mass fraction, temperature and reaction rate fields: the reaction rate field confirms that the flame is lifted and stabilized in the vicinity of the recirculation zone. This zone contains hot

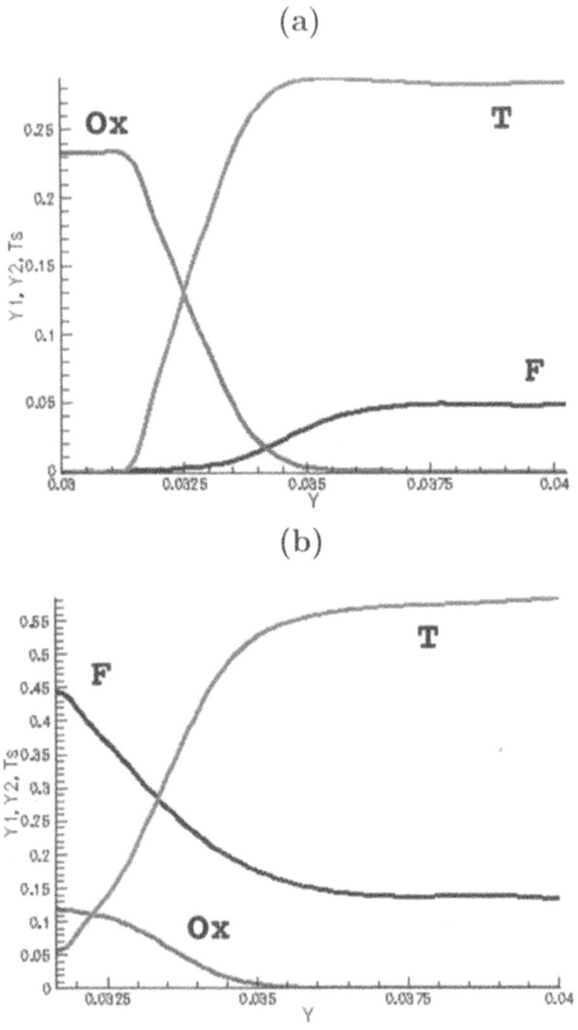

Figure 27.6. Lifted flame (case D): fuel, oxidizer and temperature profiles across the flame front in locations A (left) and B (right) displayed in Fig. 27.5.

gases and acts as a heat tank providing the energy required to stabilize the flame. The fuel mass fraction field shows that the leakage of fuel towards the recirculation zone, and due to the previous burning of rich mixtures, appears even in the mean flow map.

A quite different result is achieved when using the Magnussen eddy dissipation model which assumes that the reaction takes place when fuel and oxidizer meet. As shown in Fig. 27.8, the flame is found to be anchored at the injector lips whatever are the mass flow rates.

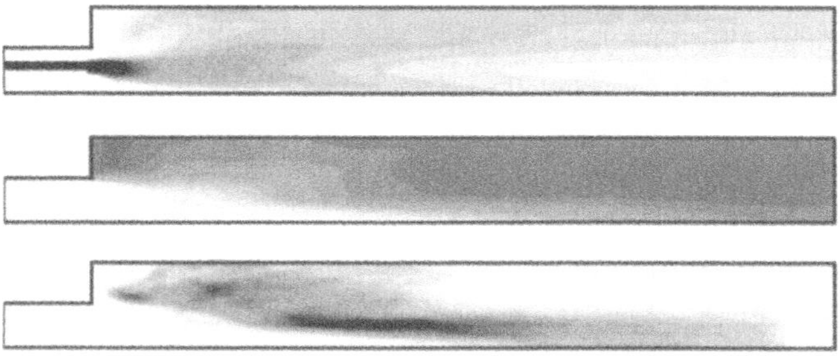

Figure 27.7. From top to bottom, averaged fuel mass fraction, temperature and reaction rate fields for case D (lifted flame).

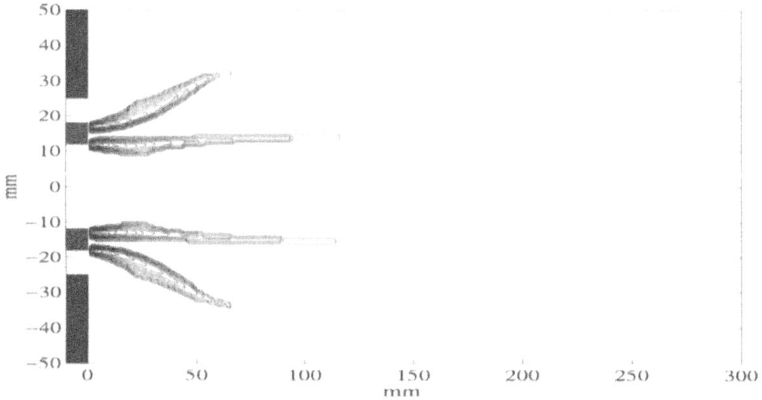

Figure 27.8. Mean reaction rate as predicted using the Magnussen eddy dissipation model (Mean flow simulations).

4. Conclusions

A dynamic thickened flame (DTF) model is developed for large eddy simulations of turbulent reacting flows and is tested against experimental data. The thickening factor F is larger than unity only in reaction zones and diffusion processes without chemical reactions are not affected. Accordingly, the DTF model is expected to be suited to situations where non-premixed, partially premixed and perfectly premixed flames are encountered such as lean premixed prevaporized (LPP) combustors developed for gas turbines.

Numerical results are compared to experimental data obtained in a turbulent propane/air non-premixed burner. This burner exhibits various regimes ("anchored", "lifted" and "extinction") which are recovered in numerical simulations. In the so-called "lifted" flame regime, the numerical simulations show that combustion mainly occurs in rich premixed flames stabilized by the recirculation zone acting as a hot gases tank. This finding seems surprising but is in agreement with a simple physical analysis: when propane and air are mixed by molecular diffusion without combustion, most of the premixing corresponds to rich mixtures (fuel in excess) because the stoichiometric iso-surface is strongly shifted towards pure oxidizer for usual stoichiometric flames. Some diffusion flames are also observed but do not correspond to usual flamelets because cold oxidizer burns with a hot mixture of fuel and combustion products. Accordingly, usual flamelet models are not adapted to correctly predict such lifted flame regimes.

Numerical results are promising but further validations, against precise instantaneous and averaged experimental data, not yet available, are required.

Acknowledgments

J.P. Légier, T. Poinsot and D. Veynante would like to gratefully acknowledge the support provided by Center for Turbulence Research (Stanford University and Nasa-Ames Research Center) during the 2000 CTR Summer Program.

References

Angelberger, C., Veynante, D., Egolfopoulos, F., and Poinsot, T. (1998). Large eddy simulation of combustion instabilities in turbulent premixed flames. In *Proceedings of the Summer Program*, pages 61–82. Center for Turbulence Research.

Boger, M., Veynante, D., Boughanem, H., and Trouvé, A. (1998). Direct numerical simulation analysis of flame surface density concept for large eddy simulation of turbulent premixed combustion. In *Twenty-seventh Symposium (International) on Combustion*, pages 917 – 925. The Combustion Institute, Pittsburgh.

Butler, T. and O'Rourke, P. (1977). A numerical method for two dimensional unsteady reacting flows. In *Sixteenth Symposium (International) on Combustion*, pages 1503 – 1515. The Combustion Institute, Pittsburgh.

Colin, O., Ducros, F., Veynante, D., and Poinsot, T. (2000). A thickened flame model for large eddy simulations of turbulent premixed combustion. *Phys. Fluids A*, 12(7):1843 – 1863.

Colin, O. and Rudgyard, M. (2000). Development of high-order Taylor-Galerkin schemes for LES. *J. Comp. Phys.*, 162:338 – 371.

De Bruyn Kops, S., Riley, J., Kosaly, G., and Cook, A. (1998). Investigation of modeling for non-premixed turnbulent combustion. *Flow, Turbulence and Combustion*, 60(1):105–122.

DesJardin, P. and Frankel, S. (1999). Two-dimensional large eddy simulations of soot formation in the near field of a strongly radiating acetylene-air jet flame. *Combust. Flame*, 119(1):121 – 133.

Ducros, F., Nicoud, F., and Schnfeld, T. (1997). Large eddy simulations of compressible flows on hybrid meshes. In *Eleventh Symposium on Turbulent Shear Flows*, Grenoble, France.

Im, H., Lund, T., and Ferziger, J. (1997). Large eddy simulation of turbulent front propagation with dynamic subgrid models. *Phys. Fluids A*, 9(12):3826 – 3833.

Légier, J., Veynante, D., and Poinsot, T. (2000). Dynamically thickened flame les model for premixed and non-premixed turbulent combustion. In *Proceedings of the Summer Program*. Center for Turbulence Research.

Moin, P., Pierce, C., and Pitsch, H. (2000). Large eddy simulations of a turbulent nonpremixed combustion. In Dopazo, C., editor, *Advances in Turbulence VIII*, pages 385 – 392. CIMNE, Barcelona, Spain.

Piana, J., Ducros, F., and Veynante, D. (1997). Large eddy simulations of turbulent premixed flames based on the g-equation and a flame front wrinkling description. In *Eleventh Symposium on Turbulent Shear Flows*, Grenoble, France.

Pierce, C. and Moin, P. (1998). A dynamic model for subgrid-scale variance and dissipation rate of a conserved scalar. *Phys. Fluids A*, 10(12):3041 – 3044.

Veynante, D. and Poinsot, T. (1997). Large eddy simulation of combustion instabilities in turbulent premixed burners. In *Annual Research Briefs*. Center for Turbulence Research.

Discussion

Brian Helenbrook, Clarkson University "I was interested to know how you feel about adding more realistic chemistry to that model and how that would go? It is one step chemistry now."

Reply: The chemistry in fact is quite simple. This is a one step, Arrhenius scheme as proposed by Egolfopolous. However, in a narrow range of parameters, it recovers the important properties of a real flame, like strain rate and quenching and this is enough for this case.

Brian Helenbrook, Clarkson University "Do you have any plans to try and implement anything a little bit more complex or is that something that you don't think is worth it?"

Reply: We are looking further at more complex chemistry. But in fact it is not so simple. This issue is essential for all turbulent combustion models. LES is not different and there are methods to implement more complex chemical schemes.

Gaby Ciccerelli, Queen's University "In this flamelet region the flame advances not only at the laminar burning velocity but it is affected by curvature straining and stuff like that. Now when you artificially thicken the flame, how do you compensate because now the curvature relative to the flame thickness is completely irrelevant."

Reply: We compensate with the efficiency functions at the subgrid scale level. In fact the subgrid wrinkling of the flame, which is not resolved in the numerical simulation, is accounted for at the subgrid level. In this, there is some curvature effect that I have shown, but there is no direct effect of the curvature on the flame speed, just on the flame surface.

Chapter 28

CHARACTERISATION OF AN AIR-BLAST INJECTION DEVICE WITH FORCED PERIODIC ENTRIES

Fabrice Giuliani *, Pierre Gajan *, Olaf Diers † & Michel Ledoux ‡
Fabrice.Giuliani@onecert.fr, olaf.diers@dlr.de

Abstract This research project is related to the control of combustion instability occurring intermittently in aeroengines. The unsteadiness of the flow is induced by a particular instability called Combustion-Driven Oscillations (CDO).

The experimental set-up at the ONERA Centre de Toulouse was designed in order to simulate the modulation exerted by CDO on the injection process. By pulsating both gas and liquid entries with specific actuators it is possible to reproduce the effect of the acoustic pressure on the resulting spray.

This paper presents the first measurements performed on the free jet under non reacting and atmospheric conditions. The aim is to acquire data on the two-phase flow characteristics (pressure, temperature, fuel particle size and local spray density) for one period of pulsation. These experimental data are used to understand the periodic flow behaviour. They are also required for a further validation of a numerical code.

Keywords: Combustion-Driven Oscillations, air-blast injector, conditioned LDA measurements, phase average processing, second invariant operator.

*ONERA Centre de Toulouse, DMAE, Unité Multiphasique Hétérogène
BP 4025 - 31055 Toulouse Cedex 4, France
†DLR, Institut für Antriebstechnik
Linder Höhe, 51147 Köln, Germany
‡CNRS - Université et INSA de Rouen, CORIA - UMR 6614
Place E. Blondel - 76 821 Mont-Saint-Aignan Cedex, France

A. Pollard and S. Candel (eds.), IUTAM Symposium on Turbulent Mixing and Combustion, 327-336.

1. Introduction

Since they are a major pollutant in the combustion processes of aircraft engine, the thermal production of nitrogen oxides has to be reduced. Due to the requirements of air propulsion systems (essentially the weight to thrust ratio) only primary measures can be undertaken to achieve this goal. The most promising way to suppress the production of thermal NOx is the use of LPP (Lean Premixed Prevaporised) combustion concepts. In comparison to conventional combustors more air from the compressor is needed for the combustion process and less air is available for damping impingement or film cooling air. Regarding the combustion air, local unmixedness in lean mixtures has a high influence on unsteady heat release rates. A non-homogeneous air/fuel distribution is a trigger for Combustion-Driven Oscillations (CDO) [Culick, 2001].

CDO correspond to an aero-acoustic-combustion coupling which appears depending on the local geometric and thermodynamic characteristics of the combustion chamber [Putnam et al., 1986]. The coupling enhances a sudden amplification of physical entities such as unsteady heat release rate, fluctuating pressure and acoustic velocity. If this phenomenon is permanent, or transiently appears close to the walls or moving parts of the system, it may be responsible for fatigue failure.

The most common criterion to describe a combustion instability is the Rayleigh criterion [Rayleigh, 1875], based on the analysis of the phase-shift between acoustic pressure p' and unsteady heat release rate q'. Considering these quantities over a volume V with a period of fluctuation τ, the criterion is defined as follows:

$$R = \frac{1}{\tau} \int_{\tau} \int_{V} p' \; q' \; dt \; dV \tag{28.1}$$

If the integral product R is positive, then the combustion instability is enhanced.

A test rig at the ONERA Centre de Toulouse was designed in order to investigate the driving parameters of thermoacoustic instabilities. It consists in a generic air-blast nozzle that can be supplied with pulsed air and pulsed liquid fuel. The driving exerted by the chamber acoustic pressure on the injection is simulated by forcing a partial fuel flow pulsation and by actuating the air fluctuation.

The incentive of the project is on one hand to observe into detail the mechanisms taking place in an aero-acoustic-combustion coupling, and on the other hand to collect data for the validation of numerical codes including a module with turbulent 2-phase flow combustion. Though the set-up was designed for fuel and air supply actuations (e.g. for the study of a simultaneous modulation of air and liquid on combustion [Haile et al., 1997] applied to an air-blast injection device), only the effect of pulsed air on the spray in free-jet configuration under non-reacting conditions is discussed here.

Measurement methods and data processing techniques are first described. The conditional averaging measurements of the air velocity are analysed in section 3. Emphasis is put on the convection sequence of liquid particles by periodic air entries. The motion of vortex structures is observed by making use of the second invariant λ_2 criterion. The physical parameters influencing the pulsed air jet are then discussed.

2. Basics of the experiment

2.1. Experimental setup

Figure 28.1 gives a global view of the test rig. It consists of two actuation systems. The first located behind the air inlet forces the air supply. The air passes through a sonic nozzle that is partially covered by a rotating cogged wheel. By choosing the rotation speed of the wheel by means of a frequency stabilised motor, a pulsation frequency of the air can be achieved up to 500Hz. A 1.5m long tube is installed between the air actuator and the air-blast nozzle in order to tranquillize the air pulsation in the radial direction and to build out a sinusoidal airflow. The second actuator is a fast oscillating electrovalve for liquid fuels with maximum pulsation frequency of 500Hz. This valve is situated as close as possible to the air-blast nozzle in order to minimise damping effects due to the supply pipe length. The air-blast nozzle is a coaxial swirler where the fuel is injected through a circular orifice (the film thickness to radius ratio is 1/40). The liquid film is pulverized through the mixing layer generated between the inner and the outer swirled air channel.

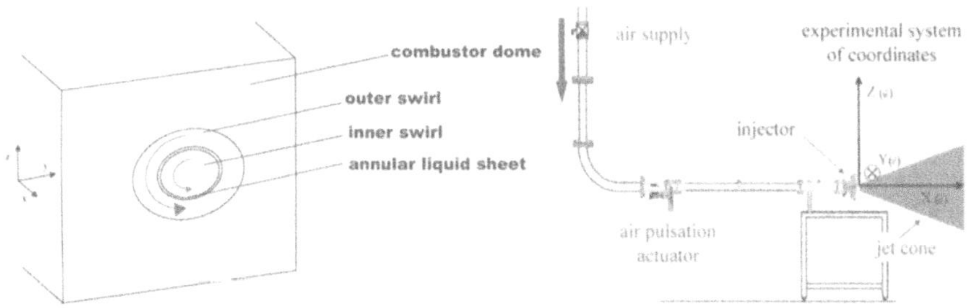

Figure 28.1. Schemes of the test rig. Left: principle of the air-blast atomiser with the injector outlet. Right: global view in free jet configuration

2.2. Measurement techniques

Feasibility tests and preliminary experiments on the two-phase flow behaviour in the pulsed case are carried out with a stroboscopic measurement arrangement, as shown in figure 28.2. With this arrangement the Mie-Scattered light of the liquid phase can be recorded [[Giuliani et al., 2001]].

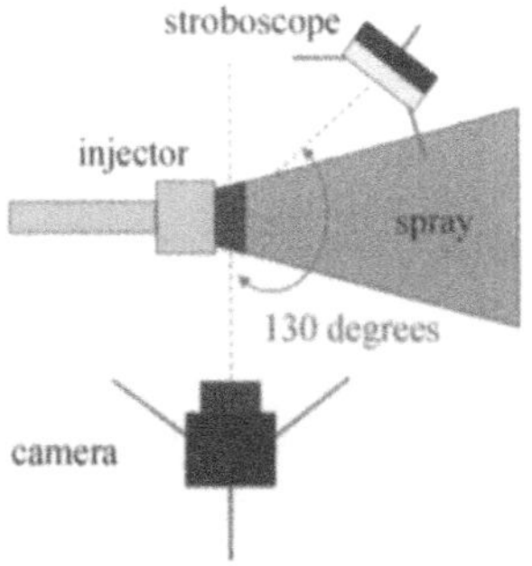

Figure 28.2. Measurement techniques. Left: visualisation setup for 2-phase flow investigation. Down: grids for 3-D conditioned LDA measurements.

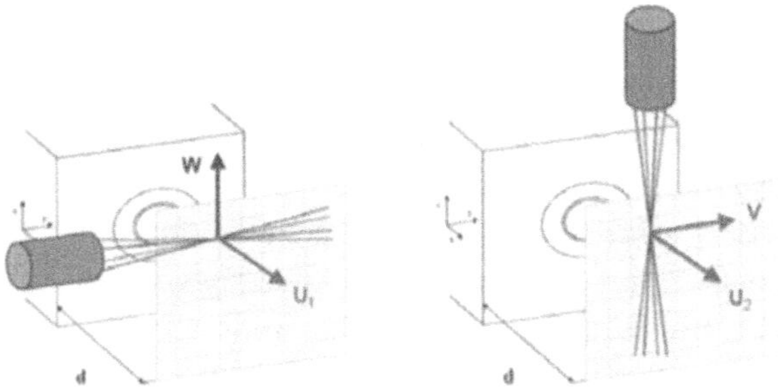

A commercial 2D-LDA system (TSI, IFA655) was conditioned in order to acquire phase averaged information about the gas velocity field. A second set of measurements was performed with the emitter/receptor arrangement rotated by 90° with respect to the first set to measure the third velocity component (figure 28.2). Taking in account the error due to the target positioning and the air supply stability, the 3D velocity measurement uncertainty can be estimated to 3%.

2.3. 2-phase flow visualisations

Figure 28.3 shows a sequence representing the evolution of the spray submitted to air pulsation. The pictures correspond to an integration of the maximum of intensity per pixel over a sequence of 256 phase-locked frames. Each frame was recorded observing the light scattered from a stroboscope synchronized and phase-shifted with respect to the pulse order. The resulting pictures were post-processed with false colours in order to subtract the intensity of light scattered by the mean flow and the background noise. This process reveals the high-density regions of the spray in terms of droplets.

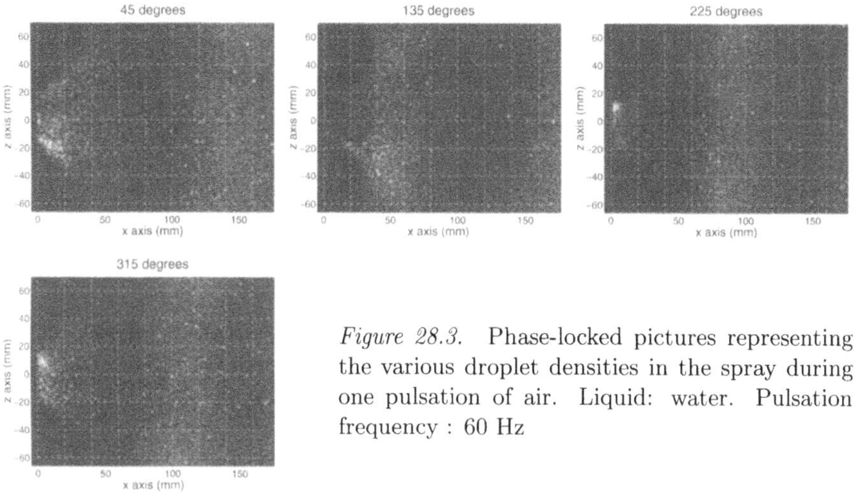

Figure 28.3. Phase-locked pictures representing the various droplet densities in the spray during one pulsation of air. Liquid: water. Pulsation frequency : 60 Hz

One sees the clear effect of convection of dense droplet packets during one cycle of air pulsation. At maximum air flow rate, the dense droplet zone is rapidly transported in the far field, whereas at minimum flow rate the dense droplet zone remains close to the injector nozzle. Therefore, the combustion zone is periodically fed with a certain quantity of fuel.

2.4. Principles of phase-average data processing

An external time information is needed in order to obtain phase averaged measurements. The LDA software receives a synchronisation trigger that resets the events clock counter at the start of each cycle of pulsation. The trigger impulse comes from an optoelectronic sensor coupled to the drive shaft of the rotating cogged wheel.

By post-processing the data, it is possible to reconstruct the cyclic evolution of the velocity with the phase averaging method [Forestier et al., 1998]. The basic principle is to sort the time resolved velocity over one period of pulsation. By subdividing this period into N regular phase intervals, one then computes "classical" statistics for the data in each interval (see figure 28.4).

3. Dynamics of the pulsed air jet

3.1. 3D velocity maps in normal jet sections

LDA phase averaged measurements as described in section §2.4 may be used to reconstruct the flow over the entire measurement grid.

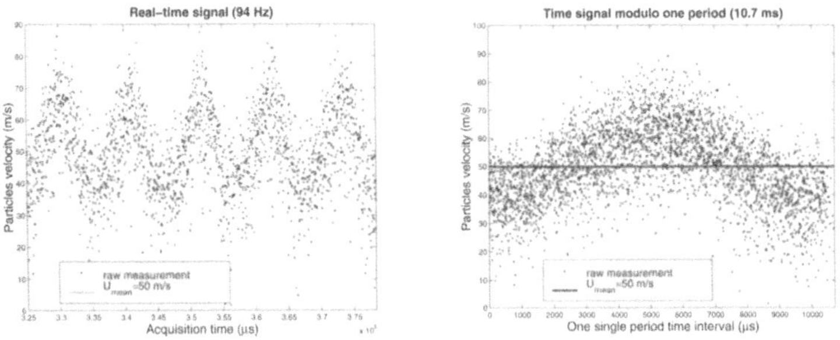

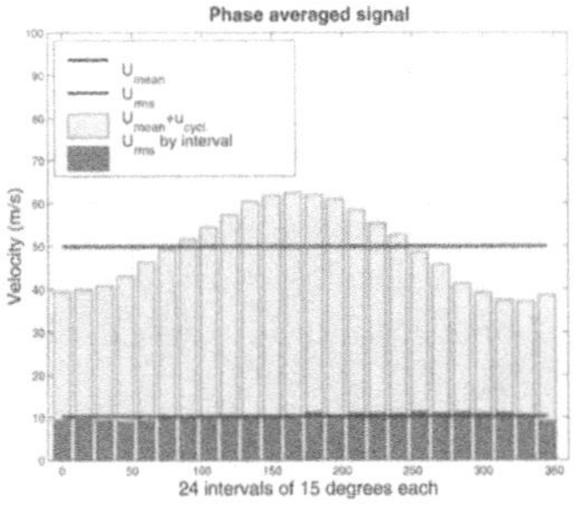

Figure 28.4. Phase average processing technique for LDA measurements

Figure 28.5 shows the evolution of the nondimensional velocity flowfields in a section situated $10mm$ downstream the injector nozzle. The background colour is related to the axial velocity, whereas the arrows show the vectorial composition of radial and tangential velocities. The convection time of the air pulse through the distribution pipe is such that the moment of greatest flow rate appears between 150 and 180 degrees of a cycle.

3.2. Vortical structure analysis

To obtain more information about the motion of the inner recirculation, and also to observe peripheral structures convected along the surface of the spray cone, one may use the λ_2 criterion [Jeong et al., 1995]. This operator (also known as "second invariant") identifies the pressure minima related only to the dynamics of the flowfield, and not to walls effects or mixing layers where vorticity is related to the fluid viscosity.

λ_2 is the second eigenvalue of the tensor issued from the vorticity transport equation. For a stationary and inviscid flow, a negative λ_2 reveals the presence of a pressure minimum, hence the core of a vortical structure. For 2-D flowfield, an incompressible λ_2 is expressed as follows:

$$\lambda_2 = \frac{1}{2}\left(\left(\frac{\partial u}{\partial x}\right)^2 + \left(\frac{\partial v}{\partial y}\right)^2 + 2\frac{\partial u}{\partial y}\frac{\partial v}{\partial x}\right) \tag{28.2}$$

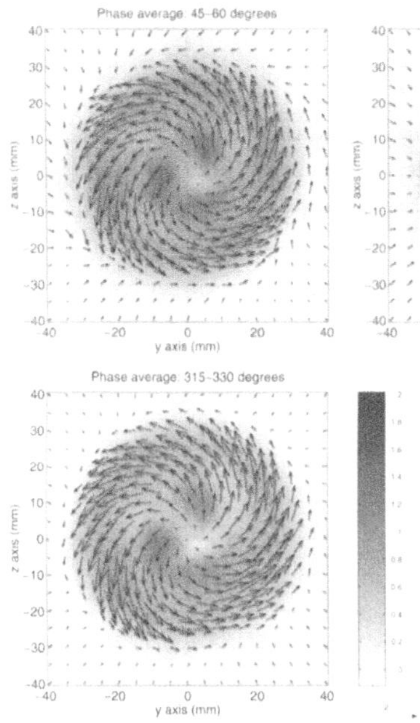

Figure 28.5. Nondimentional velocity maps. Phase averaged LDA measurement performed $10mm$ downstream the injector nozzle. Pulsation frequency: $100Hz$.

The calculation was done over a plane along the jet axis, with pulsed air. The result is shown in figure 28.6.

Inner and peripheral vortical structures appear clearly. The stop-and-go motion of the inner recirculation evolves as follows: at the moment of greatest flow rate, it is positioned far away from the nozzle exit, and the jet cone angle at the outlet is small. At the moment of smallest flow rate, the jet cone is broader while the inner structure gets close to the nozzle tip. This motion is important in the physics of the flashback phenomenon.

The peripheral vortical structures which appear at the corner of the dump outlet at the moment of smallest flow rate are detached at the moment of greatest flow rate and then conveyed at the surface of the jet expansion cone. In the case of a spray, small size droplets can be entrained by these structures, and this would induce a fluctuating supply of the combustion front.

4. Conclusions

In order to investigate the dynamics of an aeroengine fuel spray submitted to combustion oscillation, a test facility was designed and set-up. Data acquisition techniques for flow velocity measurements and spray visualisations were adaptated to high-frequency phenomena and validated.

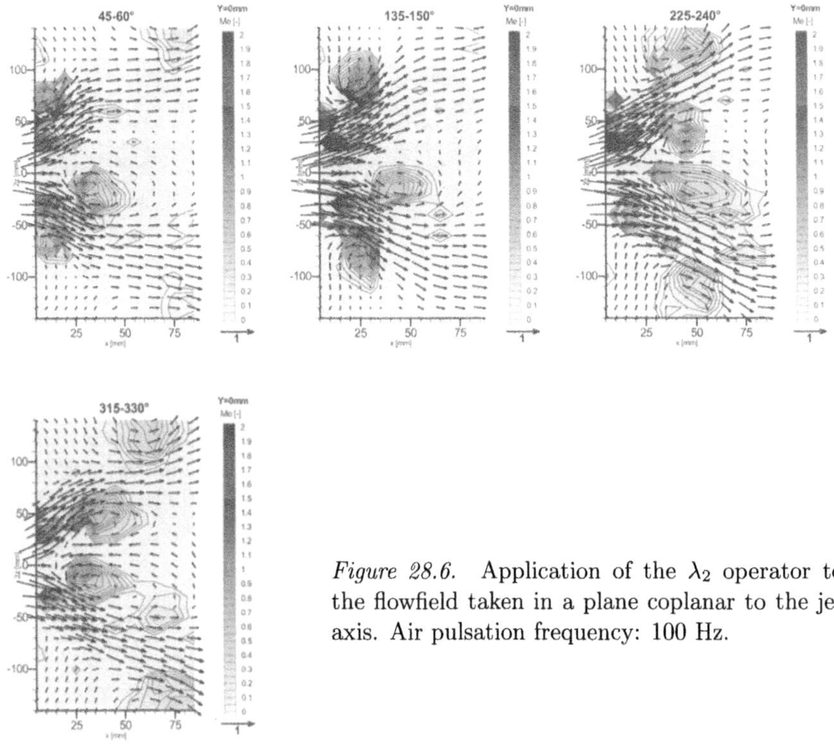

Figure 28.6. Application of the λ_2 operator to the flowfield taken in a plane coplanar to the jet axis. Air pulsation frequency: 100 Hz.

Results show a clear geometrical evolution of the global jet during a pulsation period. The expansion cone envelope deformation will lead to a fluctuating combustion zone. Inner and outer vortex motions will lead to a possible flashback phenomenon, and to a cyclic feeding of the combustion front.

The next step in this work is to observe the transported phase in terms of population density, particle size distribution and velocity. Experiments in combustion will eventually be performed in order to observe the effect of modulated entries on the flame front. Phase-averaged measurements on the flame front location and on the unsteady heat release will be required to close the loop on the aero-acoustic-combustion coupling. A local validation of the Rayleigh criterion with experimental data will therefore be possible.

Acknowledgements

This study is granted by the French General Delegation for Army Funds (DGA) in the frame of an Upstream Research Program named

TITAN. The works are done at the ONERA Centre de Toulouse in collaboration with SNECMA moteurs.

References

Culick F.E.C., *Dynamics of combustion systems: fundamentals, acoustics and control*, RTO AVT - VKI Special Course on Active Control of Engine Dynamics, VKI (Belgium), 14 - 18 May 2001.

Putnam A.A., Belles F.E. & Kentfield J.A.C., *Pulse Combustion*, Prog. Energy Combust. Sci., Vol. 12, pp. 43-79, 1986.

Rayleigh J.W.S., *Theory of Sound*, Vol. 2 (2nd edition). Dover, NYC, 1945.

Haile E., Lacas F. & Candel S., *Contrôle actif de la combustion par injection instationnaire de combustible*, Comptes Rendus de l'Académie des Sciences, Vol. 325, pp. 203-209, 1997.

Giuliani F., Gajan P., Diers O. & Ledoux M., *coulement diphasique gnr par un injecteur arodynamique en rgime pulsatoire forc*, Neuvime Congrs FLUVISU 2001, Rouen (France), June 2001.

Forestier N., Geoffroy P. & Jacquin L., *valuation de moyennes de phase dans un coulement au-dessus d'une cavit rsonante*, Sixime CFVL, Session C, No. 4, Saint-Louis (France), September 1998.

Jeong J. & Hussain F., *On the identification of a vortex*, J. Fluid Mech., Vol. 285, pp. 69-94, 1995.

Discussion

Jeff Lienau, Pratt and Whitney "Very well conceived experiment thank you for showing that. My question is, do you have any plan; this is a very well conceived what I would call a component experiment, and this kind of thing is really a system type of a problem. I mean the interaction between the compressor, the combuster and the turbine really the interaction of the three do you have any idea or method of linking these effects together for your future work. You didn't mention it on the slide I was just wondering if you do?"

Reply: First some words on the context. It is work related to SNECMA Motor. Of course it is really particular to a particular geometry. It is not a general experiment; it is really for this kind of air-blast, which is looking like an existing air-blast device. SNECMA must be provided with a big data set. Our aim is to develop also measurement techniques able to follow fluctuating quantities as I showed. So there I showed the velocity, next we will observe how the granularity (? garbled word on tape) is evolving in a variance (? garbled word on tape). So there is also a time measurement system and there will be also the visualizations of the flame front still time dependent. So finally we dont get close, real close to an existing, what exists really inside a combustion chamber, this is not the aim. The aim is to provide a data set big enough to validate later on some codes, two phase flow codes with combustion.

Also we tried to be quite close to something happening really so that they have got an idea of what happens inside a combustion chamber.

Emmanuel Villermaux, Marseille "You have indeed devised a very neat experimental set up to address important questions. However, I have a question about the principle of your experiments. You have the ability of modulating the flow rate of the air, both the flow rate of the air and of the liquid it seems clear that modulating the flow rate and the liquid will result in a spatial modulation of the density of the droplets. On the other hand, modulating the air velocity will most likely result in modulation, not on the population density of the droplets but rather on their size. This is my apriori feeling. It is true, first? My second question, in your view will these two extreme situations have the same, will they have the same consequence with respect to the Rayleigh criterion?"

Reply: To your first question, we observed that the air position is very important in the transport of the droplets so the position of air we create a denser annualize zone of droplets that are convected through the zone. Condensation of liquid is very important at the boundary layer or the boundary conditions of the nozzle. Because, as I showed, you have an all or nothing liquid flow. Later on due to the turbulence, it is not so evident to see that there is a situation of droplets we do some measurements and we did not see actually something related in terms fluctuation in size and in population. One, only the liquid is dispersed. With the average dispersion - we saw something.

Emmanuel Villermaux, Marseille "In the population density?"

Reply: Yes.

Emmanuel Villermaux, Marseille "More than in the droplet sizes?"

Reply: Also in the tube. The signal was stronger when there was a dense spray passing into the measurement. To answer the second question, I think that yeah it would have the same effect on the Rayleigh criteria. I cannot answer straight away but this is my feeling. But now we will have to observe when we are working in combustion what is happening actually because there we have to possibility to modulate and to make a delay in the position of liquid and the position of air.

Paul Dimotakis, Caltech "I wasn't sure whether the velocity measurements relied on the droplets themselves of it there was other - to go along with them. If not then the measurements are then in some way conditional on what the droplets are doing and therefore as they are modulated or bias if you will on the measurements is going to depend on the operating conditions. I was wondering if you will comment on that."

Reply: I don't know if I forgot to say all the measurements in the air were done with only air, it was not a two phased flow. It was just the air in order to get global information about the air flow rate. Now we will have of course to see the interaction between the droplets and the air transportation. I think that so far this is not a big trouble. We can see from the pictures, just from visualization we can see how they are conveyed through the jet.

Chapter 29

NUMERICAL ANALYSIS OF HYDROGEN/AIR JET DIFFUSION FLAME

Yasuhiro Mizobuchi
Shigeru Tachibana
Junji Shinjo
Satoru Ogawa
and Ryoji Takaki
CFD Research Center
National Aerospace Laboratory of Japan
mizo@nal.go.jp, tachiban@nal.go.jp, shinjou@nal.go.jp, og@nal.go.jp, ryo@nal.go.jp

Abstract The structure and stabilizing mechanism of a subsonic hydrogen/air jet lifted flame are numerically investigated using DNS. The lifted flame consists of an inner premixed flame and outer diffusion flames. The flame base is premixed, which sustains the vigorously turbulent inner premixed flame. The outer diffusion flame consist of several island flames, which are produced from the inner premixed flame. The stabilization at the flame base is maintained by the balance between the axial velocity and the laminar flame velocity. Deviation of heat release layer from hydrogen consumption is observed in the inner premixed flame.

Keywords: Lifted flame, turbulence, DNS, flame structure, flame index, flame stabilization

1. Introduction

Lifted flame is one of the most important and interesting flame configurations from the viewpoints of fundamental and practical research. The structure of the lifted flame has been investigated from the viewpoints of the flamelet extinction [Peters and Williams, 1983] and the triple flame structure [Kiori et al., 1993, Vervisch and Poinsot, 1998],

A. Pollard and S. Candel (eds.), IUTAM Symposium on Turbulent Mixing and Combustion, 337-345.

yet the details of the structure and the stabilization mechanism of the three-dimensional turbulent lifted flames have not been revealed.

The authors have simulated a hydrogen/air turbulent flame by DNS and have succeeded to capture the lifted flame. The computation has been conducted with more than 22 million grid points using the vector parallel computer Numerical Wind Tunnel (NWT) at National Aerospace Laboratory (NAL) of Japan. In this paper, the structure and the stabilization of the lifted flame are investigated mainly from the short-term observation of the unsteady flame behavior.

2. Flame configurations

The flame configurations conform to the experiment by Cheng et al. [Cheng et al., 1992]. A hydrogen jet is injected into still air from a round nozzle whose diameter D is 2mm. The jet velocity v_{jet} is 680 m/sec, the Mach number is 0.54 and the Reynolds number based on the diameter is 13600. A lifted flame was observed in the experiment and the lift-off height was 14mm.

3. Computational model

The 9-species (H_2, O_2, OH, H_2O, H, O, H_2O_2, HO_2, N_2) and 17-reaction model by Westbrook [Westbrook, 1982] is employed. The air is assumed to be composed of 22% O_2 and 78% N_2 in volume. The diffusion of chemical species is evaluated using Fick's law with binary diffusion coefficients. The transport coefficients of each chemical species are estimated using the Lennard-Jones intermolecular potential model [Chapman and Cowling, 1970]. The enthalpy of each chemical species is derived from the JANAF thermochemical table [JANAF, 1965]. Gravitational effects are ignored because the jet speed is high enough. No turbulence model is used.

4. Computational method

The governing equations are the compressible three-dimensional Navier-Stokes equations coupled with the conservation equations of chemical species. The equation of total mass conservation is solved additionally.

The governing equations are discretized in a finite-volume formulation and convective terms are evaluated using a TVD numerical flux based on Roe's approximate Riemann solver [Roe, 1981, Wada et al., 1989]. The higher-order flux is constructed extrapolating the characteristics using two types of flux limiters [Wada, 1995]. The accuracy of this flux is third-order in smooth regions and second-order around regions where the sign of characteristics gradient changes. The viscous terms are evaluated with standard second-order difference formulae. The diffusion fluxes are evaluated by Fick's law with binary diffusion coefficients and, therefore the diffusion fluxes are modified so that the total mass is conserved [Mizobuchi and Ogawa, 2000]. The time integration method is the explicit Runge-Kutta multi-stage method. The second order time integration is used.

The surfaces of the nozzle tube are assumed to be slip walls. At the jet exit the axial velocity is imposed. The total pressure and the total temperature are fixed to the values which realize a 1/7 power law boundary layer when the exit pressure is the atmospheric pressure. No artificial disturbance is imposed. At the outer boundaries the non-reflection condition [Thompson, 1987] is applied. After the cold flowfield is established, heat is added for ignition.

The grid system is rectangular. The y-direction is the jet axial direction, x- and z-directions are normal to the y-direction and the origin is the jet exit center. The computational region is $-15D \leq x, z \leq 15D$ and $-3D \leq y \leq 20D$. The grid spacing is 0.05mm in $-1.25D \leq x, z \leq 1.25D$, $0 \leq y \leq 8D$. This size is 2.5 times as large as the Kolmogorov scale measured in the experiment around the ignition point. The grid spacing is coarser with increasing distance from the above mentioned region. About 22.8 million grid points are used.

5. Results and discussion

5.1. Structure of the lifted flame

A lifted flame is obtained in the numerical simulation in the same way as in the experiment. The temperature distribution in a cutting plane including the nozzle center is shown in fig.29.1. The lift-off height in simulation is about 5.5 diameters which is slightly shorter than the experimental one.

The flame structure is analyzed using the flame index (F.I.), which is proposed by Takeno [Yamashita et al., 1996]. The definition of F.I. is,

$$\text{F.I.} = \nabla Y_{\text{H}_2} \cdot \nabla Y_{\text{O}_2}, \tag{29.1}$$

where Y_s is the mass fraction of chemical species s. The flame is premixed when F.I. is positive and diffusion when F.I. is negative. Figure 29.2 presents the normalized F.I. distribution in a cutting plane. The positive F.I. is normalized by the F.I. of the laminar premixed flame at corresponding mixture fraction, and its iso-level contours are drawn with solid lines from 0.4 to 10.0 by 0.4. The negative F.I. is normalized by the F.I. at the extinction limit of the counter diffusion flame and its contours are drawn with dashed lines from -0.02 to -0.1 by 0.02. F.I. is set to zero where the temperature is less than 600K for better observation. The stoichiometric lines are drawn with thick lines.

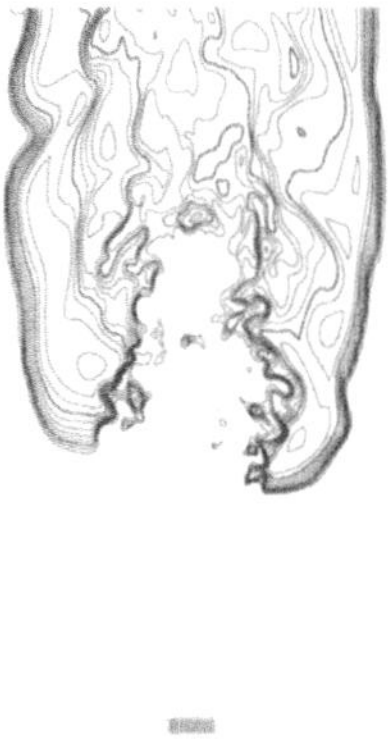

Figure 29.1. Instantaneous distribution of temperature.

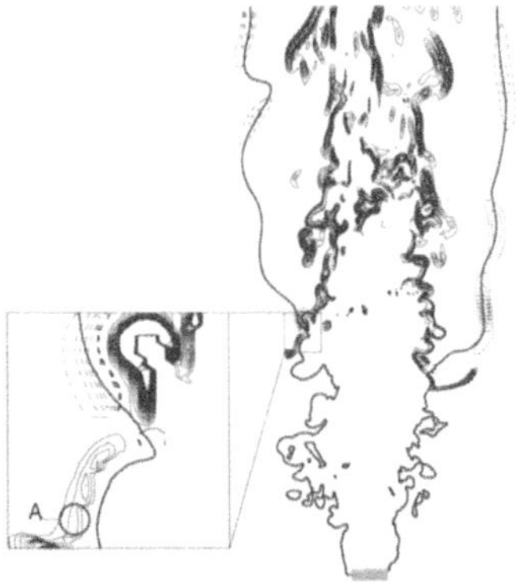

Figure 29.2. Instantaneous distribution of normalized flame index.

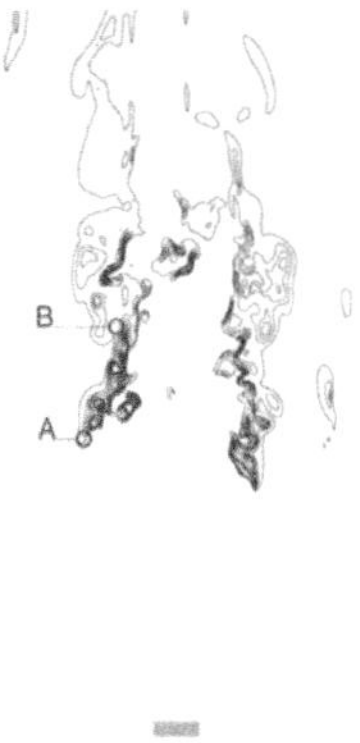

Figure 29.3. Instantaneous distribution of heat release rate.

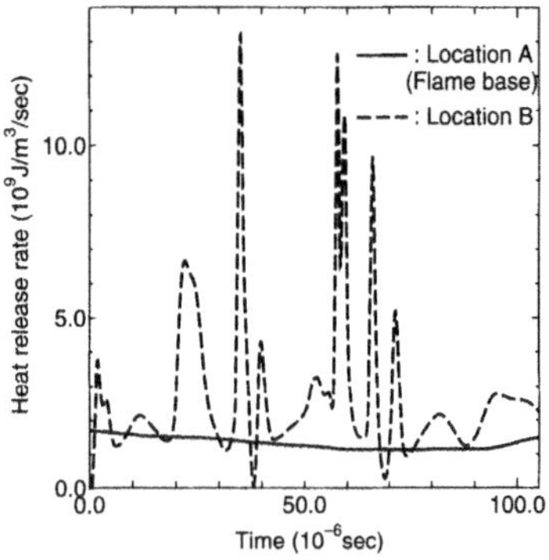

Figure 29.4. Time histories of heat release rate at locations A and B in fig.29.3.

The F.I. distribution shows that the lifted flame consists mainly of the inner fuel-rich turbulent premixed flames and the outer diffusion flames. The outer diffusion flames develop along the stoichiometric lines. An enlarged view around the flame base is also shown, where the location A is the most stable location in the flame during the observation period. The flame is a little bit fuel-lean premixed around the location A and the normalized F.I. is about unity, which indicates that the flame base is stable enough to maintain combustion. The base flame looks like a triple-flame, but more careful investigations by three-dimensional and long-term observations are needed to determine the base flame structure.

Figures 29.3 shows the instantaneous distribution of heat release rate. The main heat release sources are the flame bases and the inner premixed flames. The time histories of the heat release rate are shown in fig.29.4 at the flame base (location A in fig.29.3) and at a position in the inner premixed flame (location B). The flame base is rather stable while the succeeding premixed flame layers are vigorously turbulent. This turbulent behavior of the inner premixed flame originates from the strong instability of the hydrogen jet, which will be shown later. The inner turbulent premixed flames are sustained by the stable base flames.

The outer diffusion flames are not stabilized at the fixed positions and consist of several island flames slowly shed downstream. The time history of heat release rate around the right hand side flame base in fig.29.3 is shown in fig.29.5. The flame islands are produced by the break-off or the local extinction of the inner premixed flames near the flame base.

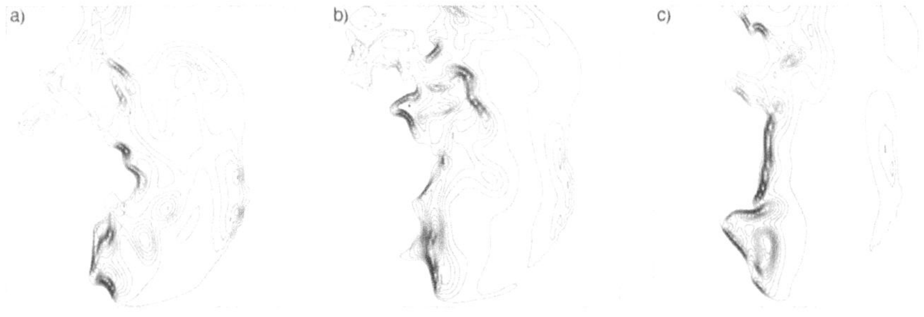

Figure 29.5. Production of flame islands. Heat release rate iso-lines are drawn around the right flame base in fig.29.3 at sequential time stages, Δt=0.03msec.

5.2. Stabilization mechanism of the lifted flame

The entrainment of air into the burnt region is one of the dominant phenomena for the stabilization of the lifted flame. The iso-surface of hydrogen mole fraction at 60% at some moment is shown in fig.29.6, where the burnt region (temperature is higher than 600K) is painted dark. The surface is strongly three-dimensional and turbulent, and has a very small scale structure especially in the unburnt region. The fineness decreases in the burnt region due to the increase of diffusivity caused by the temperature rise. Figure 29.7 shows the fluctuations of hydrogen mole fraction in the unburnt mixing region at 4.7 diameters downstream from the nozzle exit. This strong instability is the Kelvin-Helmholtz instability due to the very high speed and the very low density of the hydrogen jet. The fluctuations have quite high frequencies as shown in fig.29.8 and the dominant frequencies correspond to the length scales smaller than 1/5 diameters. The large scale entrainment, which is often observed in two-dimensional computations, is not observed. The entrainment of air into the burnt region is mostly provided by three-dimensional small scale turbulence.

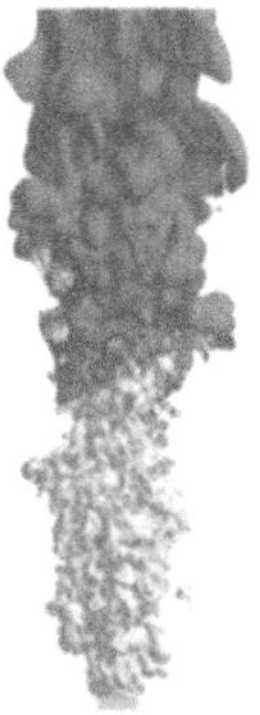

Figure 29.6. Instantaneous iso-surface of hydrogen mole fraction at 60%. The burnt region is painted dark.

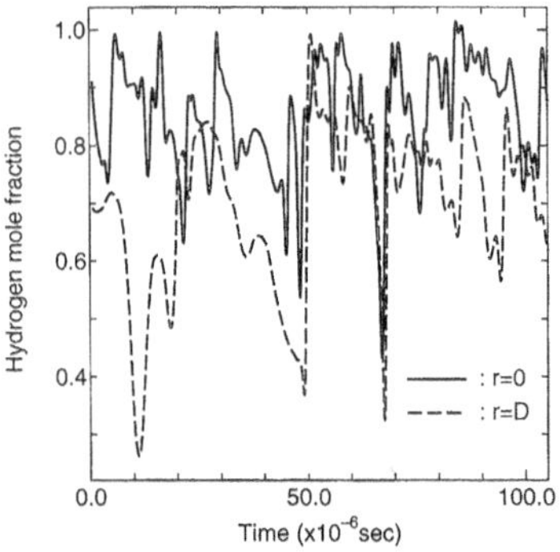

Figure 29.7. Time histories of hydrogen mole fraction in unburnt mixing region.

The lifted flame appears to be stable during the short-term observation period, although the inner premixed flames are vigorously turbulent. The time histories of the axial velocity at the locations A and B in fig.29.3 are shown in fig.29.9. The velocity in the inner premixed flame and its fluctuation are large. On the other hand, at the flame base, the axial velocity is quite small, the fluctuation is rather small and the velocity sometimes becomes negative. The averaged axial velocity is about 2.9m/sec at the location A. The laminar flame velocity near stoichiometric condition is calculated to be about 2.4m/sec. The estimation is quite rough, but it can be concluded that the axial flow velocity and the laminar flame velocity balance at the base flame.

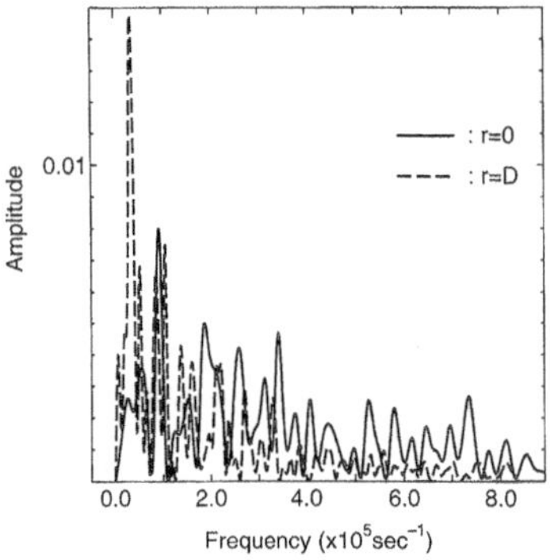

Figure 29.8. Power spectrum of hydrogen jet fluctuation in unburnt mixing region.

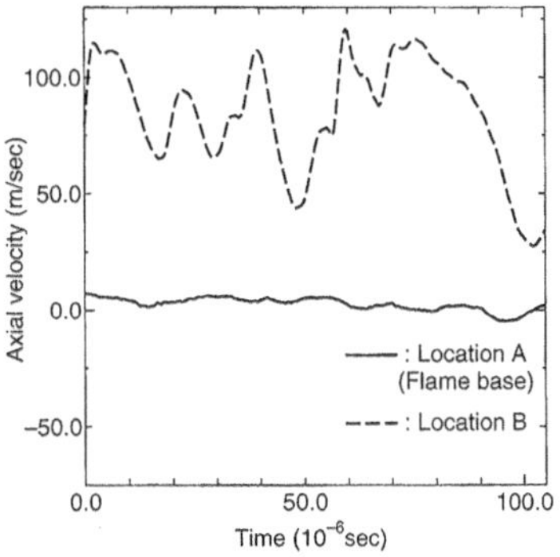

Figure 29.9. Time histories of axial velocity at the locations A and B in fig29.3.

5.3. Deviation of heat release layer from hydrogen consumption layer

Figures 29.10a and 29.11b are enlarged views of hydrogen consumption layers and heat release layers, respectively, in the inner premixed flames at the same time step. The hydrogen consumption layers are rather thin and continuous. However, the heat release layers are rather thick and disrupted. The heat release layer deviates from the fuel consumption layer. In this simulation, relatively detailed kinetics are employed. The major reactions of the hydrogen consumption and the heat release are different, and therefore there can be a displacement between the two layers. The displacement is not uniform. This is due to the vigorous turbulence that originates from the high frequency instability of the hydrogen jet. The reaction layers are disturbed by the turbulence. This observation raises questions to the validity of the laminar flamelet concept to describe the vigorously turbulent premixed flames in the simulated lifted flame.

Figure 29.10a. Heat release rate distribution in the inner premixed flames.

Figure 29.11b. Hydrogen consumption rate distribution in the inner premixed flames.

6. Conclusions

A hydrogen/air turbulent jet lifted flame is simulated by DNS and the following aspects of the lifted flame are revealed.

(1) The lifted flame consists of the inner turbulent premixed flames and the outer diffusion flames. The bases of the lifted flame is rather gentle premixed flames which sustain the inner vigorous turbulent premixed flames. The inner premixed flames and the base flames are the main heat release sources. The outer diffusion flames consist of several island flames shed downstream which are produced from the inner premixed flames.

(2) The entrainment of air into the burnt region is mostly provided by three-dimensional small scale turbulence. The stabilization of the lifted flame is maintained by the balance between the axial flow velocity and the laminar flame velocity at the flame base.

(3) In the inner turbulent premixed flames, the heat release layers deviate from the hydrogen consumption layers, which leads to a question to the validity of the laminar flamelet concept in the vigorous turbulent premixed flames of the lifted flame.

In this paper, the analyses are mainly based on the short time scale and two-dimensional observations. For the future detailed investigations, long time scale and three-dimensional analyses are indispensable.

Acknowledgments

The authors would like to thank Professor Tadao Takeno of Meijou university for his valuable suggestions on the aspect of this work.

This research was carried out as a research activity at the Center for Smart Control of Turbulence funded by MEXT of Japan.

References

Chapman, S. and Cowling, T. G. (1970). *The Mathematical Theory of Non-Uniform Gases.* Cambridge University Press.

Cheng, T. S., Wehrmeyer, J. A., and Pitz, R. W. (1992). Simultaneous temperature and multispecies measurement in a lifted hydrogen diffusion flame. *Combustion and Flame*, 91:323–345.

JANAF (1965). *JANAF Thermochemical Tables.*

Mizobuchi, Y. and Ogawa, S. (2000). Numerical analysis of fractal feature of hydrogen-air jet flame. *AIAA paper 2000-0184.*

Kiori, P. N., Rogg, B., Bray, K.N.C. and Liñán, A. (1993). Flame spead in laminar mixing layers: The triple flame. *Combusiton and Flame*, 95:276–290.

Peters, N. and Williams, F. A. (1983). Lift-off characteristics of turbulent jet diffusion flames. *AIAA Journal*, 21(3):423–429.

Roe, P. (1981). Approximate riemann solvers, parameter vectors, and difference scheme. *Journal of Computational Physics*, 43:357–372.

Thompson, K. W. (1987). Time dependent boundary conditions for hyperbolic systems. *Journal of Computational Physics*, 68:1–24.

Vervisch, L. and Poinsot, T. (1998). Direct numerical simulation of non-premixed turbulent flames. *Annual Review of Fluid Mechanics*, 30:655–691.

Wada, Y. (1995). *Numerical Simulation of High-Temperature Gas Flows by Diagonalization of Gasdynamic Matrices.* PhD thesis, the University of Tokyo.

Wada, Y., Ogawa, S., and Ishiguro, T. (1989). A generalized roe's approximate riemann solver for chemically reacting flows. *AIAA paper 89-0202.*

Westbrook, C. K. (1982). Hydrogen oxidation kinetics in gaseous detonations. *Combustion Science and Technology*, 29:67–81.

Yamashita, H., Shimada, M., and Takeno, T. (1996). A numerical study on flame stability at the transition point of jet diffusion flames. In *Proceedings of Twenty-Sixth Symposium (International) on Combustion*, pages 27–34, Pittsburgh.

Discussion

Parviz Moin, Stanford University "You mention at the beginning that you do not agree with the experiment on the height of the lifted flame. What is the reason for that do you think?"

Reply: The reason for discrepancy - I think that there are some reasons. The burst size is small enough around the ignition point but it is not small enough around the nozzle exit.

Parviz Moin, Stanford University "No, no the grid size. What is the typical diameter of the nozzle?"

Reply: 2 mm

Parviz Moin, Stanford University "And the typical velocity was?"

Reply: 680 m/s

Thierry Poinsot, CERFACS, Toulouse "Did you check the effect of the velocity profile in the tube of hydrogen and do you modulate it or is it constant? Do you inject turbulence or ... ?"

Reply: No I didn't check that so the profile one seventh power law profile

Thierry Poinsot, CERFACS, Toulouse "It's constant..So you don't inject...No modulation, nothing ..."

Reply: I fixed it total temperature, total pressure so depending on the pressure at the nozzle exit the speed changes.

Chapter 30

TURBULENT MIXING WITH SPRAYS

Rafik BEN DAKHLIA, Julien de CHARENTENAY
ONERA, BP 72 - 29 Av de la Division Leclerc,
92322 Chatillon Cedex, FRANCE
and, Laboratoire EM2C, Ecole Centrale Paris and CNRS,
Grande Voie des Vignes, 92295 Châtenay-Malabry, FRANCE
r.bendakhlia@onera.fr, marescha@onera.fr

Nasser DARABIHA
Laboratoire EM2C, Ecole Centrale Paris and CNRS,
Grande Voie des Vignes, 92295 Châtenay-Malabry, FRANCE
darabiha@em2c.ecp.fr

Vincent GIOVANGIGLI
Centre de Mathématiques Appliquées,
CNRS Ecole Polytechnique,
91128 Palaiseau Cedex, FRANCE
giovangi@cmapx.polytechnique.fr

Abstract

In this paper, we study turbulent mixing with sprays using a low Mach number DNS code coupled with a Lagrangian solver. We investigate the impact of turbulence on the spray dispersion, and its consequence on the global vaporization rates of the whole spray. In particular, we show that in two-dimensions the droplets are convected in low vorticity regions, which delays the vaporization process.

Keywords: Spray, Droplets, DNS, Turbulent Mixing

A. Pollard and S. Candel (eds.), IUTAM Symposium on Turbulent Mixing and Combustion, 347-355.

1. Introduction

Two-phase combustion appears in many practical devices, such as propulsion systems in which fuel is injected in its liquid phase. Liquid-fuel injection has a strong influence on the system performance and the combustion quality. There are many complex phenomena involved in these flows, such as sprays, atomization, droplet vaporization, combustion, and turbulence. This motivates further studies of multiphase and turbulent flows.

During the last decade, numerous Direct Numerical Simulation (DNS) of gaseous flames have been performed but the influence of liquid phase in such simulations was taken into account only very recently [Reveillon and Vervish, 2000]. More generally, recent advances in gaseous flames DNS and combustion diagrams should be extended to the situation of multiphase flows. As a step in this direction, we numerically investigate, in this paper, sprays evaporating in a turbulent mixing.

2. Mathematical formulation

The two phase flow equations are derived in a kinetic framework, in which the droplets' distribution function is discretized as the sum:

$$f = \sum_{l\in[1,N]} N_{p,l}\delta\big(\boldsymbol{x} - \boldsymbol{x}_{p,l}(t)\big)\delta\big(r_p - r_{p,l}(t)\big)\delta\big(\boldsymbol{v}_p - \boldsymbol{v}_{p,l}(t)\big)\delta\big(T_p - T_{p,l}(t)\big), \tag{30.1}$$

where N is the total droplet number, $\boldsymbol{x}_{p,l}$, $r_{p,l}$, $\boldsymbol{v}_{p,l}$, $T_{p,l}$, denote the droplet position, radius, velocity, temperature respectively and where the constant $N_{p,l}$ is the droplet number located at $\boldsymbol{x}_{p,l}$. The droplet equations are obtained by taking moments of the Boltzmann-type spray equation [Williams, 1985, Williams, 1958, Ben Dakhlia and Giovangigli, 2000]. The gas phase equations are coupled to the droplets equations through various integral sources terms. The coupling terms involved in the droplet model can be found for instance in Williams [Williams, 1985]. We have also taken into account interactions between the droplets by adding a simplified coalescence model.

Since a kinetic description is used for the spray [Williams, 1985], we note that the droplets are punctual so that, in particular, the flows around and inside the droplets are not solved. Therefore, our calculations are not model free and cannot be termed DNS *stricto sensu*. However, a full DNS taking into account droplet internal flow, transient evaporating kinetics and details of droplets surface deformations is today cost prohibitive.

3. Numerical techniques

We simulate numerically turbulent mixing with sprays by coupling a Lagrangien solver with a low-Mach number DNS code. Spatial derivatives are computed with sixth-order central finite-difference scheme. Time integration is performed using a fourth-order Runge-Kutta formulation. The low-Mach approximation leads to a two-stage integration of the momentum equation with a pressure solver using Poisson's equation.

In the code, the numerical tools that solve the problem equations and the thermochemistry and transport properties librairies are clearly separated. For the thermodynamics properties and chemical source terms, we used a vectorized version of the CHEMKIN II library [Kee et al., 1989,Giovangigli and Darabiha, 1988]. For the transport properties, such as multicomponent diffusion coefficients, thermal conductivity and shear viscosity, the vectorized EGLIB library was used [Giovangigli, 1999].

4. Turbulent mixing with sprays

4.1. Configuration

We simulate turbulent mixing with sprays in a two dimensional periodic domain. The mesh is Cartesian and uniform with 101×101 nodes representing a physical domain of $5\text{mm}\times5\text{mm}$. The initial gaseous field is hot air at a temperature T = 1000K in which methanol droplets, at a temperature T_d = 300K, are randomly distributed. We initialized a turbulent field using the Von Karman spectrum with a Pao correction for small scales. The turbulent parameters are the integral length $\lambda = 6.2\ 10^{-2}$ cm, that corresponds to the most energetic eddies, and the rms velocity is u' = 488 cm s^{-1}. Starting from this initial gas conditions, the droplets evaporate and interact with the gas while the turbulence decays.

4.2. Results

Figure 30.1 represents vorticity iso-lines with droplets positions and velocity vectors at different eddy turnover time τ. This figure clearly shows the influence of vorticity on the droplets' dispertion. In particular, the droplets are convected in the low vorticity region by the gas. Thus, one can notice some coalescence phenomena. This is confirmed by analysing figure 30.2(a) in which only few droplets are located in the vortices. Finally, in figure 30.2(b), almost the whole spray has vaporized. Note that in figure 30.1 and 30.2 and , the size of the scatter representing the droplets has been arbitrarily rescaled and multiplied by a factor of 1000 in order to visualize where the coalescence appears. Considering this, it is clear that even starting from a monodisperse distribution, we rapidly are in presence of a polydisperse spray.

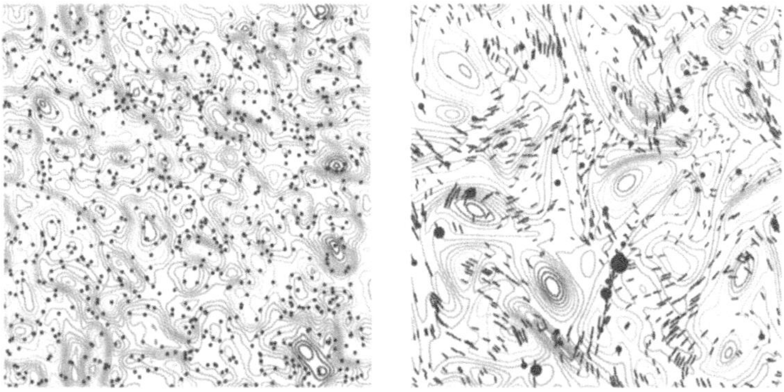

(a) t = 0 τ (b) t = 1 τ

Figure 30.1. Spray dispersion at different eddy turnover time τ : vorticity iso-lines with droplets positions and velocity vectors.

(a) t = 2 τ (b) t = 3 τ

Figure 30.2. Spray dispersion at different eddy turnover time τ : vorticity iso-lines with droplets positions and velocity vectors.

On figure 30.3 and 30.4 we plot the droplets' distributions. Starting from a monodisperse spray of 5 μm (fig. 30.3(a)), the distribution shape is modified (fig. 30.3(b)) due to multiple coalescences and due to the vaporization process. The whole histogram is slightly modified by coalescence but we clearly see (fig. 30.4(a)) that the spray behavior is driven by the vaporization process. The last histogram (fig. 30.4(b)), can be corelated with figure 30.2(b), in which only the coalesced droplets still evolves.

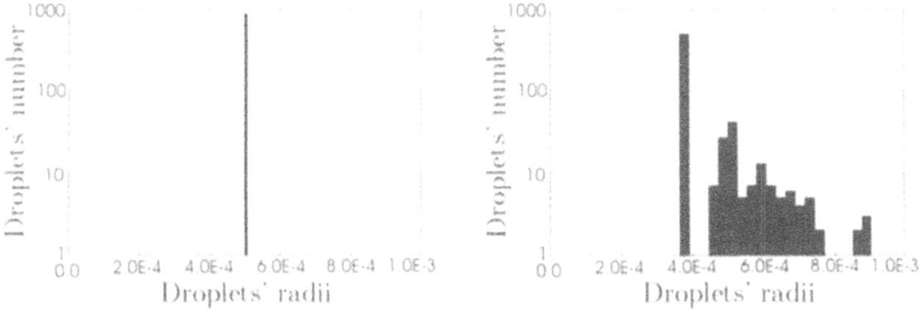

(a) t = 0τ

(b) t = 1 τ

Figure 30.3. Droplets' distribution at different eddy turnover time τ

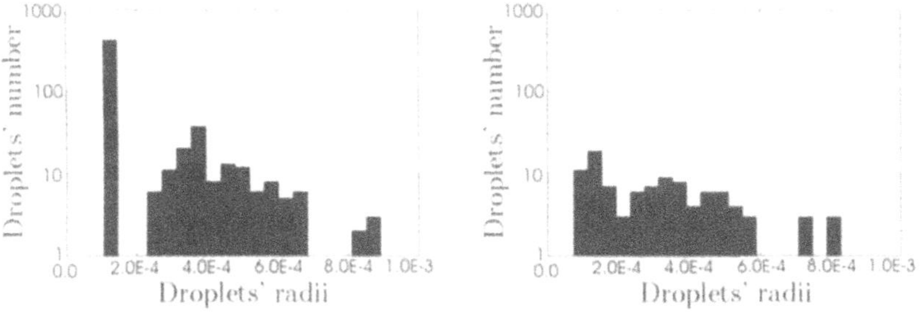

(a) t = 2 τ

(b) t = 3 τ

Figure 30.4. Droplets' distribution at different eddy turnover time τ

On figure 30.5 we have plotted the square of the droplets mean radius as a function of time. We have evaluated this radius by considering the whole mass of liquid composing the spray. On this figure, the dashed line represents a calculation of the spray in a quiescent atmosphere. One can note that the results correspond to a d^2 law. The solid line represents the squared mean radius evolution in a turbulent flow field. In this situation, the the squared mean radius behavior is different compared to the d^2 law, and the global vaporization process of the whole spray is delayed. This is due to the fact the gaseous mixing evolves during time. In particular, since the droplets are convected in the low vorticity regions, the fuel gaseous mass fraction appears between the vorticies (fig. 30.6), changing the mixing around the droplets. Thus, the atmosphere surrounding the droplets is more humid, reducing the vaporization rates. Note that calculations were conducted without any coalescence model, giving similar results.

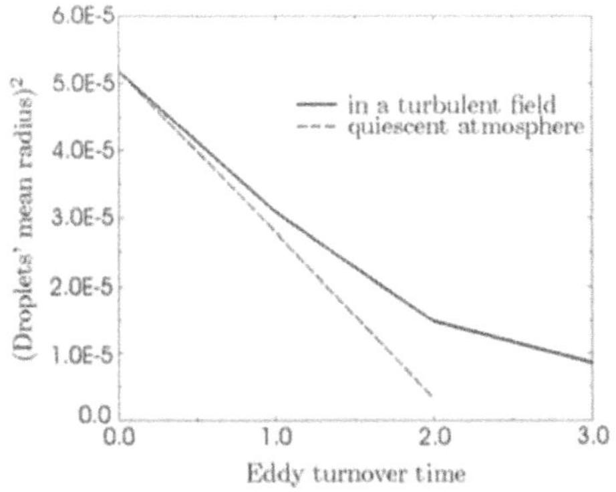

Figure 30.5. Droplets' mean radius as a function of τ

Figure 30.6. Fuel mass fraction at t = 3 τ

5. Conclusions

In this paper we have investigated turbulent mixing with sprays. In particular, we have shown the impact of turbulence on the spray distribution, and how it delays the global vaporization process of the whole spray. Futur work could include turbulent spray flame calculations.

References

Ben Dakhlia, R. and Giovangigli, V. (2000). Multiradii modeling of counterflow spray diffusion flames. In *Proc. Combust. Inst.* 28 : 1039-1045.

Giovangigli, V. (1999). *Multicomponent Flow Modeling.* Birkhäuser Boston.

Giovangigli, V. and Darabiha, N. (1988). *Vector Computers and Complex Chemistry Combustion.* Mathematical Modeling in Combustion and Related Topics, C. Brauner and C. Schmidt-Laine Eds., M. Nijhoff Pub., NATO ASI Series 140, 491-503.

Kee, R., Rupley, F., and Miller, J. (1989). Chemkin-ii : A fortran chemical kinetics package for the analysis of gas-phase chemical kinetics sand89-8009.

Reveillon, J. and Vervish, L. (2000). Spray vaporization in nonpremixed turbulent combustion modeling: A single droplet model. *Combustion and Flame*, 121:75–90.

Williams, F. (1958). Spray combustion and atomization. *Phys. of Fluids*, 1:541–545.

Williams, F. (1985). *Combustion Theory, Second ed.* Addison-Wesley Publishing Co.

Discussion

Josette Bellan, JPL "I have two questions. The first one is you said your model for the drop of evaporation was a D2 law isn't that invalidated by your last conclusion? So is there an inconsistency in the model since you show distinctively non D squared law behavior?"

Reply: Do you agree with me that the redline corresponds to the D2 level right?

Josette Bellan, JPL "The D2 law is obviously a constant slope. I don't know what your calculation shows. That is what you show there, yeah?"

Reply: So this is the model involved for each droplet the D2 squared level. If you check the vaporization rate for the global spray - you take the global mass and the corresponding radius of this measurement of the time it is not the D2 level globally because as you know the D2 level is just for a single droplet evaporating.

Josette Bellan, JPL "Is that a fair comparison then? That is what I am asking. Because in fact you are just comparing apples and oranges. So I don't know that is why I am asking so is there something there. Anyway that's my first question. My second is what are your conclusions in a two dimensional setting will be transferable to a three dimensional turbulent setting. Obviously the drops are inhibited in their motion in your two dimensional domain and certainly you don't see even if you start with a turbulent situation you don't see the stretching and tilting phenomena that you see in a three dimensional setting."

Reply: Of course this kind of study has to be done or future work within three dimensional calculation. You are right.

Johannes Janicka, Darmstadt "I also have two questions. One is related to this part. I didn't understand it really what you were going to say. We have two classic mechanisms. One is you put a single droplet in a field, evaporation will grow and you get deviation from the ? law. In the other direction if you have more than one droplet you have an interaction between droplets. What did you show over there interaction between droplets or influence of turbulence on the field? What is the message of this part?"

Reply: The message is the impact of turbulence on the spray because when the turbulence ...

Johannes Janicka, Darmstadt "Both interactions of droplets and turbulence?"

Reply: Just a second. The impact of turbulence on the spray. What I want to say here is since the droplets are convected in low region, low vorticity region you have the mixing of the droplet is completely different. For example, if you check at the fuel map iso-lines, here is the fuel iso-lines. Okay, the droplet will go there in the low vorticity region and the surrounding gas will be kind of saturated in fuel. That means, that's why the vaporization process is completely, yeah.

Johannes Janicka, Darmstadt "Okay, that's clear. Then the second question - could you perhaps specify your conditions a little bit more. What is the droplet size and what is the distant between droplets, compared to Kolmogorov scale, length scales and what is the droplet density lets say in kilograms per square metres or something like that. To get an idea about your configuration."

Reply: Okay, so the initial droplet radius was 5 micron. The Kolmogorov scale was 25 microns. The distance - since the droplets were specially randomly distributed were varying. Oh, the average, if I remember correctly, was around 15-20, at least 15-20 diameters.

Johannes Janicka, Darmstadt "And the concentration"

Reply: On this case we had 900 droplets.

Emmanuel Villermaux, Marseilles "Well I may try to help you to answer the previous questions. The departure from the D2 law that you observed is yes is some sense linked to the existence of stirring motions, but the prime ingredient is that this is a result of coalescences - you have bigger and bigger droplets as time evolves as your pdf's show. This is by the way interesting because you never get diameters smaller than your initial diameter meaning that you are in a limit where coalescence is dominating on evaporation. All the diameters as time evolves are always larger than your initial diameter, am I wrong?"

Reply: Yes.

Emmanuel Villermaux, Marseilles "Yes."

Reply: Yes, I'm sorry. Because here - due to the vaporization process the initial droplets were at 5 microns.

Emmanuel Villermaux, Marseilles "This is not the same scale."

Reply: Yeah, I'm sorry, this is not clear.

Emmanuel Villermaux, Marseilles "Yeah, you should be sorry."

Reply: I would like to answer about your comment the difference in the vaporization process due to turbulence is not in my mind completely due to coalescence. Just because of the inhomogeneous mixing, inhomogeneous distribution of the gas in the low vorticity region and the calculation where done without any coalescence and the same phenomena was involved, was observed.

Emmanuel Villermaux, Marseilles "Okay I have a second question or comment regarding your input in your model. The D2 law assumes a single droplet evaporating in a quiescent environment. The Stokes law assumes a single droplet in relative motion with a quiescent environment. Isn't there a problem using both laws at the same time?"

Reply: I think that one can use both at the same time as a first approximation because of course you can always use a more accurate model. For example concerning the vaporization rate you can use the concept of effective conductivity and for the Stokes law the same thing for the drag. I think your right, you can always use a more accurate model.

Paul Dimotakis, Caltech "It is nice to see in your simulation the depletion of the vortical regions and the accumulation in the convergence regions of lines these are problems that plague or should plague people that use particles to measure velocity by any methods what so ever. I was wondering if you could speculate how the coalescence might be different because the flow is three dimensional because there is a topological difference here that is interesting."

Reply: Yeah you are right. This kind of calculation has to be extended to 3D calculation. Right now I didn't do any 3D calculations.

Paul Dimotakis, Caltech "I was asking for speculation. My second question was actually related the Emmanuel Villermaux 's. I was surprised that even though you have coalescence and that you show in your size distribution an enhancement at large numbers that one does not see a depletion at small numbers. Considering how small particles it takes to make a large one I would have expected a depletion at small diameters, some depletion. But you show everything to be just bigger. Thank you."

Lorenz Sigurdson, University of Alberta "One quickly, the question is, you had coalescence did you consider at all that droplets might impact and produce smaller droplets?"

Reply: No, I did not take into account the collision that can introduce break up or droplet deformation. It is a very, very basic coalescence model.

"Yes, I understand. But do you think that is much of a factor, an approximation, does it happen very much."

Reply: For the general understanding of the phenomena I don't think that it is a real problem. I think that we have to use an accurate one.

Chapter 31

LARGE EDDY SIMULATIONS OF COMBUSTION INSTABILITIES IN A SWIRLED COMBUSTOR

Sébastien Ducruix
Laboratoire E.M2.C., CNRS and ECP
Grande Voie des Vignes
92295 Chatenay-Malabry, France

Thierry Poinsot
I.M.F.T.
Alle du Professeur Camille Soula
31400 Toulouse, France

Sébastien Candel
Laboratoire E.M2.C., CNRS and ECP
Grande Voie des Vignes
92295 Chatenay-Malabry, France

Abstract Modern gas turbine combustors feature pressure oscillations with relatively high amplitude levels. This dynamical behavior may lead to flashback or flame blow-out and it is generally unacceptable. This article is concerned with the simulation of this type of unstable behavior. Realistic geometrical data are used and unsteady computations are carried out and compared to experimental data gathered on a model scale system developed by Cambridge University. Calculations described in this article indicate that it is possible to simulate the dynamical behavior of the flame in this type of combustor.

Keywords: Combustion, Instability, Acoustics, Coupling, Swirl.

A. Pollard and S. Candel (eds.), IUTAM Symposium on Turbulent Mixing and Combustion, 357-366.

1. Introduction

It is well known that coupling of acoustics and combustion may lead to instabilities in many practical systems. The numerical simulation of these dynamical phenomena is of considerable technological interest but it also constitutes a difficult problem. The aim of this study is to develop unsteady flame computations in a realistic geometry. Data obtained on a gaseous fuel swirled burner operated by the University of Cambridge are compared with simulations of unstable behavior of the burner performed with a Navier-Stokes large eddy simulation code.

The experimental geometry and some observations of the combustion instabilities are first briefly described (section 2). Assumptions used to represent the experimental system and numerical models are then presented in section 3. Numerical simulations and comparisons with experimental data are finally discussed in section 4.

2. Experimental Configuration

A schematic diagram of the experimental configuration developed at the University of Cambridge is shown in Fig. 31.1. The injector was designed to be close to industrial injector characteristics.

The airflow enters the injector through an annular section and is then divided into two streams which flow through concentric channels. Vanes are placed in these channels to create in both streams a rotating component of the velocity ("swirl") with an opposite sign. The fuel (ethylene, C_2H_4) is injected in the airflow through cylindrical bars fitted with two 0.45 mm diameter holes. The swirler unit is connected to the combustion chamber, which is a 70 mm diameter quartz tube. The second end of the tube is opened to atmospheric pressure.

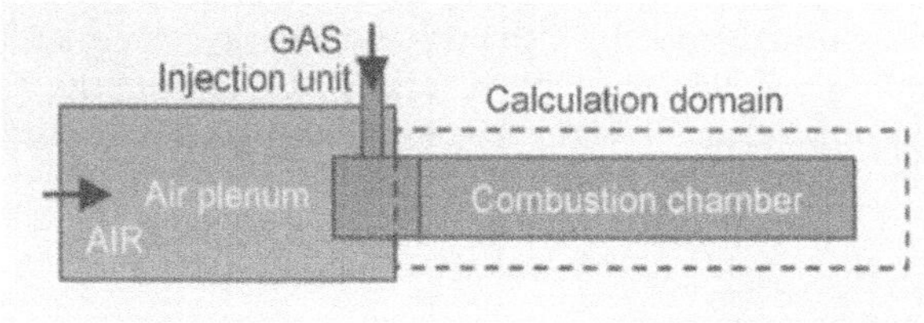

Figure 31.1. Scheme of combustion chamber.

Different types of combustion oscillations can be found, depending on the operating point. For low airflow rates and low equivalence ratios ($\Phi \leq 0.7$), the pressure power spectrum exhibits a main peak at 200 Hz. Other peaks may also be seen, but their amplitude is smaller. This behavior will be retrieved in the calculations described below.

3. Assumptions and models

Simulations of the complete system (air and fuel supplies, injector and combustion chamber) would require relatively large computer capabilities. The simulation of the mixing region between air and fuel streams constitutes a difficult problem which will determine to some extent the dynamical behavior of the system.

Three assumptions are introduced to make the simulations more tractable. These assumptions are strong, but the aim is here to show that calculations are possible in complex configurations. First, the flow beyond the injection vanes is assumed to be perfectly premixed. With this assumption, one avoids the simulation of the mixing zone, but one cannot consider partially premixed effects like instabilities driven by equivalence ratio inhomogeneities. Secondly, it is assumed that u_θ, rotating component of the flow created by the vanes, is proportional to the axial flow velocity and to the vane angle and that it is a constant in a given section. Thirdly, the system is treated in a first approximation as axisymmetric. This assumption is probably adequate for the mean flow but it is certainly not suitable with respect to the turbulence fluctuations. One may reason however that low frequency dynamics will be mainly driven by coherent axially symmetric turbulent structures. In what follows, a portion of the system corresponding to a 10° angle around the symmetry axis is considered.

3.1. Meshing of the domain

The inlet boundary condition is chosen at the outlet of the swirler unit, just below the vanes. This choice makes it possible to characterize the flow at the inlet of the combustion chamber by the three following items. The equivalence ratio Φ of the premixed gases is chosen in the range [0.5,0.8]. The airflow rate is in the range [0.05,0.09] $kg.s^{-1}$. The rotating component of the velocity is deduced from the mean flow velocity and the vane angle.

The combustion chamber, which is one meter long is also meshed and an exit zone is added at the outlet of the chamber in order to represent the reflection of acoustic waves impinging on the combustor exhaust. The downstream boundary of the computation domain is treated with non reflecting conditions.

Figure 31.2 presents an enlargement of the mesh at the outlet of the swirler unit. A hybrid mesh, which is structured in the injection channels and unstructured everywhere else, is generated. This makes it possible to resolve the swirl structure and the flame front on a fine mesh in the

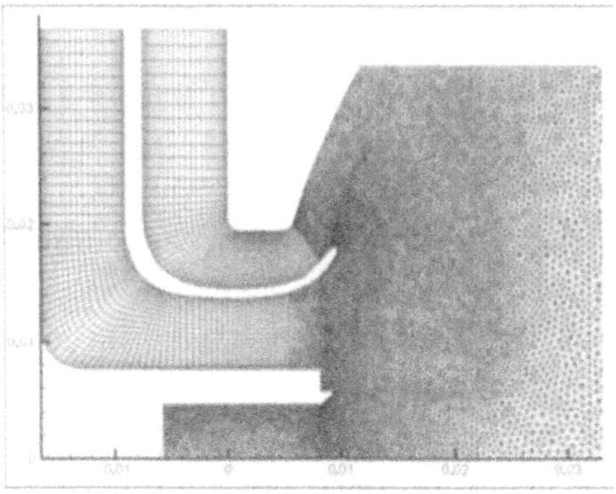

Figure 31.2. Enlargement of the mesh at the outlet of the swirler unit.

vicinity of the swirler unit. There are about 49000 nodes and 45000 cells in the entire domain.

3.2. Combustion modeling

The combustor is fed with a mixture of ethylene and air. The combustion kinetics of this mixture will be described with a single step reaction: $C_2H_4 + 3O_2 \rightarrow 2CO_2 + 2H_2O$ and an Arrhenius law given by:

$$\dot{\omega} = A\,[C_2H_4]^a[O_2]^b \exp(-E_a/RT)$$

where $[C_2H_4]$ and $[O_2]$ are the reactants molar concentrations, a and b the corresponding concentration exponents, A the pre-exponential factor, E_a the activation energy, R the gas constant, and T the absolute local gas temperature. These various coefficients were adapted to obtain reasonable approximations of the flame speed and thickness for lean mixtures.

A combustion model (artificial thickening approach) is implemented in AVBP, a Navier-Stokes flow solver developed by Cerfacs and the University of Oxford [Schönfeld et al., 1999]. This method is based on recent developments in turbulent combustion modeling. A complete description of the thickened flame (TF) technique may be found for RANS models [Butler and O'Rourke, 1976] and for LES models [Veynante and Poinsot, 1997, Thibaut and Candel, 1998]. The key idea is to thicken the flame while maintaining its propagation speed.

Following classical premixed laminar flame theories, this is achieved simply by multiplying the thermal and molecular diffusivities a by a thickening factor F and dividing the preexponential constant A by the same factor F. With this tranformation, the flame is thickened by a factor F and may be explicitely resolved on the LES mesh for sufficiently large values of F. The thickening factor F must be large enough to resolve the flame front on the LES mesh, but small enough to let the flame be wrinkled by the vortices calculated in the simulation.

The underestimation of the flame front wrinkling by the thickened flame approach must be corrected by increasing the flame speed by an efficiency factor E. This could be achieved by multiplying both the pre-exponential factor and the molecular and thermal diffusivities by a factor E, keeping the flame thickness constant. Details on the determination of the efficiency factor can be found in the literature [Angelberger et al., 1998, Colin et al., 2000].

4. Reactive flow simulations

Fresh gases flow through the internal and external channels of the injector (Fig. 31.2). The flow velocity is chosen equal to 15 $m.s^{-1}$, corresponding to an airflow rate of 60 $g.s^{-1}$.

In what follows, calculations are performed for an equivalence ratio $\Phi = 0.7$. The flame thickening technique, as described in section 3, is used. A thickening factor $F = 8$ is chosen.

4.1. Fields of temperature

Figure 31.3 shows the field of temperature in the vicinity of the swirler unit. First, it can be seen that the recirculation zone in the central part of the chamber plays an important role in the stabilization of the flame. Burned gases are in contact with fresh gases, providing a continuous source of ignition to the flame.

The stabilization region is near the lower border of the internal channel. This region is characterized by low velocities. It is difficult to determine whether the flame is lifted. This point will be discussed in the next paragraph. Snapshots of the temperature field show the development of vortices in this region which move the flame stabilization position. These perturbations cause fluctuations in the flame surface which may lead to combustion instabilities. Moreover the flame may sometimes be able to propagate in the internal channel. This behavior can lead to flashback (this phenomenon has not been observed yet).

The vortices created in the region between the external jet and the chamber wall (Fig. 31.3) are convected in the chamber. They clearly wrinkle the flame front and produce from time to time large pockets of fresh gases which are conveyed into the recirculation region and destabilize the flame. Pockets of fresh gases are also created near the flame extremity. These pockets are convected and burned with a certain delay. Depending on the delay, this phenomenon may participate in the development of acoustic instabilities.

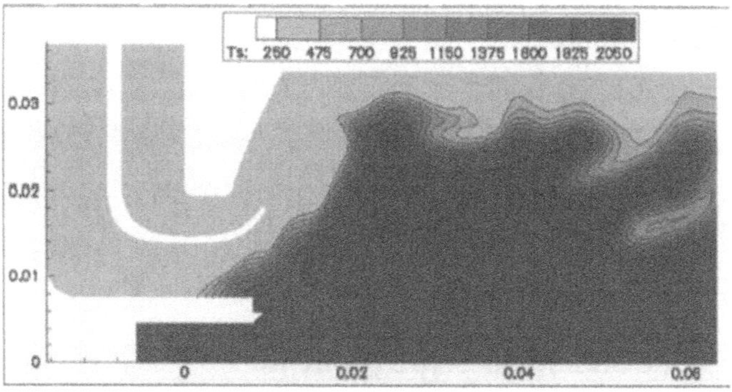

(a) $t = t_0$

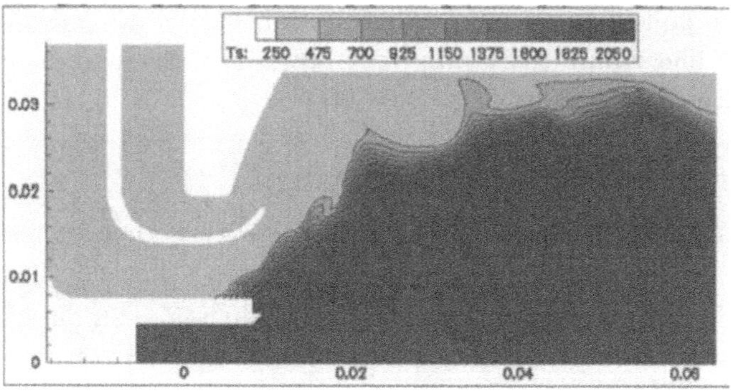

(b) $t = t_0 + \Delta t$

Figure 31.3. Snapshots of the field of temperature near the injector outlet. $\Delta t = 1.$ ms.

At $t = t_0$, the flame length is about 12 cm. This length varies between 8 and 12 cm during the calculation. It can be noticed that, in the first part of the flame front (0.1< x <0.3), the amplitude of the motion is relatively small and the flame remains close to its mean position, whereas large movements can be observed in the second part of the flame. Averaged images of the flame would then show a short flame, corresponding to the integration of the flame emission in the first part of the chamber and similar to experimental visualizations.

4.2. Fields of reaction rate

Figure 31.4 displays the reaction rate in the vicinity of the swirler outlet. The wrinkling of the flame surface due to large vortices can clearly be seen, showing that the value chosen for the thickening factor F is not too large. Moreover, the field of efficiency factor E shows that the combustion modeling plays an important role, "transforming" the wrinkling effect that cannot be resolved due to the size of the mesh cells into a larger consumption speed.

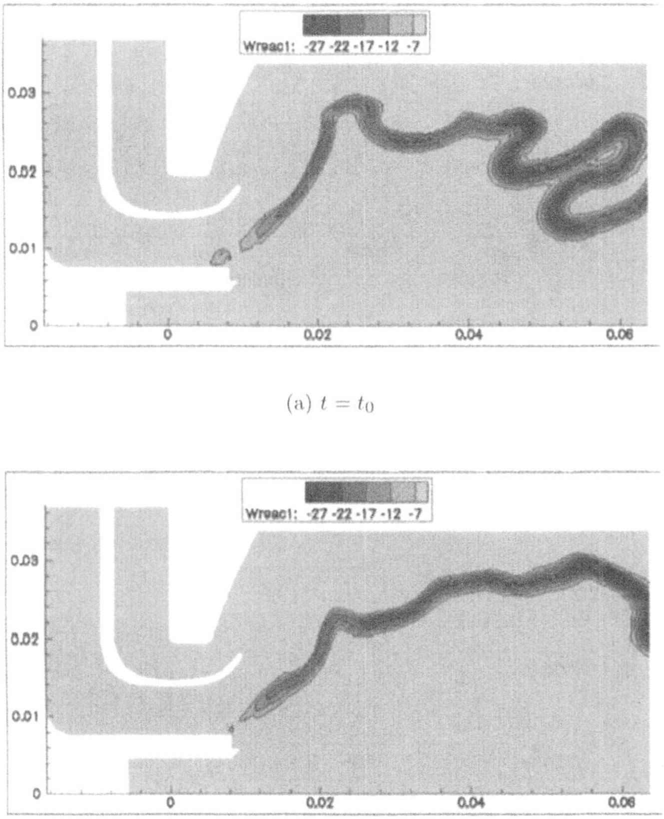

(a) $t = t_0$

(b) $t = t_0 + \Delta t$

Figure 31.4. Snapshots of the field of reaction rate. $\Delta t = 1.$ ms.

The flame is slightly lifted with respect to the injector channel, even though a small reaction rate exists near the wall of the internal channel. It is interesting to note that the stabilization position is not fixed in time. Longer simulations would make it possible to determine frequencies associated with this phenomenon.

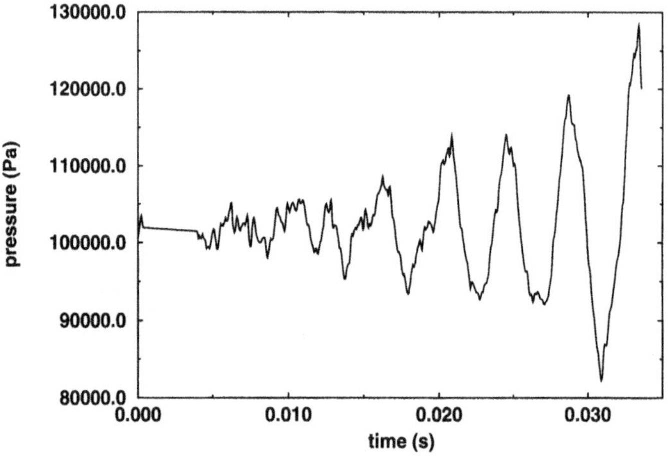

(a) Temporal evolution of pressure.

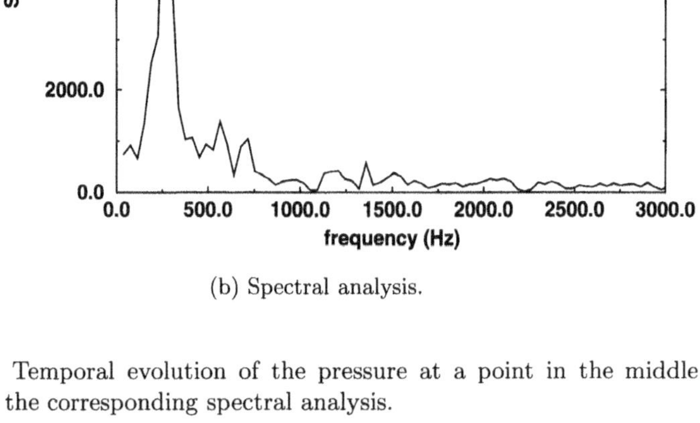

(b) Spectral analysis.

Figure 31.5. Temporal evolution of the pressure at a point in the middle of the chamber and the corresponding spectral analysis.

4.3. Combustion instabilities in the chamber

Figure 31.5 presents the temporal evolution of the pressure at a point in the middle of the chamber and the corresponding spectral analysis, for a time period of 35 ms. After a initial transient period, the signal is quasi periodic, almost sinusoidal. The growth rate is positive at the end of the simulation indicating that no limit cycle has been reached at that time.

Using the spectral analysis, it is possible to determine the main peaks associated with the pressure signal. The signal is almost sinusoidal with a frequency close to 250 Hz. The conditions which were chosen for the simulation (flow rate, equivalence ratio) correspond to the development of an instability around 200 Hz in the experiment. Unfortunately, the duration of the simulation is not long enough because the flame is rapidly blown out.

5. Conclusion

Calculations described in this article indicate that it is possible to simulate the dynamical behaviour of turbulent flames spreading in a gas turbine combustor. Numerical results are qualitatively close to observations but the validation of the method would require extensive calculations for a range of injection conditions to see if it is possible to predict the boundary of the instability domain.

Acknowledgments

Support for this research has been provided by the European Union in the frame of the Brite Euram contract ACIACOC (BE97-4324).

References

Angelberger, C., Veynante, D., Egolfopoulos, F., and Poinsot, T. (1998). Large eddy simulation of combustion instabilities in premixed flames. Annual research briefs, Center for Turbulence Research.

Butler, T. and O'Rourke, P. (1976). A numerical method for two dimensional unsteady reacting flows. In *Sixteenth Symposium (International) on Combustion*, pages 1503–1515. Combustion Institute.

Colin, O., Ducros, F., Veynante, D., and Poinsot, T. (2000). A thickened flame model for large eddy simulations of turbulent premixed combustion. *Phys. Fluids*, 12(7):1843–1863.

Schönfeld, T., Angelberger, C., Légier, J.-P., and Ducruix, S. (1999). Numerical simulation of compressible reactive flows on unstructured grids. In *AIAA 37th Aerospace Sciences Meeting and Exhibit*, Reno, NV, USA.

Thibaut, D. and Candel, S. (1998). Numerical study of unsteady turbulent premixed combustion: Application to flashback simulation. *Combust. Flame*, 113:53–65.

Veynante, D. and Poinsot, T. (1997). *New Tools in Turbulence Modeling*, chapter Reynolds-Averaged and Large Eddy Simulation Modeling for Turbulent Combustion. O. Métais and J. Ferziger, Les Editions de Physique.

Chapter 32

AN EXPERIMENTAL INVESTIGATION OF MIXING AND COMBUSTION CHARACTERISTICS ON THE CAN-TYPE MICRO COMBUSTOR WITH A MULTI-JET BAFFLE PLATE

Hang-Seok CHOI
Kazuyoshi NAKABE
Kenjiro SUZUKI
Department of Mechanical Engineering, Kyoto University, Kyoto 606-8501, Japan
t60y0915@ip.media.kyoto-u.ac.jp
nakabe@mech.kyoto-u.ac.jp
ksuzuki@mech.kyoto-u.ac.jp

Yoichi KATSUMOTO
Engineering Department. Sony Corporation, Tokyo, 141-0001, Japan.
Youichi.Katsumoto@jp.sony.com

Abstract The present research is related to the development of a hybrid system, which is composed of solid oxide fuel cell (SOFC) and micro gas turbine (MGT) assisted with recuperator. The main problems in small-size combustors for MGT are related to mixing enhancement and flame stability because the Reynolds number is very low. To solve these problems, we developed a multi-jet type baffle plate for the combustor. The reacting flow fields in this device were investigated by using flow visualizations and blow-out experiments. As expected this kind of baffle plates enhance the turbulence mixing in the low Reynolds number flows and maintain the flame stability in the small combustor.

Keywords: flame stability, turbulent mixing, micro combustor, flow visualization

A. Pollard and S. Candel (eds.), IUTAM Symposium on Turbulent Mixing and Combustion, 367-375.

1. Introduction

The micro machining technology has been rapidly developed and it is now possible to fabricate micro power generators and their additional devices. Using micro fabrication technologies, we can develop the micro power system for power generation, which may be used by small community or individual households. The present research is related to the development of a hybrid system, which is composed of solid oxide fuel cell (SOFC) and micro gas turbine (MGT) assisted with recuperator.

The aim of the present research is, in particular, to develop the micro combustor for MGT. Considering the operating conditions of the micro combustor, the main problem is related to the very low Reynolds number flow conditions compared with those of other larger commercially-available combustors. Combustion in such small combustors may be influenced by lower mixing, producing longer and sooty luminous flames. It is then necessary to establish the appropriate methods to enhance mixing between fuel and oxidizer and simultaneously keep the flame stable in the micro combustor.

The phenomena of jet mixing have been studied by many researchers [[Dahm (1987)]] in the absence or presence of combustion. For turbulent diffusion flames, research has concerned on flame stability, transition from laminar to turbulent flow, lift-off and final blow-out, [Vanquickenborne (1966)], [Broadwell (1984)], [Chen (2000)], [Dahm (1988)], [Eickhoff (1984)], [Peters (1984a)], [Peters (1984b)], [Pitts (1988)], [Pitts (1989)], [Takahashi (1982)], [Yule(1980)]. The physics of flame stability is not yet fully understood. Considering the general characteristics of non-premixed turbulent jet flame, we have developed a can-type micro combustor with a multi-jet baffle plate to enhance mixing and maintain flame stability.

According to the development concept of the present combustor, the temperature range of fuel and air supplied to MGT from SOFC is expected to be from 300K to around 1,000K. This preheating condition can affect the flame stability [Takeno (1975a)], [Takeno (1975b)]. In the present experimental study, also attention is focused on combustion stability enhancement by the preheat of the supplied air and fuel, and on the flame patterns induced by the baffle plate geometrical configuration.

2. Experimental System

2.1. Experimental apparatus

The experimental apparatus comprises a combustor section, a preheating unit, a pre-heater temperature controller and flow visualization equipment. Figure 32.1 shows the experimental apparatus for cold (non-preheated) and hot (preheated) experiments.

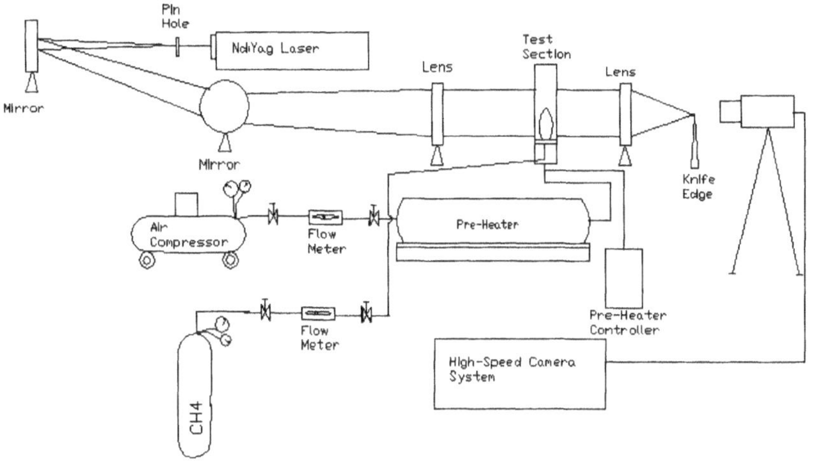

Figure 32.1. Experimental apparatus.

The combustor section consisted of an air pre-conditioning unit, a baffle plate, and a Pyrex glass tube combustor. The tube length and its inner diameter were 300 mm and 22 mm respectively. The air is delivered by a compressor, to the air pre-conditioning unit, and supplied to the combustor test section through the baffle holes. The baffle plate configuration will be explained in the subsection 2.3. The air pre-conditioning unit consists of water separator and honeycomb to reduce unexpected moisture and flow irregularities that were generated through the air feed line. Methane is discharged from a pressurized bottle, through a pressure regulator and finally supplied to the combustor test section through a smooth stainless steel pipe. The methane flow in the fuel feed line, may be assumed to be laminar for all experimental conditions considering the entrance length L_e and Reynolds number Re ($\frac{L_e}{d} \cong 0.06Re$) [White(1986)].

The gas temperature at the combustor inlet was kept constant with a pre-heater temperature controller. Between the pre-heater and combustor sections, the fuel feed stainless steel tube was located in the center of the air feed tube. Before starting the preheating experiment, the initial temperature at the combustor inlet was monitored for both air and fuel with K-type thermocouples. Based on these measurements, the maximum temperature of air and fuel was found to be 450K at the combustor inlet in the pre-heated flow conditions.

2.2. Optical system

An Nd:YAG pulse laser (Sectra-Physics, T80-YHP70) and a high-speed video camera (Kodak, Ekta-Pro) were used to capture the flow and flame images. Shadowgraph and Schlieren photographs of the flames were taken with the high-speed camera and with a digital still camera (Nikon, D-1). The frame frequency of the high-speed camera and the shutter speed of the still camera are 1/4500 s and 1/8000 s, respectively. Although the exposure period was not sufficiently fast to capture the instantaneous eddy development motion, the overall flame behavior was clearly visualized in the video frames. The direct snapshots of the flames were also taken to compare with the shadowgraph and Schlieren photographs.

2.3. Baffle plates

Four kinds of baffle plates were designed to modify the inlet flow conditions and enhance mixing more in the combustor. Figure 32.2 shows the geometrical configurations of the baffle plates. The diameters of fuel and air jets, the location of the air jets surrounding the fuel jet and the number of air jets were changed in various manners.

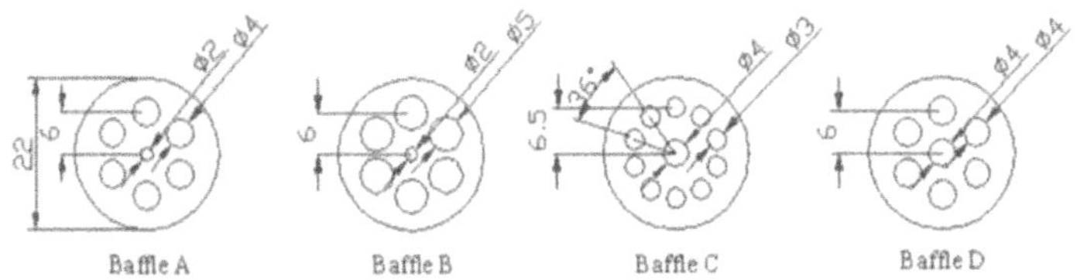

Figure 32.2. Configuration of baffle plates. (unit : mm)

3. Results and Discussion

3.1. Unconfined combustion

Before performing the confined combustion in the baffle plate-installed combustor, the mixing and stability characteristics were examined without the confining tube of the combustor. Through this experiment, the flame patterns of the multi-jet flow fields were identified.

In Figure 32.3, (A), (B) and (C) represent cases of fixed fuel mass flow rate with varying air mass flow rate, and (D), (E) and (F) cases of fixed air mass flow rate with varying fuel mass flow rate. With an increase in the air mass flow rate, (A) through (C), instabilities occur in the mixing layer located between the air and fuel jets, and the mixing layer becomes unstable from the downstream of the multi-jet flow fields. The flow field eventually features complex turbulence structures visualized in the image (C). In the direct images, (A), (B) and (C) of the right-hand side in the figure, the flame changes its luminous color to blue one, and its length becomes shorter as the air mass flow rate is increased. This means that the mixing between fuel and air is enhanced.

For fixed air mass flow rate experiments((D), (E) and (F)), the shadowgraph image (D) and high-speed video images of the flow visualizations reveal two layers. One may identify an inner layer and an outer one as suggested by [Yule(1980)]. But, the behavior of these layers is governed by the interactions among the jets unlike in Yule's experiments. With an increase in the fuel mass flow rate, the inner layer instability appears downstream, (D), and moves upstream of the fuel jet, (E) and (F). Then, the flame formed near the fuel jet exit is eventually extinguished. Just before extinction, the flame is lifted off. Also, small-scale turbulent eddies can be observed in the lifted flame between the jet exit and the stabilized base of the lifted flame.

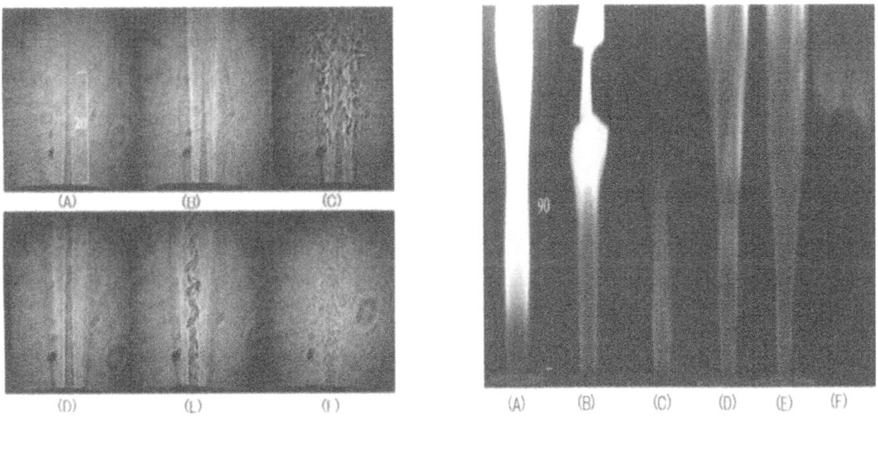

Figure 32.3. Shawdowgraph(left) and direct(right) pictures increasing air mass flow rate or fuel mass flow rate in baffle plate A; [Fuel mass flow rate: 5.43×10^{-6} kg/s, Air mass flow rate : (A) 0 kg/s, (B) 3.91×10^{-4} kg/s and (C) 9.74×10^{-4} kg/s], [Air mass flow rate: 1.95×10^{-4} kg/s, Fuel mass flow rate : (D) 2.67×10^{-5} kg/s, (E) 3.64×10^{-5} kg/s and (F) 4.06×10^{-5} kg/s], (mm).

Figure 32.4 shows the shadowgraph images of flow fields in unconfined combustion cases. The left image corresponds to baffle plate A and the right to baffle plate C. Air and fuel mass flow rates were 7.84×10^{-4}

kg/s and 5.43×10^{-6} kg/s, respectively, for both baffle plates A and C. In this figure, different turbulent length-scales can be observed between baffle plates A and C. In baffle C, the turbulence length-scale is smaller than that of baffle A. Small scale eddies appear closer to the fuel jet in the case of baffle C than for baffle A. According to the direct images, in the case of baffle C, the flame length is also smaller than that in the case of baffle A. Even if the fuel jet exit velocity in baffle C case is lower than that in baffle A case, a shorter and blue flame can be obtained by reducing the size of the air jet hole and increasing the number of the holes. This means turbulent mixing between fuel and air is augmented in the case of baffle plate C.

Figure 32.5 corresponds to the preheating combustion case with the same flow conditions as in Figure 32.4. In this case, the occurrence of transition to turbulence is observed further downstream region compared with the case shown in Figure 32.4. This indicates that the flow characteristics are affected by the increased kinematic viscosity due to preheating. Thus, the increased viscosity suppresses the early occurrence of the flow instability, delays transition and influences the mixing characteristics. In terms of flame stability, however, the effect of preheating is favorable.

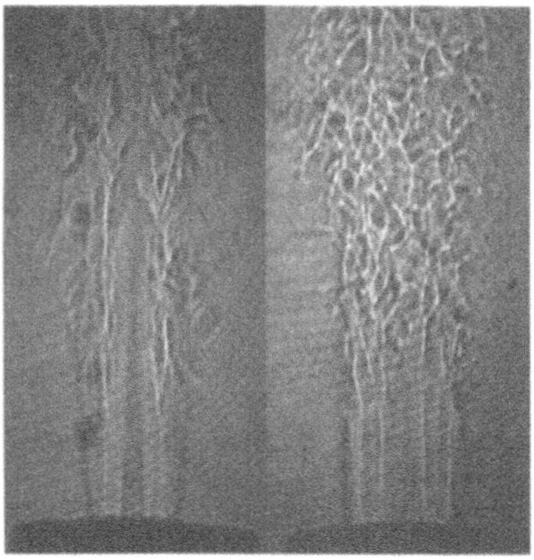

Figure 32.4. Flow structure comparison between baffle plates A and C in fixed fuel mass flow rate with varying air mass flow rate (cold case).

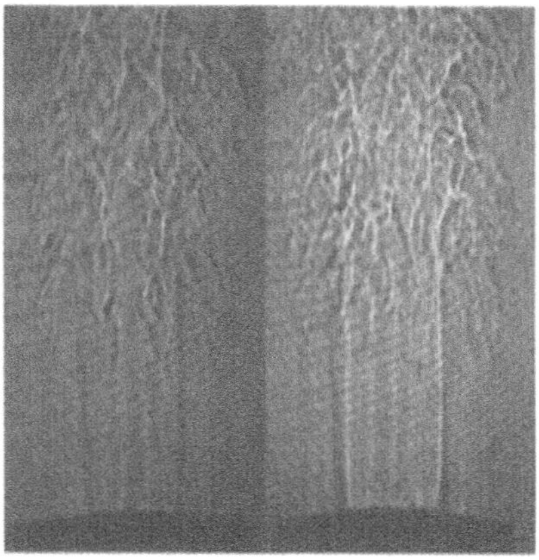

Figure 32.5. Flow structure comparison between baffle plates A and C in fixed fuel mass flow rate with varying air mass flow rate (hot case).

Figure 32.6 shows the results for the case of fuel mass flow rate 3.74×10^{-5}kg/s and air mass flow rate 1.95×10^{-4}kg/s. This flow condition is close to the lift-off condition in the case of baffle plate A. In the latter case (Figure 32.6 on the left), the flame is already unstable. In the case of baffle plate C (Figure 32.6, right image), no instability is observed in the upstream of inner layer. Only the slow wavy motion of the outer layer can be seen in the video, and the flame remains stable because of lower fuel flow velocity than that of baffle plate A. As explained later in the section of confined combustion case, the jet flow from the fuel jet of smaller diameter is relatively more unstable than that of larger diameter for the same fuel and air mass flow rates.

Figure 32.7 shows the effect of preheating with the same flow condition as in Figure 32.6. For the baffle plate A, the instability is suppressed and the starting position of instability is located further downstream than in Figure 32.6. There is no sign of inner layer instability near the jet exit in the case of baffle plate C in distinction with observations made with baffle plate A.

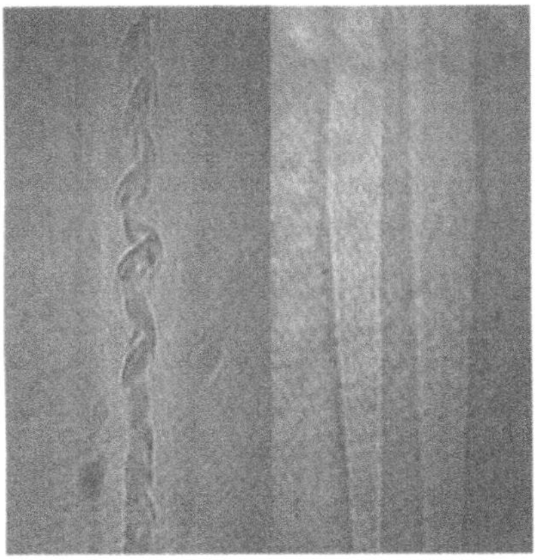

Figure 32.6. Flow structure comparison between baffle plates A and C in fixed air mass flow rate with varying fuel mass flow rate (cold case).

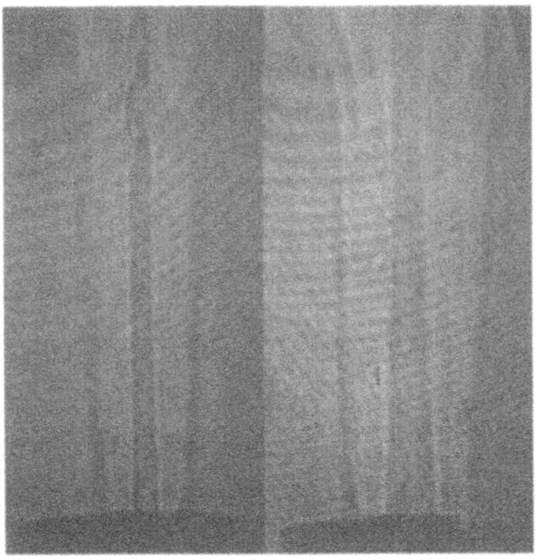

Figure 32.7. Flow structure comparison between baffle plates A and C in fixed air mass flow rate with varying fuel mass flow rate (hot case).

3.2. Confined combustion

Figure 32.8 shows results of flame stability experiments for confined combustion cases. In Figure 32.8, the abscissa and ordinate of each graph indicate air velocity and air to fuel momentum ratio, respectively. Flame stability characteristics can be seen for non-preheated (left graph) and preheated cases (right graph). Comparing the blow out limits of preheated and non-preheated cases indicates that flame stability is enhanced by preheating. For both cases, baffle plate C is much more effective under fuel-lean conditions, compared with a conventional coaxial confined jet flame (dotted line).

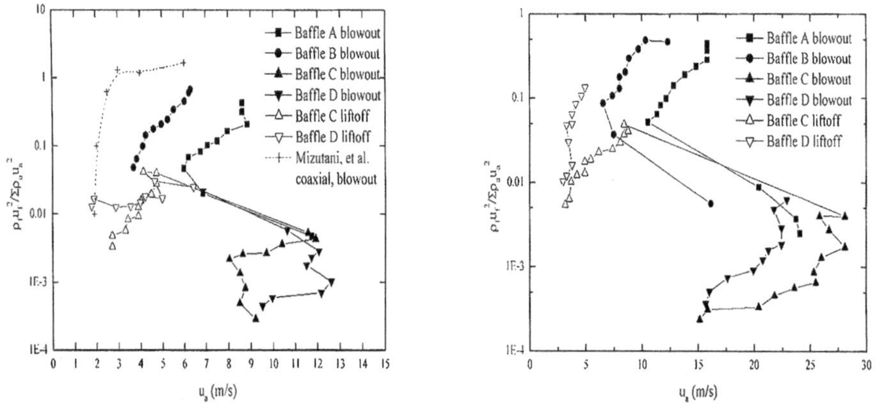

Figure 32.8. Flame stability comparison among baffle plates in cases of hot and cold flows.

4. Concluding Remarks

In the present research, several multi-jet type baffle plates are introduced to enhance the mixing rate of fuel and air in a small-size combustor. It is expected that the baffle plate can enhance the turbulent mixing in low Reynolds number flows due to the interaction between the low-velocity fuel jet and high-velocity air multi-jet. For preheating combustion, it is verified that the temperature increase of initial fuel and air jets can suppress the transition and stabilize the flame for a certain range of flow rates. In this case, the increased temperature is not high enough compared with actual operation temperature of methane in the hybrid system. But, the stability is increased.

Acknowledgement

This work was supported through the research project on ' Micro Gas Turbine / Fuel Cell Hybrid-Type Distributed Energy System ' by the Department of Core Research for Evolutional Science and Technology (CREST) of Japan Science and Technology (JST).

References

Broadwell J. E. *et al.* (1984), *Blowout of turbulent diffusion flames*, 20th Symposium (International) on Combustion, 303.

Chen Y. C. and Bilger R. W. (2000), *Stabilization mechanisms of lifted laminar flames in axisymmetric jet flows*, Combustion and Flame, 122, 377.

Dahm W. J. A. and Dimotakis P. E. (1987), *Measurements of entrainment and mixing in turbulent jet* , AIAA Journal, 25, 1216.

Dahm W. J. A. and Dibble R. W. (1988), *Coflowing turbulent jet diffusion flame blowout*, 22nd Symposium (International) on Combustion, 801.

Eickhoff H. *et al.* (1984), *Experimental investigation on the stabilization mechanism of jet diffusion flames*, 20th Symposium (International) on Combustion, 311.

Peters N. (1984a), *Partially premixed diffusion flamelets in non-premixed turbulent combustion*, 20th Symposium (International) on Combustion, 353.

Peters N. (1984b), *Laminar diffusion flamelet models in non-premixed turbulent combustion*, Prog. Energy Combust. Sci., 10, 319-339.

Pitts W. M. (1988), *Assessment of theories for the behavior and blowout of lifted turbulent jet diffusion flames*, 22nd Symposium (International) on Combustion, 809.

Pitts W. M. (1989), *Importance of Isothermal mixing processes to the understanding of lift-off and blowout of turbulent jet diffusion flames*, Combustion and Flame, 76, 197.

Takahashi F. *et al.* (1982), *Transition from laminar to turbulence free jet diffusion flames*, Combustion and Flame, 48, 85.

Takeno T. and Kotani Y. (1975a), *A study on the structure of turbulent jet diffusion flames*. Combustion and Flame, 10, 45.

Takeno T. and Kotani Y. (1975b), *An experimental study on the stability of jet diffusion flame*, Acta Astronautica, 2, 999.

Vanquickenborne L. and Van Tiggelen A. (1966), *The stabilization mechanism of lift off diffusion flames*, Combustion and Flame, 10, 59.

White F. M. (1986), *Viscous flow in ducts*, Fluid Mechanics, Second edition.

Yule A. J. *et al.* (1980), *Combustion-Transition interaction in a jet flame*, AIAA Journal ,19, 752.

Chapter 33

EFFECTS OF ONE, TWO AND QUADRUPLE TABS ON THE MIXING AND ENTRAINMENT IN THE NEAR FIELD OF ROUND JETS

S. McIlwain *, T. Holme, S. Waterman, and A. Pollard

Dept. of Mechanical Engineering, Queen's University, Kingston, ON, K7L 3N6, Canada

Abstract Large eddy simulation was used to calculate the effects of vortex-generating tabs on the turbulence structures in the near field of a round jet. The coherent structures within the turbulence were identified with plots of the discriminant. Two or four tabs compressed the flow regions containing structures, increasing the strength of the vorticity. These results demonstrate the potential for using large eddy simulation to aid in the design of efficient jet nozzles.

1. Introduction

Vortex-generating tabs are used in turbulent round jets to increase mixing and flow entrainment, improve combustion efficiency, and reduce noise. The tabs are small protrusions that extend into the flow from the nozzle wall, often placed at the nozzle exit. They distort the round shape of the exit velocity profile and manipulate the vorticity in the flow field, thereby controlling the evolution of the large-scale turbulence structures.

A relationship between the presence of large-scale structures of vorticity and the rate of entrainment in a turbulent jet flow was observed in the large eddy simulations of [McIlwain and Pollard, 2002]. The structures form due to flow instabilities downstream from the nozzle. The primary Kelvin-Helmholtz instability mechanism causes the shear layer to roll up into vortex rings. Secondary instabilities also develop that

*now at the Institute for Aerospace Research, National Research Council, Ottawa, ON, K1A 0R6

A. Pollard and S. Candel (eds.), IUTAM Symposium on Turbulent Mixing and Combustion, 377-385.

create finger-like pairs of streamwise vortex structures that emerge from the region between two vortex rings and stretch around the jet core ([Liepmann and Gharib, 1992]).

The effect of the shape and tilt (angle of attack) of the tabs, and the combination of several tabs at the nozzle exit, have been previously investigated through experiments. A tab generates a pair of counter-rotating streamwise vortices at the nozzle, largely due to the formation of an upstream "pressure hill" ([Zaman *et al.*, 1994]). Enhanced mixing within the flow and an increase in the number of small-scale structures has been observed ([Hunag and Ho, 1990]); consequently, an increase in the rate of entrainment of the surrounding fluid is also expected. However, it is not known whether the enhanced mixing in the flow is caused by the increased surface area of the mixing layer that is exposed to the surrounding fluid, or the observed vortex breakdown farther downstream. The effect of a combination of tabs remains controversial: when too many tabs are used, the flow field may settle back into a steadier configuration with a less profound distortion ([Zaman *et al.*, 1994], [Mi and Nathan, 1999]). [Zaman, 2000] reports that a double-tab configuration produces the best results.

Although tabs do increase mixing by intensifying the interaction between azimuthal and streamwise turbulence structures, they must be used with a good understanding of their mechanisms to obtain a positive and efficient effect ([Foss and Zaman, 1996]). Therefore, numerical studies of the effects caused by vortex-generating tabs are desirable since the time-dependent evolution of the turbulent structures can be observed. As a first step in this direction, the present study uses Large Eddy Simulation (LES) and flow visualization techniques to investigate the near field effects of three configurations of tabs in a turbulent round jet.

2. Numerical Procedure

In LES, the smallest turbulence scales are approximated using a subgrid scale model while the larger scales are resolved using the Navier-Stokes equations. Therefore, the velocity and pressure fields in the governing equations must be filtered to remove the subgrid scales of motion. Applying the gradient-diffusion hypothesis and representing filtered quantities using $\hat{}$, the governing equations of motion are ([Ciofalo, 1994]):

$$\frac{\partial \hat{u}_i}{\partial x_j} = 0 \ , \tag{33.1}$$

$$\frac{\partial \hat{u}_i}{\partial t} + \frac{\partial \hat{u}_i \partial \hat{u}_j}{\partial x_j} = -\frac{1}{\rho}\frac{\partial \hat{P}}{\partial x_i} + \frac{\partial}{\partial x_j}\left[(\nu + \nu_s)\left(\frac{\partial \hat{u}_i}{\partial x_j} + \frac{\partial \hat{u}_j}{\partial x_i}\right)\right] , \quad (33.2)$$

for $i = 1, 2, 3$. Here, u_i contains the components of the velocity vector in the x_i direction, t is time, ρ is density, P is a modified pressure quantity, ν is the kinematic viscosity, and ν_s is the subgrid viscosity. The effect of removing the subgrid scales of turbulence is included through ν_s, which must be modelled. In the present research, the dynamic Smagorinsky model, first proposed by [Germano *et al.*, 1991], is used.

The governing equations were discretized using second-order accurate central differencing, and they were advanced in time using the [Kim and Moin, 1985] fractional-step method combined with a Runge-Kutta semi-implicit scheme. Complete details can be found in [McIlwain and Pollard, 2002]. The round jet computations were performed on a rectangular grid of $73 \times 56 \times 56$ points over a domain size of $4.25D \times 3.0D \times 3.0D$, where D is the jet diameter at the nozzle. The edge of the round nozzle was approximated on a square grid. The jet inlet was prescribed using an analytical time-averaged turbulent velocity profile, upon which was superimposed a small sinusoidal disturbance with a maximum value of 2.8% of the mean jet velocity. The frequency of the disturbance was set to the preferred mode of the jet, as determined from several short test simulations. The standard convective outlet condition was used at the jet outlet:

$$\frac{\partial u_i}{\partial t} + U_{conv}\frac{\partial u_j}{\partial x_1} = 0 , \quad (33.3)$$

where U_{conv} is the convection velocity and x_1 is in the downstream direction. A zero gradient condition was used at all other boundaries. The Reynolds number of the flow based on D and the velocity at the nozzle was 68,000. A previous round jet study of flow at the same Reynolds number using the same grid demonstrated a reasonable level of grid independence, [McIlwain and Pollard, 2002]. The tabs were modelled by blocking off appropriate control volumes at the inlet. The dimension of each tab was chosen to induce a significant perturbation in the jet while maintaining a similar characteristic length scale. Cases with no tabs, one tab, two tabs, and four tabs were considered.

3. Structures of Small-Scale Turbulence

Structures of small scale turbulence were observed using plots of the discriminant, as defined by [Chong *et al.*, 1998]. The characteristic equation of the velocity gradient tensor, $A_{ij} = \partial u_i / \partial x_j$, is

$$\lambda^3 + \mathcal{P}\lambda^2 + \mathcal{Q}\lambda + \mathcal{R} = 0 , \quad (33.4)$$

where λ represents the eigenvalues, and $\mathcal{P}$, $\mathcal{Q}$, and $\mathcal{R}$ are the tensor invarients,

$$\mathcal{P} = -tr\mathbf{A} \ , \tag{33.5}$$

$$\mathcal{Q} = \frac{1}{2}(\mathcal{P}^2 - tr(\mathbf{A})^2) \ , \tag{33.6}$$

$$\mathcal{R} = -det(\mathbf{A}) \ . \tag{33.7}$$

For incompressible flow, $\mathcal{P} = 0$. The discriminant d is defined as the surface that divides flows with complex eigenvalues from those with all real eigenvalues,

$$d = \frac{27}{4}\mathcal{R}^2 + \mathcal{Q} \tag{33.8}$$

Chong *et al.* classified regions of the flow in which $d > 0$ as foci, and demonstrated that they correspond to vortices.

A contour plot of the jet flow with no tabs in which regions where $d > 0.2s^{-6}$ are shaded light grey is shown in figure 33.1. The plot corresponds to a single instant in time, and is representative of the results obtained throughout the simulation. The jet nozzle is at $x/D = 0$, and the flow travels from left to right in the plot. The discriminant highlights the regions of flow in which vortices are found. Although the plot shows individual bundles with $d > 0.2s^{-6}$, bundles in close proximity to each other may be connected together to form coherent structures. The vorticity initially forms in the shear layer that surrounds the jet immediately downstream from the nozzle. Downstream of $x/D = 2$, the shear layer rolls up into coherent ring-like structures, which continue to transport the small-scale structures downstream. Rings are shown as collections of $d > 0.2s^{-6}$ bundles at $x/D = 2.75$ and $x/D = 3.5$ in figure 33.1. Regions of higher pressure exist between the rings; these are shaded dark grey in figure 33.2.

Plots of regions of $d > 0.2s^{-6}$ and of high pressure for the jet flows with one, two, and four tabs are shown in figures 33.3, 33.4, and 33.5, respectively. There is little difference between the flows with one and no tabs. As the number of tabs increases to two or four, the size of the regions of high pressure also increases. Thus, the vortices defined by the discriminant contours are compressed into a smaller space. The number of bundles with $d > 0.2s^{-6}$ increases sharply throughout the flows with two or four tabs, compared to the flows with one or no tabs. This suggests that the coherent structures in the flow contain higher levels of vorticity. This effect is compounded by the concentrating effect of the higher pressure regions. The present results support the findings of [Marcouyre *et al.*, 2001], who showed that the introduction of tabs increases the amount of flow entrainment.

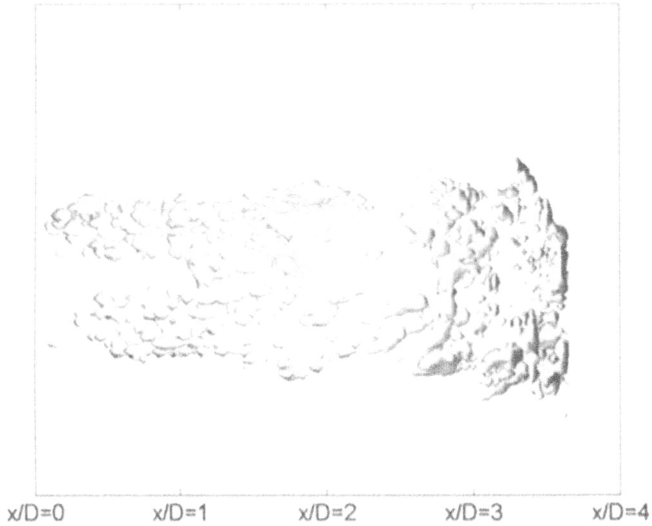

Figure 33.1. Plot of $d > 0.2s^{-6}$ in the jet flow without tabs

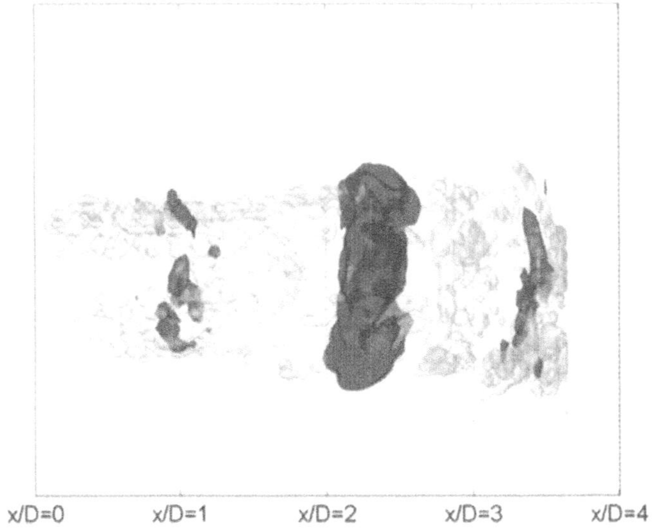

Figure 33.2. Regions of $d > 0.2s^{-6}$ (light grey) and high pressure (dark grey) in the jet flow without tabs

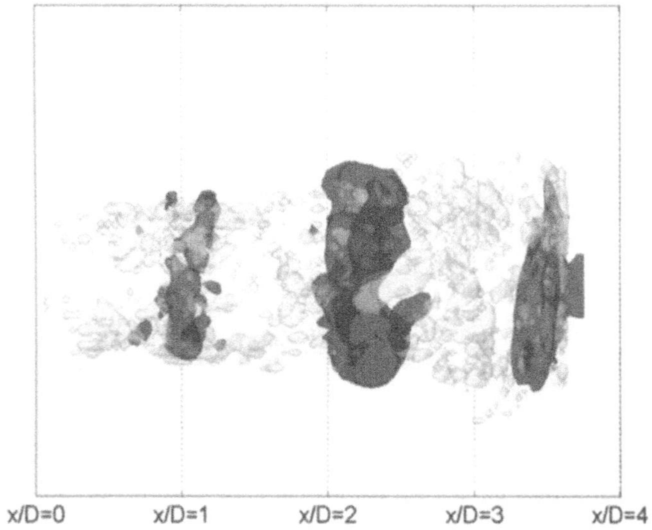

Figure 33.3. Regions of $d > 0.2s^{-6}$ (light grey) and high pressure (dark grey) in the jet flow with one tab

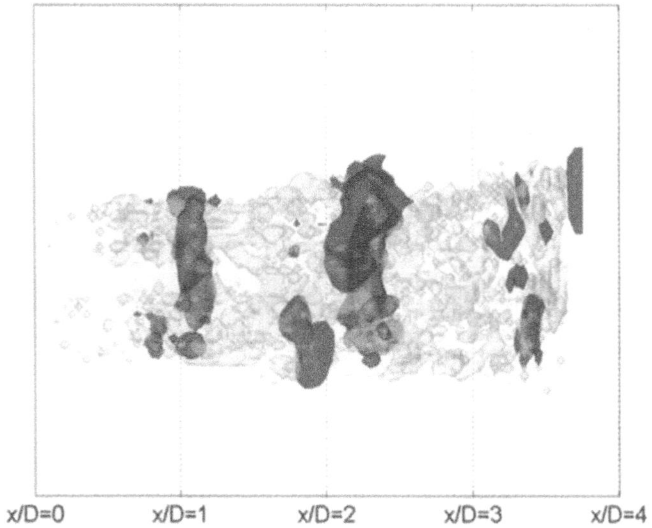

Figure 33.4. Regions of $d > 0.2s^{-6}$ (light grey) and high pressure (dark grey) in the jet flow with two tabs

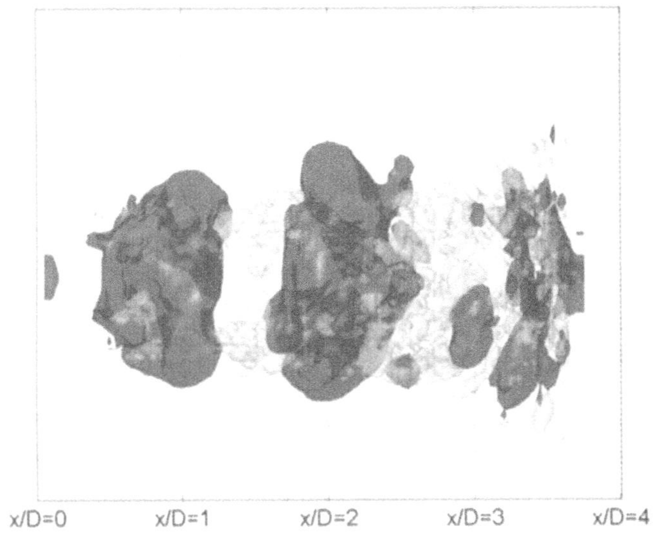

Figure 33.5. Regions of $d > 0.2s^{-6}$ (light grey) and higher pressure (dark grey) in the jet flow with four tabs

4. Conclusions and Future Work

Large eddy simulation was used to investigate the effects of vortex-generating tabs in a turbulent round jet. Turbulence structures in the flow field were identified from instantaneous plots of the discriminant. The introduction of tabs at the nozzle increased the number of positive discriminant bundles in the flow, and increased the size of the high pressure regions between the coherent ring-like structures. The rate of flow entrainment is expected to increase as a result. Two or more tabs were required at the nozzle before any significant changes were observed downstream.

These results represent a first attempt at modelling and visualizing the effects of tabs using large eddy simulation. Additional work is required to study the effects of the size and the configuration of the tabs. Also, the effect of the tabs on the inlet conditions used in the jet model needs to be investigated. However, despite the preliminary nature of these results, this study has demonstrated that large eddy simulation results can be used to illustrate the effects of vortex-generating tabs, and aid in the design of jet nozzles.

Acknowledgments

This work was supported by grants from the Natural Sciences and Engineering Research Council of Canada. Access to the computational resources of the High Performance Computing Virtual Laboratory is acknowledged. The initial computations were performed by Michel Marcourye while working as a DEA student at Queen's University.

References

McIlwain, S. and Pollard, A., 2002, "Large eddy simulation of the effects of mild swirl on the near field of a round free jet", *Phys. Fluids*, to appear.

Liepmann, D. and Gharib, M., 1992, "The role of streamwise vorticity in the near-field entrainment of round jets", *J. Fluid Mech.*, Vol. 245, pp. 643.

Zaman, K.B.M.Q., Reeder, M.F. and Samimy, M., 1994, "Control of an axisymmetric jet using vortex generators", *Phys. Fluids*, Vol. A(6), pp. 778.

Huang, L. and Ho, C., 1990, "Small scale transition in a plane mixing layer", *J. Fluid Mech.*, Vol. 70, pp. 475.

Mi, J. and Nathan G., 1999, "Effect of small vortex-generators on scalar mixing in the developing region of a turbulent jet", *Int. J. of Heat and Mass Transfer*, Vol. 42, pp. 3919.

Zaman, K.B.M.Q., 2000, "Spreading characteristics of compressible jets from nozzles of various geometries", *J. Fluid Mech.*, Vol. 383, pp. 197.

Foss, J. and Zaman, K.B.M.Q., 1996, "Effect of a delta tab on fine scale mixing in a turbulent two-stream shear layer", *AIAA Paper 96-0546*.

Ciofalo, M., 1994, "Large-eddy simulation: a critical survey of models and applications", *Advances in Heat Transfer*, Academic Press, pp. 321.

Germano, M., Piomelli, U., Moin, P. and Cabot, W.H., 1991, "A dynamic subgrid-scale eddy viscosity model", *Phys. Fluids*, Vol. A3(7), pp. 1760.

Kim, J. and Moin, P., 1985, "Application of a fractional-step method to incompressible Navier-Stokes equations", *J. Comp. Phys.*, Vol. 59, pp. 308.

Chong, M.S., Soria, J., Perry, A.E., Chacin, J., Cantwell, B.J. and Na, Y., 1998, "Turbulence structures of wall-bounded shear flows found using DNS data", *J. Fluid Mech.*, Vol. 357, pp.225.

Marcouyre, M., McIlwain, S. and Pollard, A., 2001, "Large eddy simulation of the near field of round jets with vortex-generating tabs", *Proc. TSFP-2.*

Discussion

Dan Ewing, MacMaster University "Nice work. With the tabs, are you using Delta tabs or triangular tabs?"

Reply: These are triangular.

Chapter 34

MIXING AND COMBUSTION, PERSPECTIVES

J. Janicka, A. Kempf, C. Mengler, and A. Sadiki
Energie- und Kraftwerkstechnik, Technische Universität Darmstadt,
Petersenstr. 30, 64287 Darmstadt, Germany

Abstract Besides RANS-techniques, large eddy simulation in three dimensions is nowadays used to study turbulent flow and mixing as well as combustion in various configurations. This paper firstly aims to present some results based on the conserved scalar method in connection with the concept of sub-grid scale pdf for combustion-LES. The second objective is to provide an evaluation of the combustion-LES. To illustrate these, LES results of a range of flows including DLR-flame, counterflow configuration and jet in cross flow are presented. The performance and limitations of submodels are evaluated and discussed.

It is shown that LES is able to deliver good results very close to available measured data. In spite of its high computational cost, LES represents a suitable tool for prediction of combusting flows. Efforts to extend the applicability of LES in complex configurations of technical importance under conceivable computing time must therefore be encouraged.

Keywords: Mixing, Combustion, LES, applied combustion systems

1. Introduction

In an increasing number of practical turbulent flow problems in industrial, environmental and technical applications, turbulent mixing is ubiquitous. In design process for conventional and modern gas turbine engines, both in the field of power generation applications and aircraft jet engines or for any highly turbulent confined combustion sytems as well as in many other combustion processes in open configurations, mixing and combustion are coupled. In non-premixed flames, combustion processes depend on the mixing of fuel and air stream, which in turn is

A. Pollard and S. Candel (eds.), IUTAM Symposium on Turbulent Mixing and Combustion, 387-403.

determined by the state of the flow turbulence. The importance of large scale structures of the flow in scalar mixing and combustion is now well accepted: turbulent eddies can really wrinkle the flame and even quench it; eddies improve heat transfer and mixing of species which contribute to increase the efficiency of technical combustion devices. For modeling, the use of a method that incorporates both large-scale and diffusion effects is highly recommendable. Large eddy simulation (LES) allows all scales of motion that are larger than the grid resolution to be explicitly computed whilst the unresolved small scales must be modeled, so that a very high accuracy in prediction of the flow field can be achieved. The intrinsic unsteady nature of turbulent flows then will be well taken into account. Cabot and Moin (1993) presented an extension of LES to scalar transport problems. Recently, Flohr and Vassilicos (2000) proposed a new approach to simulate passive scalar in LES of turbulence. This is particularly interesting for simulation of chemically reacting flows where an accurate description of mixing is essential. A recent overview of advances in LES methodology for complex flows is provided by Moin (2001).

With regard to combustion, the chemical reactions, in non-premixed flames for example, occur only by molecular mixing of fuel and oxidizer, which in general occurs on the dissipative turbulence scales. Thus, the combustion process occurs essentially at the smallest scales of the subfilter level, which has to be modeled. Therefore, as for RANS-techniques, the chemical source terms need to be modeled. Many recent work deal with the applicability of RANS-like combustion models in LES. Though improvements are achieved continuously, many works imply major simplifications. When using elaborate chemistry models, often the Navier-Stokes equations are solved only in two dimensions in order to reduce the computational cost. A very strong simplification for LES of reacting flows is to restrict density to be constant is space and time. While such simplifications are acceptable to study fundamental phenomena, they can not be applied to flow configurations that occur in technical applications. This fact is now under consideration (see Forkel and Janicka, 2000, Hawkes and Cant, 2000, Steiner and Pitsch, 2000). The proposed models can be classified in five categories: direct method, linear Eddy model, transported pdf method, conserved scalar method and extensions for premixed cases (see DesJardin and Franklin), 1998, Veynante and Poinsot, 2000).

From a numerical point of view, to attempt a description of combustion systems and an accurate prediction of the process evolution in these, it is necessary to formulate an appropriate global model, which includes well-tested submodels and validated against data from well-conceived experimental configurations of technical importance.

This paper firstly aims to present some results based on the conserved scalar method in connection with the concept of sub-grid scale pdf. The validation is achieved by studying turbulent flow and mixing fields as well as combustion in different configurations including DLR-flame, counter-flow configuration and a jet in cross flow.

2. Modeling and numerical procedure

The flow field is described by the filtered equation of continuity together with the three-dimensional filtered Navier-Stokes equations for variable density fluids.

The evaluation of the density requires a description of the combustion process. In this work, we use the mixture fraction formulation. It is then assumed that the chemical state and thereby the species mass fractions can be related to a conserved scalar, the mixture fraction, where the sub-grid scale pdf is presumed to follow a Beta-function distribution. Numerically, fluctuations of density in space and time are considered to only depend on chemistry, not on pressure. To represent the sub-grid scale stresses and scalar flux, a Smagorinsky model is used along with the dynamic procedure and a constant turbulent Schmidt number of 0.7, a common value used by Forkel and Janicka (2000) whilst Cook and Riley (1994) set 0.72. A flamelet approach is used to relate the filtered mixture fraction to density, temperature and species concentrations. The filtered mixture fraction is determined by solving its transport equation, and its sub-grid scale variance is given by an appropriate sub-grid scale model.

Thus, the chemical state is defined by the local mixture fraction, the subgrid fluctuation of the mixture fraction and the rate of scalar dissipation. For scalar dissipation, a δ-function is used to describe its pdf. For modeling the rate of scalar dissipation, the most basic approach is chosen, modeling χ according to

$$\chi = 2D\left(\frac{\partial \widetilde{f}}{\partial x_i}\right)^2 .$$

Following Forkel and Janicka (2000), equations of continuity, momentum and mixture fraction were transformed into cylindrical coordinates and discretized in space by finite volumes utilizing central schemes. The accuracy of approximation is 4th order for convective terms and 2nd order otherwise. The equations are integrated in time by a 3rd order low-storage Runge-Kutta method, pressure is determined by solving a Poisson equation derived from the equation of continuity.

3. Applications

Some LES results from a range of flows that includes the DLR-flame, counterflow configuration and jet in cross flow are presented and discussed. We focus on mixing and combustion, in particular with turbulence, a good overall agreement of the flow field is a prerequisite. Using LES the major features of these fields have been very well captured without any ad hoc parameter adjustement as usually needed in RANS calculation methods, before more detailed investigations could be performed. Detailed results are discussed in Kempf *et al.* (2001), Mengler *et al.* (2001) and in Düsing *et al.* (2000) for LES of a model combustion chamber using oscillating inflow conditions. Furthermore, recent results of LES of a turbulent diffusion flame (test case "H3" defined for the Second International Workshop on Measurement and Computation of Turbulent Non-Premixed Flames) can be found in Forkel and Janicka (2000).

3.1. Counter Flow Configuration

The burner configurations studied here are those investigated by Mastorakos (1999) and Sardi (1998) in which non-reactive flow and reactive case have been considered.

In the non-reactive case, both nozzles use plain air. Apart from the flow field, the mixture fraction was measured. For that purpose, one stream is heated to make the temperature an indicator of the mixture fraction. With the temperatures of the cold jet T_c, of the hot jet T_h and the local temperature T, the mixture fraction is described by $f = (T - T_c)/(T_h - T_c)$. The air investigated is of ambient temperature (300 K) and standard-pressure (0.101 MPa), resulting in a kinematic viscosity of $\eta = 1,84 \cdot 10^{-5}\mathrm{Ns/m}^2$ and a density of $\rho = 1.29\mathrm{kg/m}^3$.

For the reactive case (Mastorakos, 1999), one nozzle uses plain methane instead of air. A plane non-premixed flame is established at half the distance of the nozzles by adjusting the momentum of the methane-jet. The velocity of the air remains almost unchanged. The co-flows spend nitrogen to inhibit methane from igniting in the flue.

In the simulation, a cylindrical domain between the nozzles was investigated. Its height is defined by the nozzle distance, its radius was chosen as a compromise. The domain radius was set to approximately 1.5 times the radius of the pipe, so that part of the co-flow is simulated as well. This domain was discretized by a grid of 104,000 cells, consisting of 65 in axial, 25 in radial and 64 in tangential direction.

The nozzles and the co-flows were modeled using an inflow condition on the abutting face of the cylinder. This consists of a Dirichlet condition for both velocity and mixture fraction and a von Neumann condition for pressure. On the outer ring, laminar inflow is enforced. On the inner section, velocities corresponding to the fluctuating flow out of the nozzle are required.

As boundary condition on the cylinder shell, a simple outflow condition was applied. It consists of a Dirichlet condition for pressure and a von Neumann type condition for velocities and scalars. To ensure stability, negative outflow-velocities are "cut-off" to be zero.

The computed results are found in overall agreement with experimental data for the non-reactive case (Sardi, 1998). For the reactive case, in which a simple combustion model has been used, a satisfactory agreement with measured data by Mastorakos (1999) could be achieved as shown in Fig.34.1-34.2.

3.2. DLR-Flame

To achieve non-lifted methane flames equally suited for experimental and numerical investigation, Meier *et al.* (2000) suggested a fuel mixture of methane (33.2 % vol.), hydrogen (22.1 % vol.) and nitrogen (44.7 % vol.). This fuel will burn stable and not form soot when used for a highly turbulent jet flame, which is referred to as DLR-Flame (8 mm of nozzle-diameter, 0.3 m/s velocity in coflow, 15 200 Reynolds number, 0.167 stoichiometric mixture fraction; DLR = German Aerospace Center). This flame is even more interesting since a wealth of experimental data (see Meier *et al.*, 2000) and simulation results (RANS data by H. Pitsch *et al.*, 2000) is already available. The work presented here applies LES to this flame and examines its suitability for the simulation of such systems.

For the simulation, a cylindric computational domain of 0.38 m length and 0.48 m in diameter was chosen, just beginning at the exit plane of the nozzle. The nozzle and the coflow were modeled using an inflow condition on the abutting face of the cylinder. This consists of a Dirichlet condition for both velocity and mixture fraction and a von Neumann condition for pressure. To describe the turbulent inflow, a turbulent velocity profile computed by the preceeding LES of a pipe-flow was applied.

The computational domain was discretized with an axis-symmetric grid of $129 \times 32 \times 60$ (axial, tangential, radial) cells. To check if this grid resolution was fine enough, the flowfield at the nozzle was recomputed again with refined cells on a domain of half the previous size ($257 \times 32 \times 60$ cells). The computation in the small domain will be labelled C2 (see Fig.34.3). Assuming the highest gradients in velocity and mixture appear close to the nozzle, this method is capable to check for "grid-independence" without the need for twice as many cells in each direction (i.e. 16 times more CPU-time, approx. 5 months). This study shows that an almost grid-independent solution had been achieved, although it shall be stressed here that there is no "absolute" grid independence to be expected in LES-context since the subgrid model varies with varying filter widths (i.e. cell-size with Schumann-Filtering). With smaller resolved length scales, the modeled part of the turbulence vanishes when Kolmogorov-scales are resolved. Furthermore, the over-all error from the model will diminish with finer resolution.

Although experimental and numerical results in Fig.34.3-34.6 generally agree, it is interesting to see how turbulence develops just downstream the nozzle and influences the distribution of mixture fraction. In particular, the turbulence intensities in the self-preservation region showed discrepancies compared to measurements. These differences observed may be due to the inflow-conditions and to reasonably resolving the high gradients (mixture-fraction field in particular) right at the nozzle.

All in all, the prediction of the flow-field is satisfying, although the results suffer of inaccuracies in the nozzle-area. This well known problem in the literature is connected to the transition from the wall-affected turbulence in the nozzle to the free turbulence, along with a structural change in the turbulence. With RANS modeling, some corrections are generally considered, with regard to the determination of some model parameters (H. Pitsch *et al.*, 2000). This kind of ad hoc parameter adjustment does not exist in LES using the Germano procedure.

Since the main species are defined by the mixture fraction, they all behave similarly and thus do not need to be presented. However, flamelet chemistry introduces some interesting information on minor species, so that OH-concentration shall be given here.

3.3. Jet in cross flow

The aim is to investigate the capability of LES using the eddy viscosity approach along with the dynamic procedure to predict flow (mean velocity, Reynolds stress tensor) and a specified scalar eddy diffusivity approach to capture mixing fields (mean and variance of mixture fraction) in a jet discharging into a crossflow. This is achieved by comparing the LES results with linear and non-linear second order turbulence closure models on the one hand and with experimental data of Andreopoulos (1983), Andreopoulos and Rodi (1984) on the other hand. For the RANS simulations, results by Heinrich (2000) are used.

The filtered Navier-Stokes equations along with filtered continuity equation as well as the filtered scalar equation are considered here with constant density. As basis for the simulations the 3-dimensional CFD-Code FASTEST (by INVENT Computing GmbH, Erlangen, Germany) is used which is extended for the large-eddy simulation in complex geometries.

The computational domain is $-2 < x/D < 7$ in the streamwise (x-) direction, $0 < y/D < 4$ in jet exit (y-) direction and $-2 < z/D < 2$ in spanwise (z-) direction. The resolution is about 275000 CV's for the whole domain. As mentioned by Jones and Wille (1996) this domain is big enough. While it is clear that the jet exit is influenced by the crossflow, this fact has to be considered when choosing the inflow conditions of the jet. With regard to this influence, a length of two diameters of the jet pipe are modeled. As can be shown in the results (see Fig.34.7-34.10) the asymmetry of the jet exit is qualitatively reproduced.

For the wall boundary of the pipe and the bottom of the crossflow channel, no-slip conditions were used. The upper and the spanwise faces of the domain were treated as free-slip. The outflow condition is a simply zero-gradient condition like in Yuan et al. (1999). The inflow conditions are described by mean velocity profiles as measured by Andreopoulos and Rodi (1983, 1984) for the crossflow inlet. For the pipe inlet experimental data from Durst et al. (1995) is used. The mean profiles of mean velocities and fluctuations are in one case perturbed with white noise and in the other case calculated without pertubations.

As first part of the results, the influence of inlet pertubation is pointed out. The simulations were performed on an Alpha 21264. Due to the artificial non-realistic velocity field produced with white noise, the implicit solver needs about a factor of 4.4 the time than simulations without pertubation (for details, see Mengler *et al.*, 2001).

After these considerations, the results of the large-eddy simulation using the dynamic procedure, labelled GERM, and Smagorinskys model, labelled SMAG, are compared to the experimental data measured by Andreopoulos and Rodi (1983, 1984) and to RANS simulations by Heinrich (2000) using the models of Jones and Musonge (1988), labelled JM, and Speziale et al. (1991), labelled SSG. All data is shown in the symmetry plane $z/D = 0$ at several positions ($x/D = -0.25, 0, 0.25, 0.5, 1, 2, 4$). Taking in mind the experience with inlet conditions, the results presented in Fig.34.7-34.10 are calculated without pertubations.

The flow field shows excellent agreement with the expertimental data and emphasizes the advantage of the LES compared with RANS modeling, although there are some discrepancies compared to the measurements which cannot be explained. In particular the mixture fraction fluctuation and related quantities show strong deviation. This fact needs to be investigated in detail by using advanced subgrid models for turbulence and especially for mixing.

4. Conclusions

LES is able to deliver good results very close to available measured data. Flamelet chemistry appears well able to reproduce experimental data for species, in particular with regards to kinetic effects prediction, whereas equilibrium chemistry strongly deviates. However, a good predictability could be achieved when appropriate choice of boundary and inflow conditions is made. The use of an inflow generator (e.g. Klein *et al.*, 2001) appears to be very helpful. Combustion-LES prediction results may be improved by using unsteady flamelet concept or other advanced chemistry models.

For complex flows in practical devices, subgrid combustion models should be able to predict without ad hoc model adjustments premixed, non-premixed and partially premixed combustion, which often occur simultaneously. Consideration of liquid spray break up and vaporization brings additional complexities that must be included in advanced subgrid models. Efforts directed to extend the applicability of LES in complex configurations of technical importance under concevable computing time must be encouraged.

Acknowledgments

The authors gratefully acknowledge the financial support by the *Deutsche Forschungsgemeinschaft (DFG)* through the "Sonderforschungsbereich" SFB 568.

References

Bray, K.N.C.; M. Champion, P.A. Libby, *Combustion and Flame* 84: 391-410 (1991)

Cook,A.W.; J.J. Riley, *Physical Fluids*, 6(8), 2868-2870 (1994)

Cook, A.W.; J.J. Riley,*Combustion and Flame* 112, 593–606 (1998)

Forkel, H.; J. Janicka, *Large-Eddy Simulation of a turbulent Hydrogen Diffusion Flame*, Proceedings of 1st Symposium on Turbulent Shear Flow Phenomena, 65–70 (1999)

Germano, M.; U. Piomelli, P. Moin and W.H. Cabot, *Phys. Fluids A*, 3, S. 1760-1765 (1991)

Mastorakos, E. *Turbulent Combustion in Opposed Jet Flows, Ph. D. Thesis*, Imperial College of Science, Technology and Medicine (1999)

Möller,S.I.; E. Lundgren, C. Fureby, *Large Eddy Simulation of Unsteady Combustion*, Proc. Combust. Inst. 26: 241–248 (1996)

Pierce, C.D.; P. Moin, *A dynamic model for subgrid-scale variance and dissipation rate of a conserved scalar*, Physics of Fluids 10(12) (1998)

Sardi, K.; A. Taylor, J. Whitelaw, *Conditional scalar dissipation statistics in a turbulent counterflow*, *Journal on Fluid Mechanics*, vol. 361, pp. 1-24 (1998)

Hawkes, E.R.; R.S. Cant, *Proc. Combust. Inst. 28*, 51–58 (2000)

Kempf, A.; H. Forkel, J.-Y. Chen, A. Sadiki, J. Janicka, "Large Eddy Simulation of a Counterflow Configuration with and without Combustion", *Proc. Combust. Inst. 28*, 35-41 (2000)

Meier, W.; R. Barlow, J.-Y. Chen, "Raman/Rayleigh/LIF Measurements in a Turbulent CH4/H2/N2 Jet Flame: Experimental Techniques and Turbulence Chemistry Interaction", *Combustion and Flame* 123, pp. 326-343 (2000)

Peters, N., "Laminar Flamelet Concepts in Turbulent Combustion", *Proc. Combust. Inst. 21*: pp. 1231-1250 (1986)

Pitsch, H.; O. Kunz, J.-Y. Chen RANS Modeling of the DLR-flame, *5th Workshop on Measurement and Computation of Turbulent Non-Premixed Combustion*, Delft, Netherlands (2000)

Steiner, H.; H. Pitsch, "Large-Eddy Simulation of a Turbulent Piloted Methane/Air Diffusion Flame (Sandia Flame D)", accepted for publication in *Physics of Fluids*, (2000).

Cabot, W. and Moin, P., In Galperin, B. and Orszag, S.A., editors, p. 141-158, Cambridge Univ. Press, (1993)

Colucci, P. J., Jaberi, F. A., Givi, P. and Pope, S.B. 1998: *Phys. Fluids A*, 10: 499-515

DesJardin, P. E. and Franklin, S.H. 1998: *Phys. Fluids A*, 10:2298-2314

Düsing, M., Sadiki, A. and Janicka, J. 2000: LES of a model combustion chamber using oscillating inflow conditions. *EUROMEC Colloquium 412*, Munich

Flohr, P. and Vassilicos, J.C. 2000: *J. Fluid Mech. 407*:315-349

Forkel, H. and Janicka, J. 2000: LES of a turbulent Hydrogen Diffusion *Flame, Flow, Turbulence and Combustion 65*, 163–175

Kempf, A., Schneider, C., Sadiki, A. and Janicka, J. 2001: LES of a highly turbulent methane flame: application to the DLR Standard flame, Proceedings of *Turbulent Shear Flow Phenomena 2*, 347–352, Stockholm.

Kempf, A., Forkel, H., Chen, J.-Y., Sadiki, A. and Janicka, J. 2000: LES of a Counterflow Configuration with and without Combustion,*Proc. 28th (Int.) Combustion Symp.* 35–41

Mengler, C., Heinrich, C., Sadiki, A. and Janicka, J. 2001: Numerical prediction of momentum and scalar fields in a jet in cross flow: Comparison of LES and Second order turbulence closure calculations, Proceedings of *Turbulent Shear Flow Phenomena 2*, 425–430, Stockholm.

Veynante, D. and Poinsot, T. 1997: LES of combustion instabilities in turbulent premixed burners. *Annual Research Briefs.*

Alvarez J., Jones W. P. and Seoud R., 1993, "Predictions of momentum and scalar fields in a jet in crossflow using first and second order turbulence closure", *Computational and experimental assessment of jets in crossflow, AGARD-CP-534*, pages 24.1–24.10. AGARD

Andreopoulos J., 1983, *Phys. Fluids*, 26(11):3201–3210

Andreopoulos J. and Rodi W., 1984, *J. Fluid Mech.*, 138:93–127

Durst F., Jovanovic J., and Sender J., 1995, *J. Fluid Mech.*, 295:305–335

Heinrich C., 2000, "Numerische Berechnung gasturbinentypischer Mischungsvorgnge mit vollstndigen Turbulenzmodellen zweiter Ordnung", *PhD thesis*, TU Darmstadt

Jones W. P. and Wille M., 1996, "Large eddy simulation of a round jet in a crossflow", In *Engineering Turbulence Modelling and Experiments 3*, volume 1, pages 199–208. Elsevier

Kim, J. and Moin, P., 1989, Proceedings of *Turbulent Shear Flows 6*, Selected Papers from the Sixth International Symposium on Turbulent Shear Flows, pages 85–96, Springer Verlag

Lilly, D. K., 1992, *Phys. Fluids A*, 4(3):633–635

Schlüter J. U., Angelberger C., Schönfeld T., and Poinsot T., 1999, Proceedings of *Turbulent Shear Flows Phenomena 1*, pages 71–76

Weinberger C., Rewerts J. and Janicka J., 1997, *J. Fluid Mech.*, *11th Symp. on Turbulent Shear Flows*, Grenoble, 8-10. september 1997. 25.17–25.22

Yuan L. L., Street R. L., and Ferziger J. H., 1999, "Large-eddy simulation of a round jet in crossflow", *J. Fluid Mech.*, 379:71–104

Moin, P. 2001, Advances in large eddy simulation methodology for complex flows, Proceedings of *Turbulent Shear Flow Phenomena 2*, 1–10, Stockholm.

Klein, M., Sadiki, A., Janicka, J. 2001, Proceedings of *Turbulent Shear Flow Phenomena 2*, 401–406, Stockholm.

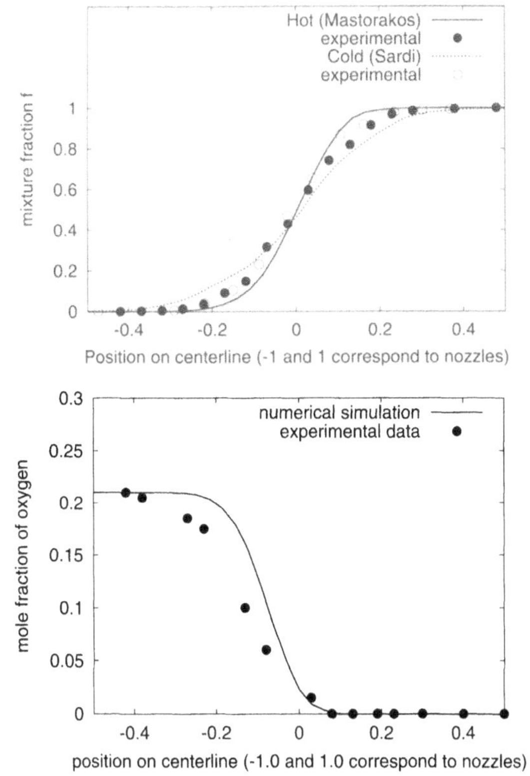

Figure 34.1. Mean of Mixture Fraction and Species fraction for oxygen

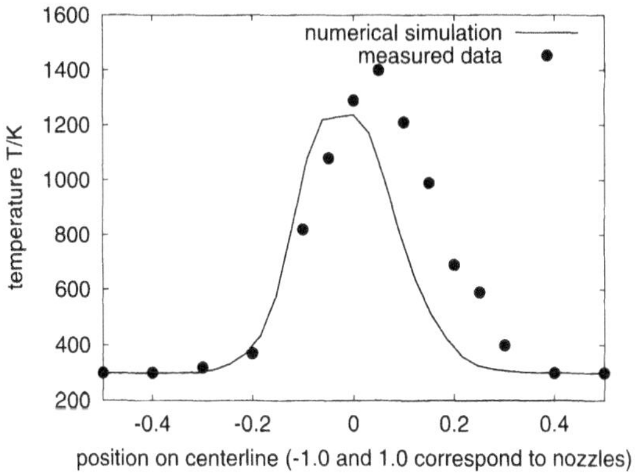

Figure 34.2. Mean of temperature, reactive case

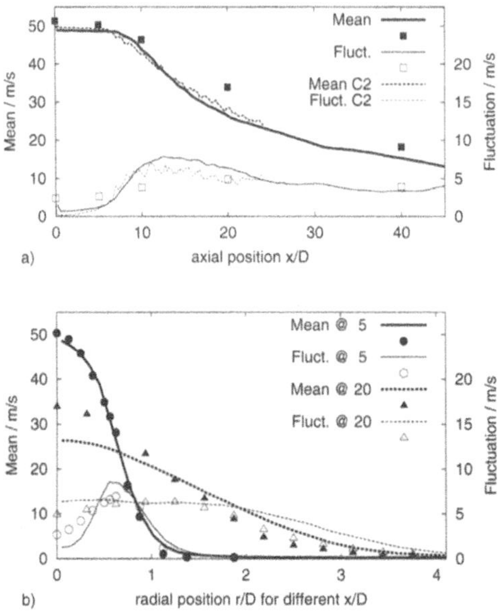

Figure 34.3. Axial velocities over axis/radius

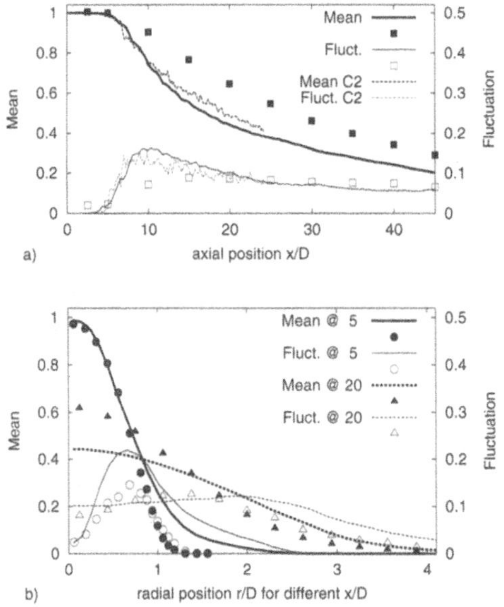

Figure 34.4. Mixture fraction over axis/radius

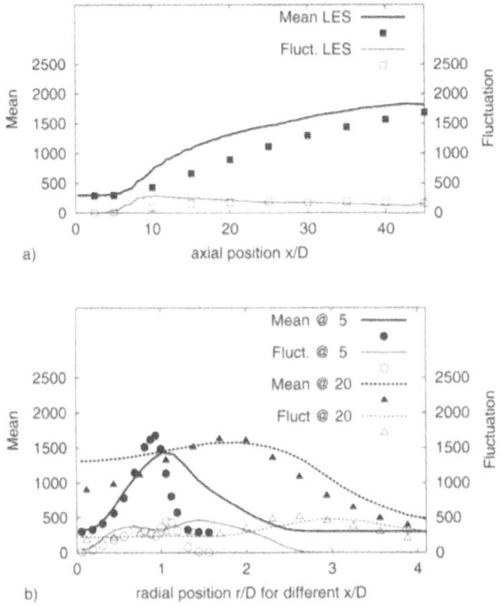

Figure 34.5. Temperature over axis/radius

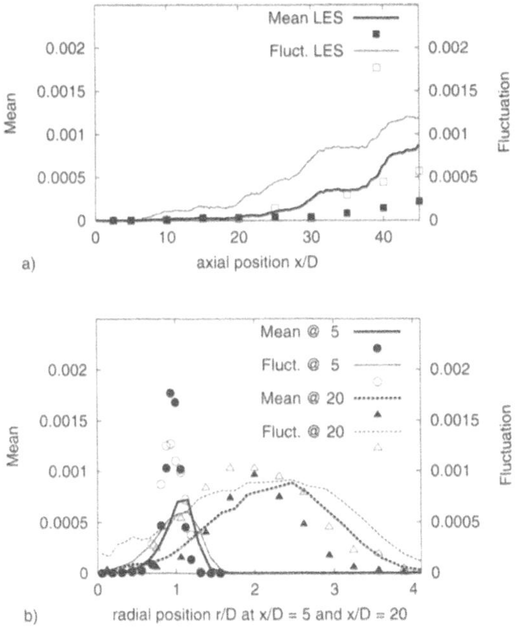

Figure 34.6. OH-Concentration over axis/radius

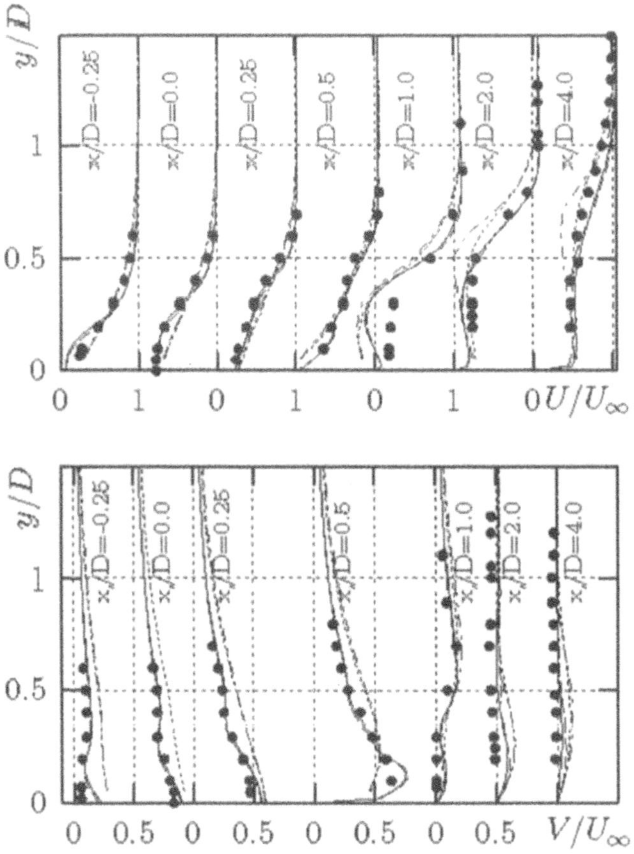

Figure 34.7. Profiles of mean Velocity U/U_∞ and V/U_∞ (Exp.:•; SMAG:– – –; GERM:——; JM:· · · ·; SSG:- · - ·)

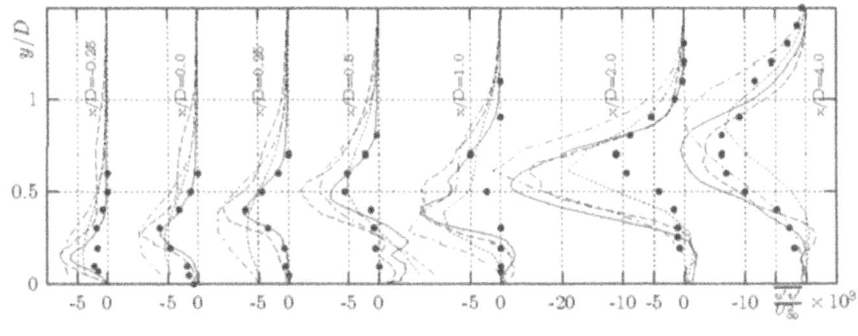

Figure 34.8. Profiles of $\overline{u'v'}/U_\infty^2 \times 10^3$ (Exp.:•; SMAG:– – –; GERM:——; JM:· · · ·; SSG:- · - ·)

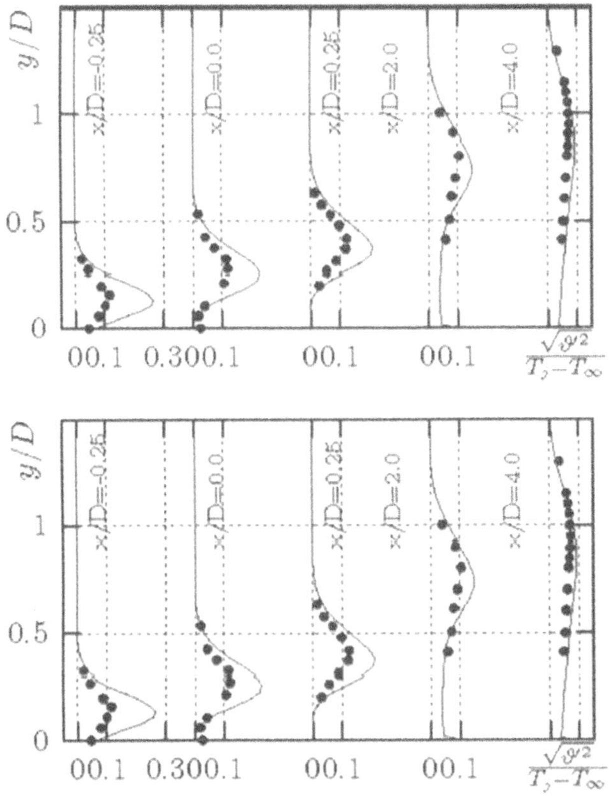

Figure 34.9. Profiles of mean temperature $T - T_\infty/T_j - T_\infty$ and rms temperature fluctuations $\sqrt{\vartheta'^2}/T_j - T_\infty$ (Exp.:•; GERM:—; JM:$\cdots$)

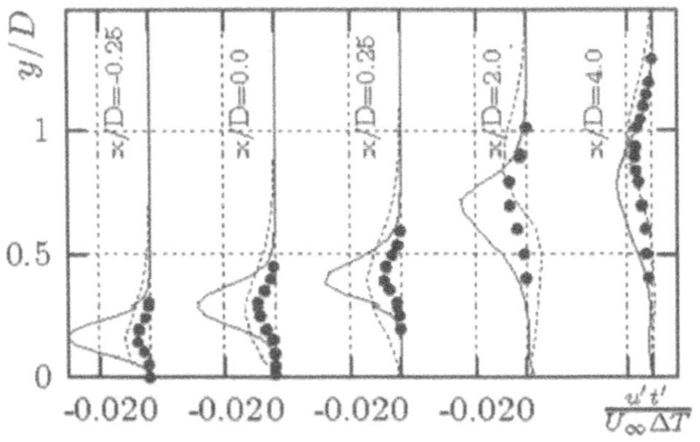

Figure 34.10. Profiles of $\overline{u'\vartheta'}/U_\infty(T_j - T_\infty)$ at various positions x/D and z/D=0 (Exp.:•; GERM:—; JM:····)

Discussion

Sébastien Candel, EMC2, Paris "Thank you for this nice lecture. I had one question concerning the G equation. You cited one of the difficulties about renormalizing the field. The other is the fact that across the g interface the temperature will jump from low to high and this is due to the heat release taking place. Many years ago we did some studies of how to handle that. And I would like to have your opinion on how you handled that."

Reply: Ah, well. Two statements. It is still a work in progress. So we are not quite sure. On the other hand we have a lot of experience in applying volume of fluid methods for liquid gas interfaces and that's a problem so we handle this up to density ratios up to thirty approximately. I have to remember - you have to apply - especially at the interfaces discretization methods where you make sure that the conservation of mass will be sustain and that a flux method where you switch between upwinding and central differences in a proper way. I think that in the last five years some methods, Soleski is a good name for volume of fluid methods, and I think we could apply this. But that is still not done, it is underway.

Norbert Peters, Stanford University "Would you also like to know how we do it?. Well there a two aspects. Laminar flow of course a steep jump and then we do in cell reconstruction methods. We really go into the grid cell and see which part is burnt and what part is unburned, Rupert Kline has developed his techniques. And then the other aspect is of course when your turbulent or LES then the flame brush is a measure of the thickness of the flame and this is much smoother. We don't run into the problem that you mentioned."

Reply: For RANS environment it is no problem. We did it and you smooth it out over ten cells or so and that is no problem.

Norbert Peters, Stanford University "But I wanted to comment on your talk. You mention that there is no problem with boundaries and combustion. I don't think that is true. We have found with diesel engine. Now a days diesel engines with high injection velocity, high pressure all the combustion takes place on the walls and soot formation comes by quenching of these reaction zones by the wall."

Reply: For internal combustion engines I agree.

Valery Zapryagaev, Novosibirsk, Russia "What you say about the use of LES method for calculation of super sonic and hyper sonic flow with shock waves?"

Reply: We don't do this. Because we feel, because hypersonic and supersonic combustion is not a big problem in Germany because we don't have a strong military industry. So we are basically interested internal combustion engines, gas turbines, and power plants. And we make the decision at the very beginning otherwise we had to switch to a compressible code. So we said we can cover 95% or 98% of the applications in our environment with this incompressible code. We made the decision that missed the other 2%. But they are not important in terms of funding in Europe. This was clearly the decision we had to make. We went on the other hand, we had a factor of 10 which is 10 or 20 times faster then compressible code. That won't be nice in certain environments I admit but that is something we will do in cooperation with other people. We would not develop alone.

A. Pollard, Queen's University "You mentioned that there is no data for jet in a cross flow."

Reply: Not so much. There is a lot of data but you don't really have data that is that complete as much as I would like to see- say scalar field, vorticity field at different positions, boundary conditions are very important. I don't think that there is such complete data on the market.

A. Pollard, Queen's University "Come and see me."

Steve Pope, Cornell University "Maybe I missed this in your talk one of the TNF target flames is the Barlow and Flame X (?) which has local extinction and reignition. Are you able to do that and if so with what combustion model?"

Reply: We did not do this so far. We will do it, but we didn't put it on the first part of the list because we know that there are a lot of other predictions and its probably not that attractive to start with flames that are done by a lot of other people so we don't do this. So we started with other flames. I expected these questions.

Paul Dimotakis, Caltech "You mentioned the military program. Right now the nuclear weapons industries are suffering under a comprehensive test ban treaty in some sense this community has imposed such a treaty on itself and I am reminded of a statement that was made by the head of the atomic weapons establishment in England many years ago before this one where in the absence of experiments our confidence went up and our competence went down so I think it is very important for this community, as you pointed out, to re-establish and reaffirm its commitment to find detailed experiments with those we are going to become increasing confident in our simulations."

Reply: Yes, I agree.

Chapter 35

USING LARGE EDDY SIMULATIONS TO UNDERSTAND COMBUSTION INSTABILITIES IN GAS TURBINES

Thierry Poinsot*
IMF Toulouse
Av.C. Soula, 31400 Toulouse Cedex
France

Jörg Schlüter, Ghislain Lartigue, Laurent Selle,
CERFACS
42 Av. Coriolis
31057 Toulouse Cedex
France

Werner Krebs and Stefan Hoffmann
Siemens AG KWU
45466 Mulheim
Germany

Abstract This paper presents a study of the stability of a swirled premixed combustion chamber both with and without reaction using Large Eddy Simulation and a numerical solver able to handle complex geometries. It is shown that cold flow instabilities, which are observed in the non-reacting case (especially precessing vortex cores), are strongly damped in the reacting case, suggesting that these natural flow instabilities are not the first source of combustion instabilities in such configurations.

Keywords: Large eddy simulation, combustion instabilities, swirling flows

*Corresponding author, email: poinsot@cerfacs.fr

A. Pollard and S. Candel (eds.), IUTAM Symposium on Turbulent Mixing and Combustion, 405-413.

Introduction

Unsteady reacting flows can not always be computed using classical Reynolds Averaged Navier Stokes (RANS) approaches. Turbulent combustion phenomena such as flame flashback, blowoff or combustion instabilities [McManus et al., 1993,Colin et al., 2000] are not well defined in a RANS framework where only time-averaged quantities are solved for. Moreover, such flows are strongly coupled with acoustic waves and their computation often requires high-order schemes and low turbulent viscosity. For such problems, large eddy simulation (LES) techniques are viewed today as a promising tool. More specifically, LES is the best tool to address a question of interest for modern lean combustors: to satisfy emission regulations, modern gas turbines operate in very lean combustion regimes. These flames are extremely sensitive to combustion oscillations but the exact phenomena leading to instabilities are still discussed. A central question for modeling approaches is to determine the phenomena inducing unsteady reaction rates, required to sustain oscillations, when an acoustic wave enters the combustion chamber. For most gas turbines, this question is linked to the development of large scale vortices at the turbine inlet. These large-scale structures may be due to different causes:

- An amplification of the natural hydrodynamic modes of the flow, especially the precessing vortex core (PVC) phenomenon well known in swirling flows

- A purely acoustically forced mechanism similar to vortices formed in impulsively started jets in which a velocity excursion at the chamber inlet creates a mushroom-shaped vortex.

In the present work, LES was used to study the instabilities appearing in a swirled configuration, close to an industrial gas turbine. Both non-reacting and reacting cases (in premixed mode) were studied and compared. For reacting cases, comparisons with experiments performed at University of Karlsruhe are presented in terms of velocity field. For non-reacting flows, no experimental data was available and a simple comparison between non-reacting and reacting cases is proposed in terms of instabilities observed in the two LES.

1. Configuration

A sketch of the configuration is given in Fig. 35.1. The premixed gases are injected into the chamber through a diagonal swirler where vanes are used to swirl the flow. For the present LES, the computation

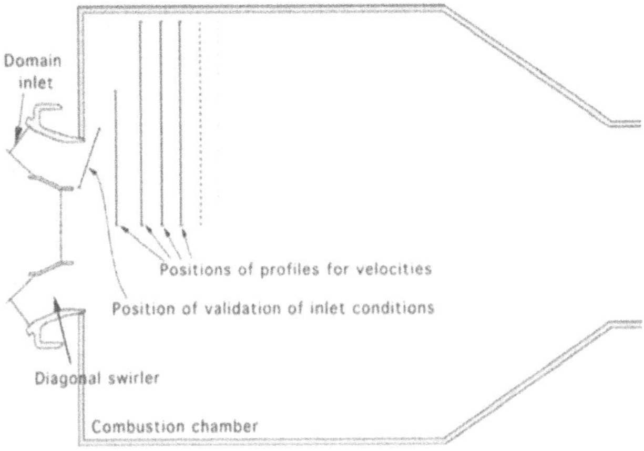

Figure 35.1. Mean reaction rate as predicted using the Magnussen eddy dissipation model (Mean flow simulations).

starts after these vanes and swirl is introduced as a boundary condition at the domain inlet to minimize the computational load.

2. Models and numerical tool

For this work, the AVBP solver (www.cerfacs.fr/cfd) was used. This solver includes the following features:

- An LES model for combustion based on the Thickened Flame model proposed by [Angelberger et al., 1998], [Colin et al., 2000] and [Légier et al., 2000]. This model thickens reaction zones but reduces to normal LES mixing models away from flame fronts. It has the capability of handling pure mixing, premixed flames as well as diffusion flames with fast chemistry.

- Acoustic boundary conditions which allow the control of waves entering and exiting the combustion chamber ([Poinsot and Veynante, 2001]) and therefore the amount of acoustic power left in the chamber.

- An hybrid mesh solver [Schönfeld and Rudgyard 1999] in which these models are implemented. This solver uses high-order schemes on unstructured grids [Colin and Rudgyard, 2000] which are an essential feature of adequate LES simulations.

- A filtered Smagorinsky model to evaluate subgrid scale stresses. This model has a sufficient accuracy for such applications and is much simpler to implement than dynamic models on this kind of meshes [Ducros et al., 1997].

3. Inlet boundary conditions

Fixing boundary conditions in complex geometries is one of the difficult tasks of LES. For the present case, profiles of velocities were measured at the entrance of the chamber (i.e. at the end of the diagonal inlet, see Fig. 35.1) and boundary conditions at the inlet of the computational domain were tuned (in terms of axial and swirling velocities) until a reasonable agreement was found at this first station between measured mean velocities and LES data [Schlüter 2000], [Schlüter et al., 2001]. This set of conditions was then conserved for all computations. All acoustic boundary conditions (inlet and outlet) were non-reflecting to limit the possibility of creating an acoustically coupled oscillation.

4. Non-reacting mean flow results

The results obtained at University of Karlsruhe include measurements (using LDA) of the axial velocity u_x, of the circumferential velocity u_θ and of their RMS values. A full discussion of these results compared to LES data is given in [Schlüter et al., 2001]. Only the main results are summarized here. Fig. 35.2 shows profiles of the mean velocity $\bar{u}_x$ obtained by LES and experiment. Fig. 35.3 shows profiles of the mean circumferential velocity $\bar{u}_\theta$. All velocities have been normalized by the bulk velocity U_b determined by the total volume flux and the surface of the burner nozzle.

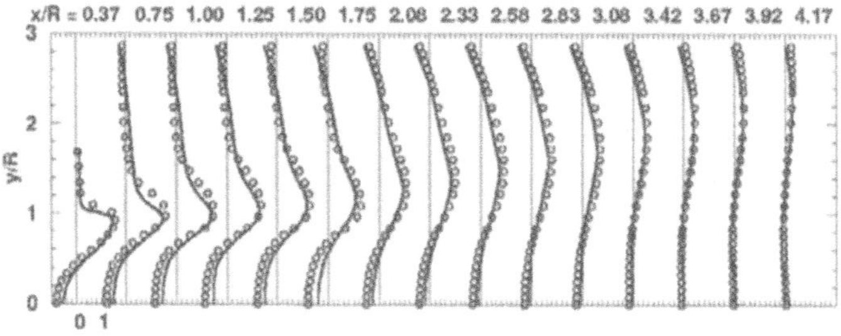

Figure 35.2. Transverse profiles of normalized mean axial velocity $\bar{u}_x/U_b$ on various stations along the burner axis. Lines: LES results. Circles: experiments (Karlsruhe).

A very good agreement is obtained between LES and experimental results. Similar agreements are obtained for RMS values, confirming the capacity of LES to predict such swirling flows correctly.

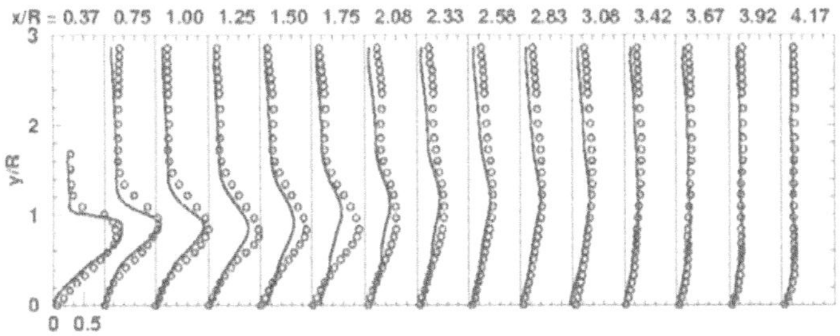

Figure 35.3. Transverse profiles of normalized mean circumferential velocity $\bar{u}_\theta/U_b$ on various stations along the burner axis. Lines: LES results. Circles: experiments (Karlsruhe).

5. Non-reacting flow instabilities

In addition to mean flow fields, LES also provides insight into the unsteady structure of the flow. In the present case, the non reacting flow is highly unsteady. Oscillation frequencies are in the range $St = fD/U_b = 0.6$ where D is the burner diameter. Instantaneous snapshots of the magnitude of the circumferential velocity are given in Fig. 35.4 to 35.6 for three instants separated by 0.8 ms.

Fig. 35.4 to 35.6 reveal that vortex shedding takes place at the chamber inlet. This shedding takes place on the inner cone of the swirler, aternatively between opposite sides of the chamber inlet and lead to the formation of a single vortex spiral oscillating inside the burner, a configuration very similar to classical PVC. Note that vortices are also formed on the outer side of the cone but with smaller intensity.

6. Reacting flow simulations

The existence of a strong PVC instability with cold flow leads to the question of its existence for reacting flows. This was investigated using LES of a premixed case with reaction activated. The main result is that combustion seems to inhibit the PVC mode observed in the cold flow. Fig. 35.7 shows a snapshot of temperature isosurfaces in two planes passing through the axis of the burner. Very limited vortex shedding is observed in this case and no PVC seems to appear in the flow, indicating that the strong unstable mode observed without combustion may be damped when reaction is turned on. This results has probably no generic character: it is well known that combustion can also destabilize flows. However, in the present burner, and for this regime, these results suggest that hydrodynamic instabilities (and especially PVC) cannot ex-

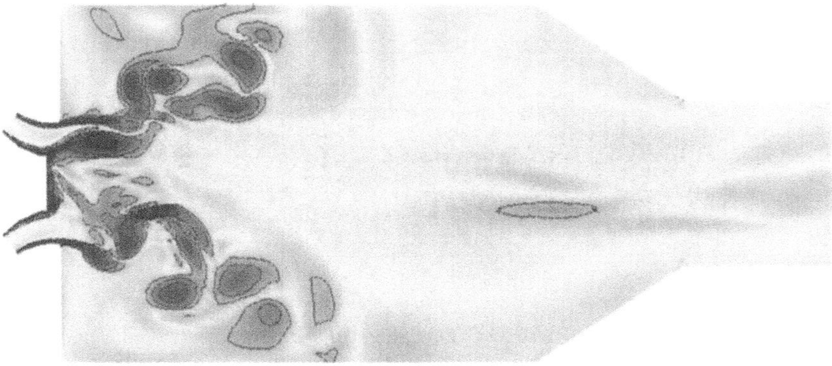

Figure 35.4. Magnitude of the circumferential velocity in the central plane of the burner. Instant t_0.

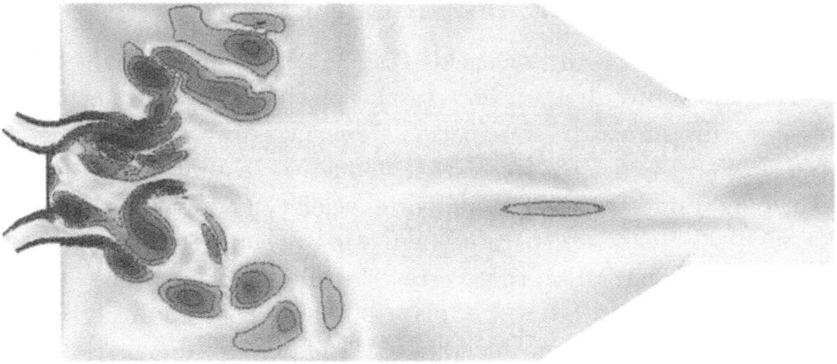

Figure 35.5. Magnitude of the circumferential velocity in the central plane of the burner. Instant t_0+0.8 ms.

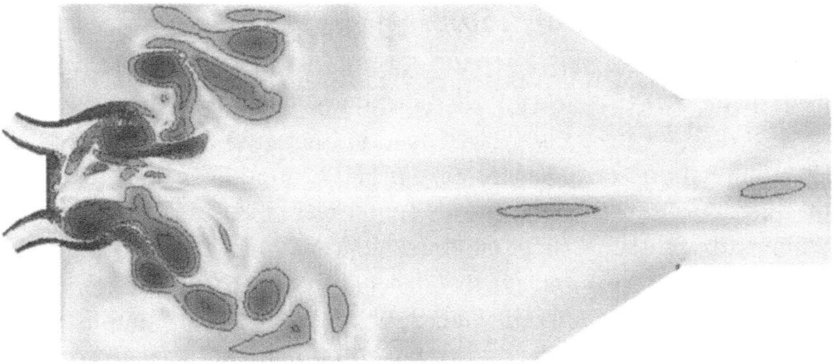

Figure 35.6. Magnitude of the circumferential velocity in the central plane of the burner. Instant t_0+1.6 ms.

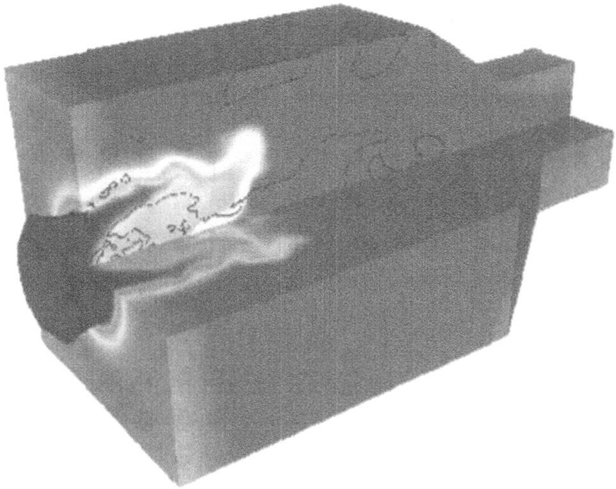

Figure 35.7. Temperature field in two planes for the reacting flow case.

plain reacting flow instabilities. Such instabilities must involve a strong acoustic forcing at the inlet and cannot appear only by hydrodynamic instabilities. Of course, both phenomena (acoustic and hydrodynamic) are probably coupled in real devices.

7. Conclusion

LES has been performed in a complex gas turbine configuration, both with and without combustion. This shows first the feasibility of such studies using available tools and computers. In the present application case, LES was first used to compute the mean flow field without combustion: comparison with experimental data of Karlsruhe showed very good agreement. LES also revealed the existence of a strong precessing vortex core in the absence of flame, which disappears when combustion is activated.

8. Acknowledgments

The support of CINES, which provided the computing power required for this study, is gratefully acknowledged.

References

Angelberger, C., Veynante, D., Egolfopoulos, F., and Poinsot, T. (1998). Large eddy simulation of combustion instabilities in turbulent premixed flames. In *Proceedings of the Summer Program*, pages 61–82. Center for Turbulence Research, Stanford.

Colin, O., Ducros, F., Veynante, D., and Poinsot, T. (2000). A thickened flame model for large eddy simulations of turbulent premixed combustion. *Phys. Fluids A*, 12(7):1843 – 1863.

Colin, O. and Rudgyard, M. (2000). Development of high-order Taylor-Galerkin schemes for LES. *J. Comp. Phys.*, 162:338 – 371.

Ducros, F., Nicoud, F., and Schnfeld, T. (1997). Large eddy simulations of compressible flows on hybrid meshes. In *Eleventh Symposium on Turbulent Shear Flows*, Grenoble, France.

Légier, J., Veynante, D., and Poinsot, T. (2000). Dynamically thickened flame les model for premixed and non-premixed turbulent combustion. In *Proceedings of the Summer Program.* Center for Turbulence Research, Stanford.

McManus, K., Poinsot, T. and Candel, S. (1993). A review of active control of combustion instabilities. *Prog. Energy Comb. Sci.*, 19: 1 – 29.

Poinsot, T. and Veynante, D. (2001). "Theoretical and numerical combustion" R.T. Edwards. ISBN 1-930217-05-6.

Schlüter J. (2000). Large Eddy Simulations of Flow and Mixing in Jets and Swirl Flows : Application to a Gas Turbine. *PhD thesis*, INP Toulouse.

Schlüter J., Schonfeld, T., Poinsot, T., Krebs, W. and Hoffmann, S. (2001). Characterization of confined swirl flows using Large Eddy Simulations. *Proc. of ASME Turbo Expo 2001, New Orleans*, Paper 2001-GT-0060.

Schönfeld, T. and Rudgyard, M. (1999). Steady and Unsteady Flows Simulations Using the Hybrid Flow Solver AVBP. *AIAA Journal*, 37: 1378-1385.

Discussion

Norbert Peters, Stanford "Well unfortunately it is very often the other way around. Instabilities appear with combustion. Are you confident that the second flame model will be able to response to pressure fluctuations in the way reality does?"

Reply: Well I tell you something. Its been the first case in which we've been able to run large eddy simulation with the second flame model that didn't go unstable. In most cases they go unstable very rapidly and again it because the details of the model here are not so important the real problem is the acoustics and we have many other simulations the one that was on the poster outside for example the flow couples very rapidly. So my feeling is that again that if you get the acoustics right the details of the model will not matter. But there is only one way to do that and it is to try with another model.

Parviz Moin, Stanford University "Non deflecting bonding conditions. Do you really have real non reflecting boundary conditions in the sense that you non linear waves and perturbation going in and out of the domain. Have you been able to come up with a true non reflecting boundary conditions."

Reply: No. It can not be truly non reflecting it is partially non reflecting because you need to keep the main pressure constant and the other thing is if the wave is totally non-linear it is partially reflecting and the answer to the last part of this question it is only non reflecting for the outlet here but there are walls here. So there are reflections for these walls and you can get coupling, even if you set up non-reflecting conditions.

Parviz Moin, Stanford University "I agree with you 100% about the accuracy of numerical techniques. What makes you think that your numerical method is very accurate?"

Reply: Well, it depends what you call very accurate. When we do a lot of tests on this flow we do that for canonical cases like acoustic waves, 1D flames and diffusion flames and we are sure for those cases that we are getting a quite accurate method.

Parviz Moin, Stanford University "But do you have numerical dissipation."

Reply: No. No, very little. It is very small. We have a fourth order, artificial viscosity.

Parviz Moin, Stanford University "Our experience has been even with the fifth order, upwinding, by the time you get half way through your combuster you annihilate all the fluctuations. You pretty soon you don't deal with turbulent eddies you just deal with a glue."

Reply: For the core flow we got a lot of measurements. We did the swirl flow the case of Delenback, and we did the jet and cross flow. We got excellent statistics all over, even downstream, I think we also used the third order scheme, seems to be doing a good job now I think that it is also a matter of mesh. We found that we could do a good job. But I agree with the you a better scheme gives you a better result.

Parviz Moin, Stanford University "But if we lower statistics in general they do fine. But if you look at say power spectra which I am wondering for problems as sensitive as combustion instabilities to the power spectrum of turbulence if it might be important to look at that."

Reply: Yes, I agree.

Chapter 36

LARGE-EDDY SIMULATION OF TURBULENT SPRAYS

S. Menon and V. Sankaran
Georgia Institute of Technology
Atlanta, Georgia, 30332, USA
suresh.menon@ae.gatech.edu

Abstract Large-eddy simulations (LES) of high Reynolds number spray with and without combustion in a full-scale gas turbine combustor are carried out to investigate the interaction between the liquid and gas phases under realistic conditions. Effect of changing the inlet swirl on droplet dispersion and transport is first studied followed by an investigation of effect of heat release for the same initial swirl conditions. Results show that for the non-reacting spray, increase in swirl dramatically increases both spray dispersion and breakdown of the initially coherent azimuthal vortices shed from the dump plane. Vortex breakdown and formation of the centerline recirculation bubble is shown to be primarily a function of initial swirl for this device. With heat release, the shape and extent of the recirculation bubble (when it exists) changes significantly from the equivalent non-reacting case due to changes in the velocity field caused by thermal expansion in the near field of the dump plane.

Keywords: Turbulent sprays, swirling flows, Large-eddy simulations

Introduction

Modeling of two-phase reacting flows is currently carried out using steady-state methods where only the time-averaged flow field is simulated. However, for complex flows, the turbulence models that are employed are not adequate, and yield poor agreement between calculations and experiments. In particular, combustion in realistic gas turbine engines occurs in a highly swirling, turbulent flow which is intrinsically unsteady. Therefore, conventional steady-state models are incapable of dealing with the physics of these types of flow. In the present study,

A. Pollard and S. Candel (eds.), IUTAM Symposium on Turbulent Mixing and Combustion, 415-425.

we employ large-eddy simulations (LES) to study unsteady mixing and combustion in a gas turbine engine. Earlier studies ([Menon and Calhoon, 1996,Kim et al., 1999,Kim and Menon, 2000]) have established the ability of the present LES model. Extension to two-phase mixing layers flows was also demonstrated earlier ([Menon and Pannala, 1997, Pannala and Menon, 1998]). The present study focuses on the simulation of turbulent mixing and combustion in a liquid fueled gas turbine engine.

1. Simulation Model

Although the inflow velocity into a realistic gas turbine combustor (e.g., General Electric LM6000) is subsonic it is not necessarily small (typically, it is around 100 m/s) and significant interaction between the various modes of oscillation (e.g., acoustic, flow, unsteady heat release) occurs, which under certain conditions can lead to large-amplitude, pressure oscillations often identified with combustion instability. To resolve and understand the dynamics of these interactions a full compressible formulation is essential. The compressible LES equations are obtained by spatial filtering of the conservation equations of motion. Details are given elsewhere and therefore, avoided here for brevity ([Menon et al., 1996,Kim et al., 1999,Kim and Menon, 2000]). The closure of the subgrid flux terms is achieved using a one-equation transport model for the subgrid kinetic energy ([Menon et al., 1996]). A localized dynamic approach has been developed using a scale-similarity approach to obtain the coefficients of this model ([Kim et al., 1999,Kim and Menon, 1999]).

In the present study, a gradient closure for the subgrid scalar flux is employed along with infinite rate kinetics to close the scalar LES equations. Note that, since large-scale motion is resolved in a LES, associated counter-gradient processes are resolved (even when a gradient closure is employed for the subgrid scalar flux). A more accurate closure approach based on a subgrid simulation (as opposed to subgrid modeling) of the reaction-diffusion equations ([Menon and Calhoon, 1996, Chakravarthy and Menon, 2001]) has also been developed and is currently being used for finite-rate kinetics. These results will be reported elsewhere.

For spray combustion, the LES gas-phase equations contain volume averaged inter-phase source terms which are obtained using standard relations for dilute sprays, as described elsewhere ([Faeth, 1987, Pannala and Menon, 1998]). The droplets in the spray are tracked using a Lagrangian method which is fully coupled to the gas phase Eulerian model. In this method, instead of each droplet, droplet groups (that represent droplets of same size, location, temperature and velocities) are tracked. The mass, momentum, energy and species transfer between the continuum and the dispersed phase is included at every time step. The gas phase LES velocity field and the subgrid kinetic energy are used to estimate the instantaneous gas velocity at the droplet location. Drag effects on the gas phase is included using semi-empirical models [Faeth, 1987]. To include stochastic dispersion of the droplets, a random velocity component is added to the gas phase fluctuating velocity. Heat transfer from gas phase to the liquid phase aids in the vaporization and the subsequent mass transfer to the gas phase. Thus, full coupling is achieved between the two phases.

2. Numerical Approach

The LES equations are solved using a finite-volume scheme that is nominally second-order accurate in space and time (a fourth-order spatially accurate scheme is also available). No-slip, adiabatic wall conditions are used along with non-reflecting inflow/outflow boundary conditions. Clustering is employed near walls and in shear layers to better resolve large scale fluctuations. A computational grid of 141 x 75 x 81 is employed in all these simulations.

The configuration under study is a cylindrical dump combustor which consists of a converging inlet section with a central injector cone. Figures 27.1(a) and (b) shows respectively, the geometry and the computational grid. The inlet pipe simulates the region downstream of the swirler. A swirl velocity profile (along with a turbulent fluctuating field of (7%) intensity is imposed at the inlet with a mean inlet mass flow rate of 0.435 kg/sec at a temperature and pressure of 816 K, and 1.38 MPa, respectively. The Reynolds number based on inlet bulk velocity and inlet diameter is $\approx$ 260,000.

The n-heptane droplets are injected from the circumference of the central cone. Particle displacement is integrated using a fourth order Runge-Kutta method. Elastic wall collisions are assumed and a total of (approximately) 100,000 droplet groups are simulated in the non-vaporizing case. In the present study, particles of a constant diameter (40 microns, in the non-vaporizing case) are injected and when they become smaller than 5 microns (in radius) they are assumed to become fully vaporized. Effect of this cut-off size on the combustion process will be addressed later using a new subgrid two-fluid model ([Pannala and Menon, 1998]). In the present study, combustion of n-heptane is modeled using infinite rate kinetics resulting in a flame temperature of 2100K.

Inclusion of spray introduces new time scales in the flow. To accurately calculate the particle trajectories, size and temperature, the

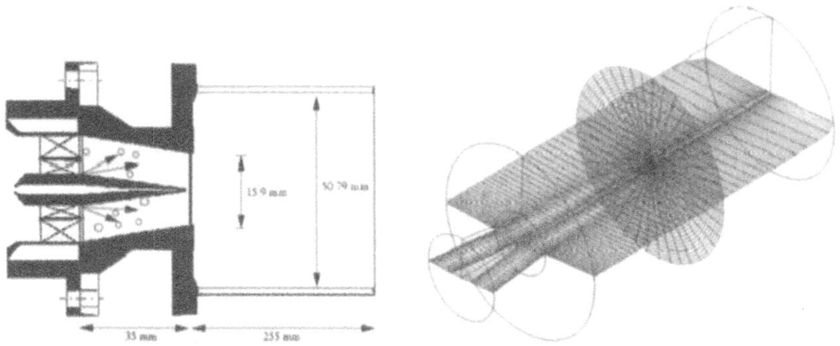

(a) Combustor configuration (b) Computational grid

Figure 36.1. Schematic of the simulated device. Only the domain downstream of the swirl vanes is simulated.

Lagrangian time step used for integration has to be the smallest of the time scales involved. The spray time step at any instant is determined as the minimum of the droplet velocity relaxation time, the droplet life time, the droplet surface temperature constraint time, the local grid time scale, and the turbulent eddy droplet interaction time. The definitions and the expressions for calculating these time steps are given elsewhere ([Faeth, 1987]).

3. Results and Discussion

3.1. Spray Combustion Dynamics

This study focuses on comparing the effect of swirl and heat release in particle laden flows. Two different swirl flows with inlet swirl numbers of 0.75 and 1.5 are simulated here with and without heat release. Here, the swirl number is defined as the ratio of axial momentum due to azimuthal velocity to the axial momentum due to the axial velocity itself. The actual swirl numbers at the dump plane (hence, the swirl conditions entering the combustor) are approximately 0.5 and 0.8, respectively due to flow relaxation in the inlet. For all these simulations, the Stokes number (the ratio of characteristic particle time to characteristic flow time) is of the order 1.

General observation shows that the larger particles do not follow the gas phase due to their large inertia. But, as they get smaller due to vaporization, they equilibrate with the gas phase. Smaller droplets are observed (not shown, for brevity) in the recirculation bubble near the dump plane. On the other hand, without vaporization (also not shown), fewer particles are seen in the recirculation bubble because of the larger Stokes number for these particles. Analysis of the flow features showed that the droplets tend to accumulate in regions of low vorticity. This type of preferential accumulation of droplets in regions of low vorticity was also observed in our earlier studies of mixing layer ([Pannala and Menon, 1998]).

The injection from the centered cone in the inlet and its subsequent interaction with the swirling air flow impacts the spray transport and dispersion considerably. Figures 36.2(a) and (b) show the vorticity distribution for the low and high swirl, non-reacting cases, respectively. Formation and shedding of near-circular, azimuthally coherent vortical rings from the inlet dump plane is clearly seen in the low swirl case, where as for the high swirl case, these coherent rings quickly breakdown. It can be seen that droplet dispersion has increased in the high swirl case. Droplet motion is modulated by the vortical rings (with drops tending to accumulate in low vorticity regions) but as the droplets evaporate, the gaseous fuel is entrained into the vortical regions where they mix with air and burn. Increase in swirl accelerates the breakdown of these rings forming relatively smaller regions of concentrated vorticity. The dominant component of the vorticity quickly changes from the azimuthal to the streamwise with the formation of small-scale structures. This vortex breakdown and the subsequent formation of the recirculation bubble enhance the dispersion of the droplets. Fuel-air mixing and thus, combustion is also enhanced with increase in swirl.

3.2. Effect of Swirl and Heat Release

Figure **??**(a) shows the streamwise variation of the center-line axial velocity for the non-reacting swirl flows. For S=0.75, immediately downstream of the dump-plane ($x/D_0 = 1.26$) flow accelerates and reaches a maximum, and then begins to decay as a conventional jet. This is due to the combined effect of the inlet convergence (see Fig. 27.1(a)), swirl and the sudden expansion, which results in a non-trivial interaction between azimuthal and axial velocities. The convergent inlet increases the angular velocity of the flow, due to the conservation of angular momentum. This in turn induces a favorable axial pressure gradient that accelerates the flow along the axis. On the other hand, for the high swirl case, such an acceleration does not occur and in fact, forms a recirculating bubble. Analysis shows that in this case, an adverse axial pressure gradient exists along the centerline.

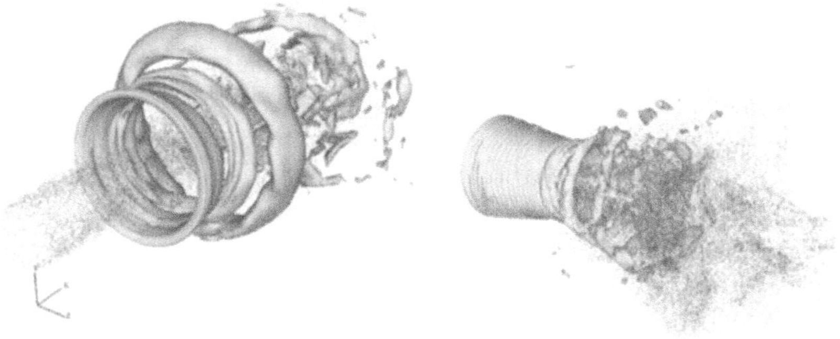

(a) Low swirl (S=0.78) (b) High swirl (S=1.56)

Figure 36.2. Effect of swirl on droplet dispersion

The inflow swirl allows for the natural development of a strong vortex core in the inlet pipe with a reduced pressure along its axis. This induces a favorable pressure gradient which tends to accelerate the flow. However, as the flow expands suddenly into the combustor, the radius of the vortex core increases, leading to a decrease of the angular velocity (by conservation of angular momentum) with the subsequent increase of pressure along its axis. This imposes an adverse pressure gradient which tries to decelerate the flow. Thus, it is the balance between the favorable and adverse pressure gradient, that determines the flow structure near the center line of the combustor. For low swirl, the favorable pressure gradient imposed by the inlet convergence accelerates the flow, whereas, the adverse pressure gradient induced by the swirl and sudden expansion decelerates the flow in the high swirl case.

Figure **??**(b) shows the centerline axial velocity for the high swirl case with and without combustion. For the low swirl case(not shown), axial variation was similar with and without combustion except that thermal expansion in a favorable pressure gradient further accelerates the flow. In the high swirl case, vortex breakdown and formation of the centerline recirculation bubble occurs in both cases. However, the shape of the bubble is quite different. Without heat release, the recirculation bubble is oriented axially and extends approximately $2D_0$ whereas with heat release, the size of the recirculation bubble is 50 percent smaller than the non-reacting case. This is due to the combination of thermal expansion

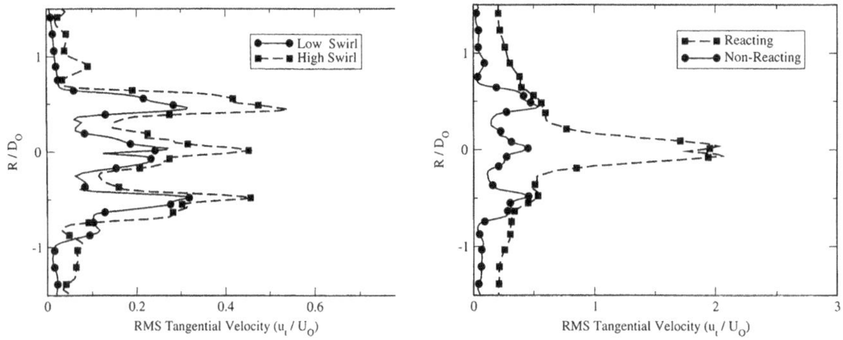

(a) Effect of Swirl

(b) Effect of Heat Release, S=1.5

Figure 36.3. Tangential rms velocity profiles at $x/D_0 = 1.26$

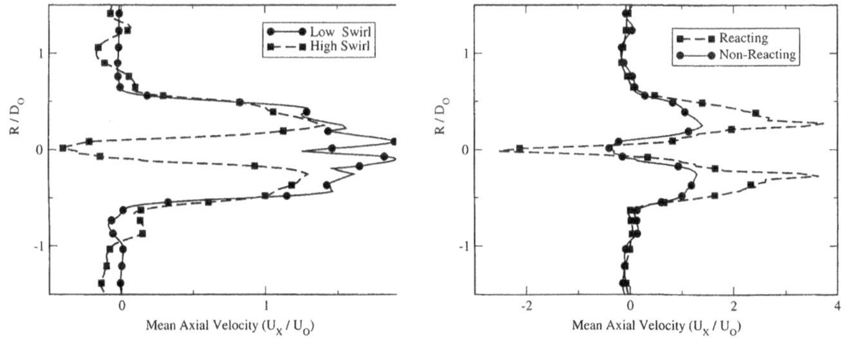

(a) Effect of swirl

(b) Effect of heat release, S=1.5

Figure 36.4. Axial mean velocity profiles at $x/D_0 = 1.26$

effect and the aforementioned adverse pressure gradient induced by swirl and sudden expansion.

Figure 36.4(a) shows the effect of swirl and Fig. 36.4(b) shows the effect of heat release on the radial profiles of the streamwise velocity. Presence of central recirculation zone is clearly visible in the high swirl case. The smaller bubble with heat release has much stronger reverse flow and is surrounded by an equally stronger axially positive flow. In all cases, the radial extent of these changes are limited to a length equivalent to the inlet diameter.

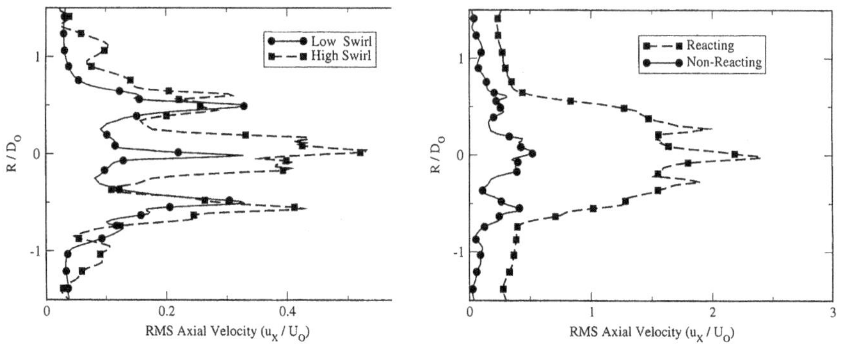

(a) Effect of swirl

(b) Effect of heat release, S=1.5

Figure 36.5. Axial rms velocity profiles at $x/D_0 = 1.26$

Figure 36.5(a) shows the effect of swirl on the radial profile of the axial velocity fluctuations and Fig. 36.5b shows the effect of heat release. Turbulent fluctuations peak in regions of shear layer and also near the center line. Augmented turbulence levels near center line is partly due the presence of shear layer on the central cone in the inlet pipe. In case of high swirl, magnitude of these turbulence levels further increase near the center line. This is due to the increased turbulence production in the central recirculation zone that has been observed for the high swirl case. With heat release, the overall level of axial velocity fluctuation further increases in the centerline region suggesting an increase in turbulence production due to a combined effect of vortex breakdown, thermal expansion and formation of recirculation bubble.

Finally, Figures 36.6(a) and (b) show the corresponding figures of the radial profiles of the azimuthal velocity fluctuations. For the non-reacting cases, turbulent fluctuations peak in regions of shear layer and also near the center line. Flow acceleration in the reacting case also seem to increase the turbulence levels at these locations. With heat release, the peaks in the shear layer region are no longer present and a much stronger tangential fluctuation intensity peak occurs near the centerline. This result when combined with the increased axial fluctuations suggests a direct impact of thermal expansion on turbulence production (since expansion occurs in all directions).

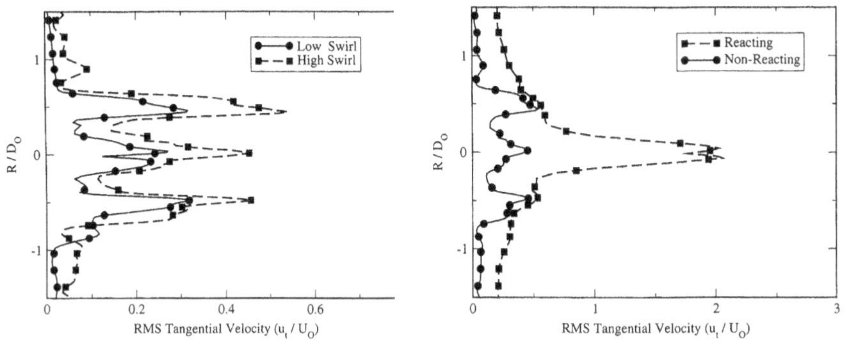

(a) Effect of Swirl (b) Effect of Heat Release, S=1.5

Figure 36.6. Tangential rms velocity profiles at $x/D_0 = 1.26$

4. Conclusions

LES of two-phase, swirling non-reacting and reacting flows in a gas turbine combustor has been carried out to study the dynamics. Effect of swirl and heat release has been addressed. Global behavior of spray mixing and combustion, such as the preferential concentration of droplets in low vorticity regions, droplet dispersion, and turbulence modification by the particles are all captured reasonably well. It is shown that vortex breakdown and formation of the recirculation bubble are most likely to occur under high swirl conditions. In the present configuration, the inlet convergence and the sudden expansion at the dump plane also play important roles in determining the vortex breakdown process. Spray dispersion and fuel-air mixing efficiency increases with increase in swirl. Heat release and thermal expansion effects change the shape and size of the recirculation bubble, and significantly enhance the turbulence intensity levels in the centerline region.

It should be noted that these results were obtained using infinite-rate kinetics. The effect of finite-rate kinetics and small scale turbulent mixing are currently being addressed using the subgrid simulation model noted earlier ([Chakravarthy and Menon, 2001]). Extensions to include droplet-droplet interactions and breakup are also being considered in order to deal with the dense spray regime.

5. Acknowledgments

This work was supported by the Army Research Office (ARO). Computational time was provided by DOD High Performance Computing Centers at NAVO (MS), and ERDC (MS) under a ARO HPC Grand Challenge Projects. GEAEC provided data for initialization.

References

Chakravarthy, V. and Menon, S. (2001). Large-eddy simulations of turbulent premixed flames in the flamelet regime. *Combustion Science and Technology*, 162:175–222.

Faeth, G. M. (1987). Mixing, transport and combustion in sprays. *Progress in Energy and Combustion Science*, 13:293–345.

Kim, W.-W. and Menon, S. (1999). A new incompressible solver for large-eddy simulations. *International Journal of Numerical Fluid Mechanics*, 31:983–1017.

Kim, W.-W. and Menon, S. (2000). Numerical modeling of turbulent premixed flames in the thin-reaction-zones regime. *Combustion Science and Technology*, 160:119–150.

Kim, W.-W., Menon, S., and Mongia, H. C. (1999). Large eddy simulations of a gas turbine combustor flow. *Combustion Science and Technology*, 143:25–62.

Menon, S. and Calhoon, W. (1996). Subgrid mixing and molecular transport modeling for large-eddy simulations of turbulent reacting flows. *Proceedings of the Combustion Institute*, 26:59–66.

Menon, S. and Pannala, S. (1997). Subgrid modeling of unsteady two-phase turbulent flows. *AIAA Paper No. 97-3113.*

Menon, S., Yeung, P.-K., and Kim, W.-W. (1996). Effect of subgrid models on the computed interscale energy transfer in isotropic turbulence. *Computers and Fluids*, 25(2):165–180.

Pannala, S. and Menon, S. (1998). Large eddy simulations of two-phase turbulent flows. *AIAA 98-0163, 36th AIAA Aerospace Sciences Meeting.*

Discussion

Sebastian Candel, EM2C, Paris "Just a comment. I noticed that in your unstable flame modulated by plane waves apparently the shape of the flame, which is looking like something we have in the lab in the laminar case, which is a pagoda flame. Have you seen those?"

Reply: This is one iso-surface.

Sebastian Candel, EM2C, Paris "The iso-surface is exactly looking like that but in the laminar case. We have experiments that show that."

Reply: This is one - the Reynolds numbers run high, but the oscillation may be modulating the flame.

Chapter 37

NUMERICAL SIMULATIONS OF PARTIALLY PREMIXED TURBULENT COMBUSTION

D.C. Haworth
The Pennsylvania State University, University Park, PA, USA

C. Jiménez, B. Cuenot, T. Poinsot
CERFACS, Toulouse, France

and R.J. Blint
GM R&D Center, Warren, MI, USA

Abstract Turbulent flame propagation through non–homogeneous reactant mixtures is investigated using high–resolution two–dimensional numerical simulation, including detailed chemical kinetics and molecular transport. Systematic variations in the distribution of reactant composition are imposed, and the effect of these variations on the primary propagating flame, secondary post–flame reaction, and pollutant formation are analyzed. In this manner, new fundamental insight into partially premixed turbulent flame propagation is generated and information relevant to modeling practical combustion systems is developed.

Keywords: partially premixed turbulent combustion, combustion simulation

1. Introduction

The propagation of a turbulent flame through a reactant mixture of non–uniform strength (e.g., spatial variations in equivalence ratio) is a problem of current fundamental interest and practical importance. Interest in partially premixed systems has been motivated by recent trends in practical combustion systems. Automotive manufacturers have been moving towards increasingly stratified combustion systems such as direct–injection spark–ignition gasoline engines [Zhao, Lai and Harrington, 1999] to reduce fuel consumption. At the same time, gas–turbine manufacturers have been increasing the level of premixing to reduce NO_x

A. Pollard and S. Candel (eds.), IUTAM Symposium on Turbulent Mixing and Combustion, 427-437.

emissions [Lefebvre, 1998]. Significant gaps in fundamental physical understanding of flame propagation, combustion stability, and pollutant formation have kept these systems from achieving their full potential.

Here a simulation–based approach is used to address these issues. The objectives are to generate new fundamental physical understanding of turbulent combustion with non–homogeneous reactants, and to develop and validate turbulent combustion models suitable for device–scale engineering computations (CFD) of practical combustion systems.

2. Partially Premixed Flame Propagation

A two–stage combustion process is hypothesized [Haworth et al., 2000]. A healthy essentially premixed flame (established by spark discharge, in the case of a gasoline direct–injection engine) propagates into a stratified reactant mixture; this flame is responsible for most of the heat release. Behind the flame in locally fuel–lean zones are products of combustion, oxidizer, and active radicals; in locally fuel–rich zones are products, fuel fragments or stable intermediates (CO and H_2), and radicals. Eventually the post–flame reactants mix and complete the reaction in a secondary combustion zone, which also is the site for formation and burnout of key pollutant species such as NO and CO.

Several competing effects determine whether the premixed burn rate is increased or reduced as a result of reactant non–homogeneity [Jiménez et al., 2001]. For example, flame area is expected to be higher for non–homogeneous reactants by virtue of increased wrinkling as the flame encounters stronger/weaker reactant pockets, while burn rate per unit flame area may be lower or higher, depending on the distribution of equivalence ratio along the active flame front. Early numerical studies of turbulent flame propagation through globally stoichiometric reactants reported modest differences in the premixed burn rate between homogeneous and non–homogeneous reactants [Hélie and Trouvé, 1998] [Haworth et al., 2000]. More recently, large variations in burn rate have been found in simulations with globally fuel–lean reactants [Jiménez et al., 2001].

These computational findings can be contrasted with the experimental measurements of Zhou et al. (1998) for a constant–volume bomb. Measurements were taken over a range of rms turbulence levels u', equivalence ratios ϕ ($\phi = 0.7$ to $\phi = 1.6$), and degrees of heterogeneity α ($\alpha = 0$ for a homogeneous mixture). For all equivalence ratios, the burn rate first increased and then decreased with increasing α, the variation being as much as a factor of three. An increase in flame corrugation was noted with increasing α. These results suggest that there is an optimum level of reactant stratification for fastest burn.

Secondary post–flame burning also is of interest. The rate of NO_x formation, for example, is affected by variations in local temperature and radical concentrations. Shih, Lee and Santavicca (1996) reported significant increases in NO and NO_x with incomplete fuel–air mixing at a fuel–lean condition ($\phi = 0.6$) for a gas–turbine combustor configuration, while at a condition closer to stoichiometric ($\phi = 0.92$), NO and NO_x decreased as the level of mixedness decreased.

3. A Model Problem

The system considered is a compressible multi–component reacting ideal–gas mixture. Propane–air oxidation is described using a 28–species 73–reaction scheme [Haworth et al., 2000]. In some cases, this has been supplemented with an extended Zeldovich thermal NO mechanism [Jiménez et al., 2001]. The governing equations and numerical methodology are described fully in Baum (1994).

The governing equations are solved in a two–dimensional rectangular domain. A steady planar laminar premixed flame initially separates reactants from products. Isotropic two–dimensional turbulence is prescribed using a two–parameter turbulence energy spectrum [Haworth and Poinsot, 1992], where the two parameters correspond to the initial rms turbulence velocity u'_{T0} and integral length scale l_{T0}. A turbulence Reynolds number is defined as $Re_{T0} \equiv \rho_u u'_{T0} l_{T0}/\mu_u$, where subscript u refers to unburned–gas properties. Non–homogeneity is introduced by varying the reactant mass fractions of C_3H_8, O_2, and N_2 in a manner that maintains the same total quantity of fuel and oxidizer as for a baseline homogeneous case, and maintains a uniform ratio of O_2 to N_2. A stratification scale is defined as $l_{\phi 0} \equiv L_y/n$, where L_y is the y–direction length of the domain (parallel to the initial flame) and n is the number of periods in $Y_{C_3H_8}$ across L_y. Both symmetrical and asymmetrical distributions of $Y_{C_3H_8}$ have been investigated (Table 37.1).

Aero–thermo–chemical conditions approach those in an automotive gasoline direct–injection engine at a low–speed light–load operating condition: $p = 4$ atm, $T_u = 700$ K (unburned gas temperature), $u'_{T0} = 7.7$ m/s, and $l_{T0} = 0.23$ mm. It is the turbulence length scale that is furthest from being typical of a practical engine ($l_T \approx 2$ mm – 4 mm near piston top–dead–center).

A chemical time scale is defined as the time required for the change in reactant stoichiometry to penetrate the primary premixed flame: $\tau_f = 2\delta_l^0/[s_l^0(1 + \rho_u/\rho_b)]$. Here ρ_u and ρ_b are the unburned– and burned–gas

Table 37.1. Simulation parameters.[a]

Case[b]	ϕ_{glob}	$\frac{\tau_f}{\tau_{T0}}$	$\frac{u'_{T0}}{s_l^0}$	$\frac{l_{T0}}{\delta_l^0}$	Re_{T0}	n	ϕ_{min}/ϕ_{max}	$\frac{l_{\phi 0}}{l_{T0}}$	*NO*
H_1.0_23	1.0	1.26	4.67	1.57	71	0	1.0/1.0	∞	no
NH_1.0_23	1.0	1.26	4.67	1.57	71	1	0.5/2.7	20.0	no
H_0.6_33	0.6	7.35	11.6	1.58	105	0	0.6/0.6	∞	yes
NHA_0.6_33	0.6	7.35	11.6	1.58	105	1	0.06/1.14	13.0	yes
NHB_0.6_33	0.6	7.35	11.6	1.58	105	1	0.03/2.06	13.0	yes
NHC_0.6_33	0.6	7.35	11.6	1.58	105	4	0.05/2.06	3.3	yes
H_1.0_43	1.0	0.84	4.67	2.36	105	0	1.0/1.0	∞	no
NHA_1.0_43	1.0	0.84	4.67	2.36	105	1	0.10/1.98	13.0	no
NHB_1.0_43	1.0	0.84	4.67	2.36	105	2	0.10/1.98	6.5	no
NHC_1.0_43	1.0	0.84	4.67	2.36	105	4	0.10/1.98	3.3	no
NHD_1.0_43	1.0	0.84	4.67	2.36	105	8	0.10/1.98	1.6	no
NHE_1.0_43	1.0	0.84	4.67	2.36	105	16	0.10/1.98	0.8	no

[a]Pressure is four atmospheres, reactant temperature is 700 K. [b]The final two numbers in the Case name refer to the size of the computational domain; e.g., *_23* is 2 mm × 3 mm.

mass density, respectively, and δ_l^0 and s_l^0 are the laminar flame thickness and speed for the initial one–dimensional steady unstrained premixed flame. For the present thermo–chemical conditions: $\delta_l^0 = 0.096$ mm, $s_l^0 = 1.65$ m/s for $\phi = 1.0$; and $\delta_l^0 = 0.142$ mm, $s_l^0 = 0.66$ m/s for $\phi = 0.6$.

Two additional quantities characterizing partially premixed combustion are a mixture fraction z and a progress variable c [Haworth et al., 2000]. Here the mixture fraction is defined as a carbon–plus–hydrogen element mass fraction; the value of z corresponding to reactants in stoichiometric proportion is $z_{st} = 0.06095$. Equivalence ratio ϕ is related to mixture fraction by $\phi = \big(z/(1-z)\big)\cdot\big((1-z_{st})/z_{st}\big)$. The reaction progress variable is defined in terms of the mixture fraction: $c = 1 - Y_{C_3H_8}/z$. In unburned reactants, $z = Y_{C_3H_8}$ so that $c = 0$; and at any point where there is no propane, $c = 1$. The secondary reaction zone is distinguished from the primary flame based on c: $c > c^*$, with $c^* = 0.9999$.

4. Results

4.1. The Primary Premixed Flame

The maximum difference in flame length between homogeneous and non–homogeneous reactants for equivalent global mixture strength ϕ_{glob} is less than 20%, while maximum differences in heat release exceed 50%. The largest differences have been for globally fuel–lean reactants having an asymmetrical equivalence ratio distribution. Global heat re-

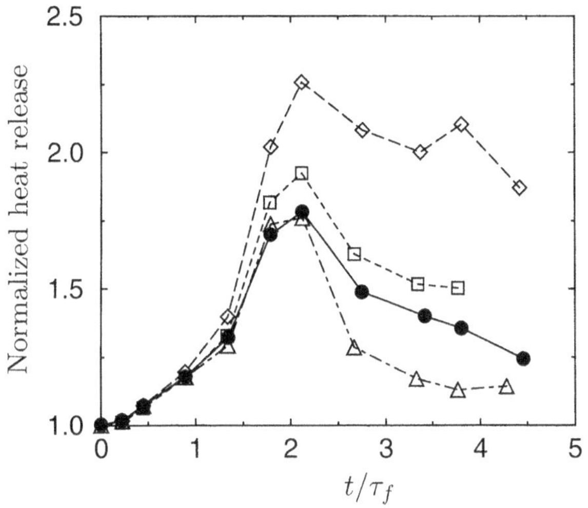

Figure 37.1. Time evolution of total heat release, normalized by its initial value, for globally fuel–lean reactants. Case H_0.6_33 (•) is a homogeneous case, and Cases NHA_0.6_33 (□), NHB_0.6_33 (◇), and NHC_0.6_33 (△) have different distributions of reactant non–homogeneities (Table 37.1).

lease for non–homogeneous reactants can be lower or higher than for an equivalent–strength homogeneous mixture (Figure 37.1). These differences result from interactions between turbulence and the propagating flame and the distribution of equivalence ratio along the active flame front [Jiménez et al., 2001].

A quantitative assessment of an engineering combustion model is provided in Figure 37.2. There the computed global heat release is compared to that obtained using a simple flamelet approximation [Jiménez et al., 2001]. Following an initial transient, agreement between the simulations and the model is surprisingly good, given the simplicity of the model (neglect of strain, unsteady effects, curvature effects, etc.). This suggests that relatively simple adaptations of laminar flamelet models developed for homogeneous premixed turbulent combustion may suffice to describe the primary heat release in highly non–homogeneous systems.

4.2. Secondary Combustion

Attempts to describe the secondary reaction zone using flamelet concepts have not been successful [Haworth et al., 2000], [Jiménez et al., 2001]. Results from a conditional moment closure (CMC) model [Bilger, 1993] have been somewhat more positive. A "first–order" CMC hypoth-

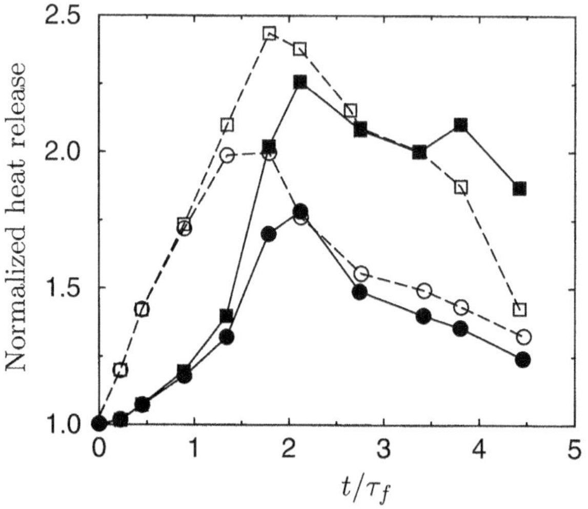

Figure 37.2. Assessment of a premixed flamelet model. Case H_0.6_33 (simulation •, model ∘), homogeneous reactants; Case NHB_0.6_33 (simulation ■ , model □), non–homogeneous reactants.

esis is that the conditional average of the species–α chemical production rate $\dot{\omega}_\alpha$ is equal to the chemical production rate evaluated using the conditionally averaged composition and temperature:

$$\langle \dot{\omega}_\alpha(\underline{Y}, T) | z = \eta \rangle = \dot{\omega}_\alpha(\langle \underline{Y} | z = \eta \rangle, \langle T | z = \eta \rangle) \quad , \tag{1}$$

and similarly for heat release. The CMC model is evaluated by comparing the left– and right–hand sides of Eq. 1 (Figure 37.3). In general, good agreement is found for heat release, key carbon–containing species (CO, CO_2), and other slow–chemistry–dominated species (e.g., NO); agreement is less satisfactory for radical species and for hydrogen–containing species. The latter discrepancy may be a result of differential diffusion.

4.3. Pollutant Emissions (NO)

Conventional engineering practice in modeling thermal NO production is to neglect reverse reactions (appropriate for low NO concentrations) and to invoke a steady–state approximation for N–atom concentration. Then the local rate of NO formation can be expressed as,

$$\frac{d[\mathrm{NO}]}{dt} = 2k_f[\mathrm{O}][\mathrm{N_2}] \quad , \tag{2}$$

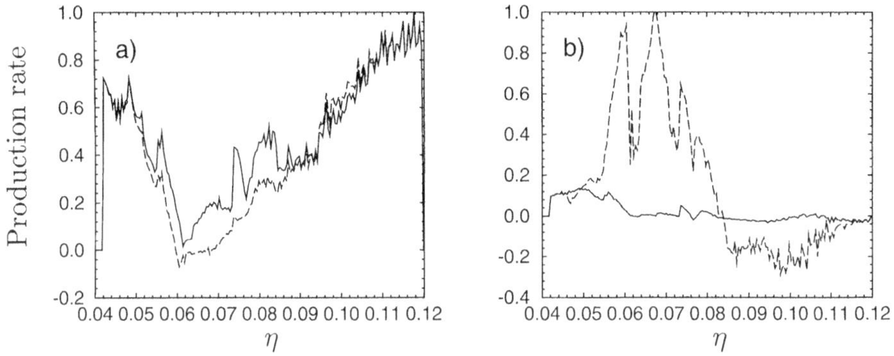

Figure 37.3. Assessment of a CMC model for secondary combustion, Case NH_1.0_23 at $t/\tau_f \approx 3.5$. Solid lines (———) $\langle \dot{\omega}_\alpha(\underline{Y}, T)|z = \eta \rangle$; dashed lines (– – – –) $\dot{\omega}_\alpha(\langle \underline{Y}|z = \eta\rangle, \langle T|z = \eta\rangle)$. The ordinate is normalized by the maximum value for each frame. a) Heat release rate. b) Chemical production rate of OH.

where square brackets indicate molar concentrations and $k_f = k_f(T)$ is the forward rate coefficient for the reaction $O + N_2 \rightleftharpoons NO + N$ [Turns, 2000]. Often it is further assumed that the O–atom concentration corresponds to chemical equilibrium. Quantitative assessment of these approximations has been made by comparing results obtained using the full extended Zeldovich mechanism with those obtained using the simplifications [Jiménez et al., 2001]. For the conditions simulated, steady–state N–atom and neglect of reverse reactions are quite good approximations, while neglecting superequilibrium O–atom severely underpredicts NO formation. These results pertain only to the initial formation of NO in and immediately behind the primary flame, because of the limited extent of the computational box.

NO formation is found to vary strongly with the length scale of the initial reactant stratification. For the conditions simulated, non–homogeneous reactants yield higher NO levels compared to an equivalent–strength homogeneous mixture. In the case of globally stoichiometric reactants, the total quantity of NO in the computational domain for large–scale non–homogeneities ($n = 1$, $n = 2$) is over 100 times greater than for homogeneous reactants (Figure 37.4). As the length scale of the non–homogeneity decreases (n increases), the homogeneous results are recovered. The high NO levels for non–homogeneous reactants are a consequence of locally high O–atom concentrations and temperatures compared to the homogeneous case [Jiménez et al., 2001]. Again, these findings pertain only to the initial formation of NO.

4.4. Heat–Release Marker Species

To extract the full potential from modern laser–based combustion diagnostics, it is necessary to correlate measurable quantities (e.g., instantaneous concentrations of single or multiple chemical species) with important but experimentally inaccessible quantities including heat release. Recent suggestions for heat–release/flame markers include the formyl radical (HCO) and simultaneous presence of formaldehyde (CH_2O) and hydroxyl radical (OH) [Najm et al., 1998]. Correlations of heat release with mass fractions of several chemical species from the present simulations have served to corroborate and to extend to highly non–homogeneous turbulent flames the marker species proposed in the literature [Haworth et al., 2000].

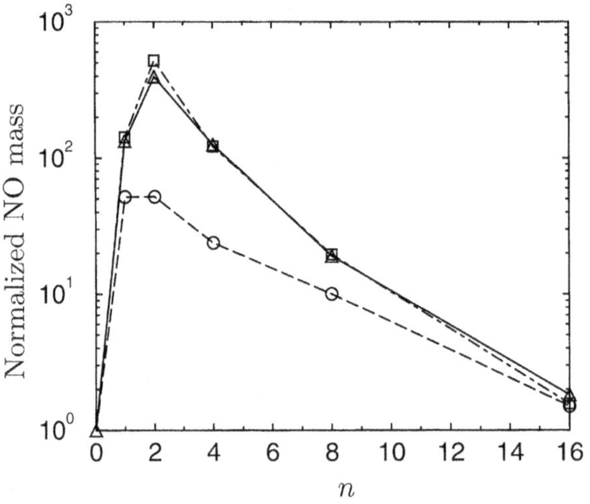

Figure 37.4. NO mass versus wavenumber of non–homogeneity n at $t/\tau_f = 4.8$, globally stoichiometric reactants (_1.0_43 series, Table 37.1). Each curve has been normalized by its homogeneous–reactant ($n = 0$) value. Primary flame ($0.5 \leq c \leq 0.9999$) ---- ; secondary combustion ($c > 0.9999$) –·–· ; total in computational box ——— .

5. Summary and Conclusions

Numerical simulations provide new insights into several aspects of partially premixed turbulent combustion. Additional results and discussion can be found in Haworth et al. (2000) and Jiménez et al. (2001). Salient findings/results include:

- Heat release and NO can be lower or substantially higher for non–homogeneous reactants compared to an equivalent–strength homogeneous mixture, depending on the nature of the non–homogeneities. This is in qualitative agreement with experimental measurements.
- Relatively small differences in flame length have been found for homogeneous versus non–homogeneous reactants, in apparent contradiction with experiment.
- Steady–state N–atom concentration and neglect of reverse reactions are excellent approximations for modeling the initial thermal NO formation. Equilibrium O–atom is a very poor approximation.
- A simple premixed flamelet model yields generally good results for the primary premixed burn.
- CMC satisfactorily represents heat release, major carbon–containing species, and NO in the secondary reaction zone, while doing less well for hydrogen–containing species and radicals.
- HCO and simultaneous presence of CH_2O and OH correlate nearly linearly with heat release in the primary and secondary combustion zones for homogeneous and non–homogeneous reactants.

The temptation to extrapolate these results to practical combustion systems must be tempered by an awareness of the limitations of the model problem. These include spatial two–dimensionality, neglect of radiation, incomplete chemical kinetics, limited integration times, limited number of realizations for statistical analysis, and the small dynamic range of length scales. In the authors' opinion, it is the last of these that is most restrictive, and which most likely is responsible for the failure to find large differences in flame length. Implications of spatial two–dimensionality are discussed in Haworth and Poinsot (1992).

Acknowledgments

This research was initiated during the 1998 and 2000 Summer Programs of the Stanford University/NASA Ames Center for Turbulence Research. Computational resources have been provided by CTR, the Institut du Développement et des Resources en Informatique Scientifique (IDRIS), and the GM R&D Center.

References

Baum, M. (1994). Etude de l'allumage et de la structure des flames turbulentes. Ph.D. Thesis, Ecole Centrale, Paris.

Bilger, R. W. (1993). Conditional moment closure for turbulent reacting flow. *Phys. Fluids A.* 5:436–444.

Haworth, D. C., Blint, R. J., Cuenot, B. and Poinsot, T. J. (2000). Numerical simulation of turbulent propane–air combustion with non-homogeneous reactants. *Combust. Flame.* 121:395–417.

Haworth, D. C. and Poinsot, T. J. (1992). Numerical simulations of Lewis number effects in turbulent premixed flames. *J. Fluid Mech.* 244:405–436.

Hélie, J. and Trouvé, A. (1998). Turbulent flame propagation in partially premixed combustion. *Proc. Combust. Institute.* 27:891–898.

Jiménez, C., Haworth, D., Cuenot, B. and Poinsot, T. (2001). Direct numerical simulation and modeling of lean stratified propane–air flames. *Combust. Flame.* To appear.

Lefebvre, A. H. (1998). *Gas Turbine Combustion*, Second Edition. Taylor & Francis.

Najm, H. N., Paul, P. H., Mueller, C. J. and Wyckoff, P. S. (1998). On the adequacy of certain experimental observables as measurements of flame burning rate. *Combust. Flame.* 113:312–332.

Shih, W. P., Lee, J. G. and Santavica, D. A. (1996). Stability and emission characteristics of a lean premixed gas turbine combustor. *Proc. Combust. Institute.* 26:2771–2778.

Turns, S. R. (2000). *An Introduction to Combustion*, Second Edition. McGraw–Hill.

Zhao, F. Q., Lai, M. C. and Harrington, D. L. (1999). Automotive spark-ignited direct-injection gasoline engines. *Prog. Energy Combust. Sci.* 25:437–562.

Zhou, J., Nishida, K., Yoshikazi T. and Hiroyasu, H. (1998). Flame propagation characteristics in a heterogeneous concentration distribution of a fuel–air mixture. *SAE Paper No. 982563.*

Discussion

Charles Meneveau, Johns Hopkins "Do you think there will be an effect of correlations between concentration with the velocity field in other words if your blobs of scalar are on the left side of the flame were they place, were they placed there uncorrelated from the velocity field, the turbulence. Do you think that there will be an additional effect if there is correlations."

Reply: Sure. Certainly the right way to do this, and of course we don't do it that way because we can't afford to would be to do many, many simulations with different initial conditions that are all the same in the mean but different initial realizations of the turbulence and so on and see how robust all of the conclusions are. So yes, it is something that we have to watch out for. These columns of fluid get nicely broken down into structures that are the initial size or width of the slab but that is a fairly qualitative thing.

Chapter 38

TURBULENT COMBUSTION OF SPRAYS UNDER SUPERCRITICAL CONDITIONS

M. Juniper
Laboratoire EM2C, Ecole Centrale de Paris, 92295 Châtenay-Malabry, France
juniper@em2c.ecp.fr

A. Tripathi, P. Scouflaire and S. Candel

Abstract Experimental results on a gaseous hydrogen / liquid oxygen spray flame are compared with a simplified model. The gasification rate, which depends on rates of turbulent mixing and vaporization, has a strong effect on flame structure. The rate-limiting process is vaporization at subcritical pressures and turbulent mixing at supercritical pressures. This is because surface tension and the latent heat of vaporization tend to zero above the critical pressure.

Keywords: cryogenic, supercritical, spray combustion

Introduction

In many combustion applications a liquid spray is injected within a gaseous reactant above the liquid's critical pressure ($p_{crit}(C_{12}H_{26}) = 18.2$ bar, $p_{crit}(O_2) = 50.5$ bar). Above p_{crit} both the surface tension and the latent heat of vaporization tend to zero. We show that this changes the mechanism governing the flame structure if this depends on a balance between turbulent mixing and vaporization.

There is little experimental data on supercritical spray combustion. The results presented here are obtained from a liquid oxygen / gaseous hydrogen injector. The injector geometry and chemical reaction are both simple, which aids comparison with theoretical and numerical results.

The flame structure is introduced in section 1. A parametric study of a one-dimensional axisymmetric model is proposed in section 2. The

A. Pollard and S. Candel (eds.), IUTAM Symposium on Turbulent Mixing and Combustion, 439-446.

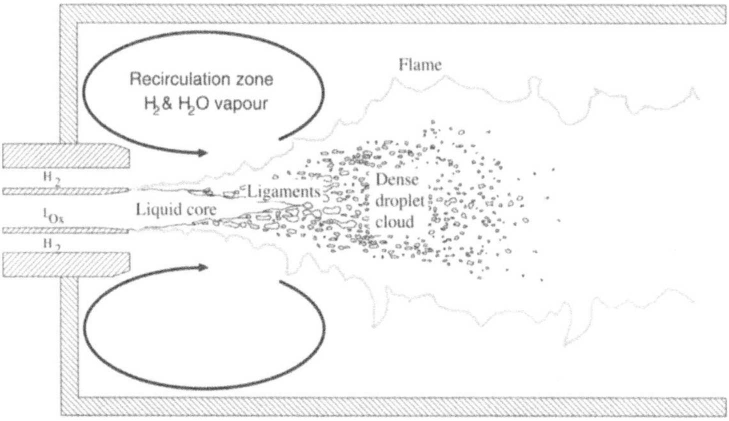

Figure 38.1. Representation of the oxygen/hydrogen cryogenic jet flame operating at subcritical pressure. The reactants are injected coaxially with liquid oxygen in the center.

experimental facility, diagnostics and results are presented in section 3 and compared with predictions of the model. The consequences of exceeding the critical pressure are explained in section 4.

1. The structure of spray flames

The flame structure is shown in figure 38.1. The oxygen jet breaks up in a fibrous regime, as defined by [Fargo, 1991]. The flame is attached to the lip of the oxygen injector, fed by gasification from the outer edge of the droplet cloud. Gasification refers to the transition of the central stream from a high density to a low density fluid. At subcritical pressures this is vaporization.

The flame can be characterized by its length and its expansion angle. Hydrogen is in excess, so the length depends on the distance after which all the oxygen has reacted. The flame's expansion angle is determined by the enclosed volume of evaporated oxygen and products.

A crucial factor is the rate of oxygen gasification, which is controlled by: (1) Heat transfer from the flame, which is mainly governed by turbulent mixing. (2) Heat and mass transfer across the liquid oxygen surface (vaporization) and its equivalent in the supercritical case. The rate of combustion is too rapid to be a limiting factor so the structure is determined by a balance between turbulent mixing and vaporization.

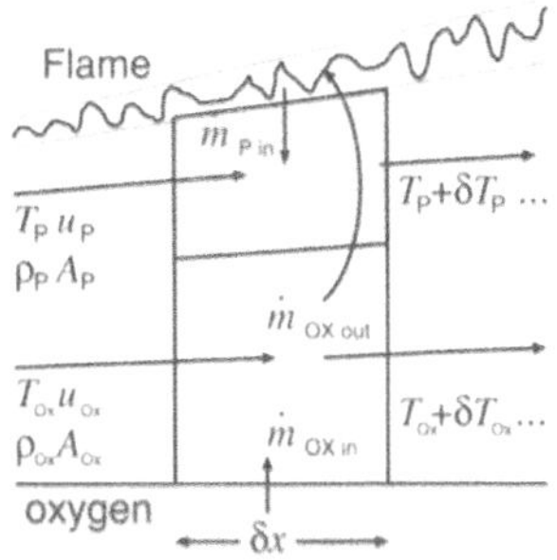

Figure 38.2. Control volumes containing product (top) and evaporated oxygen (bottom). The changes in flow variables over an incremental distance δx can be evaluated via mass, momentum and energy balances.

2. 1-D axisymmetric model of a spray flame

2.1. General description

A simple 1-D axisymmetric model (figure 38.2) is proposed. A dense oxygen core is separated from a turbulent flame sheet by two control volumes at distance x from the injector. One contains evaporated oxygen and another contains hot products. They have axial extent δx and cross-sectional areas A_{Ox} and A_P. To simulate gasification and combustion, $\delta\dot{m}_{OXin}$ is injected into the oxygen control volume and $\delta\dot{m}_{OXout}$ is removed. A proportional quantity of hot H_2O is injected into the products control volume. The velocity of the products is found from a momentum balance between the hydrogen and the oxygen. The proportion (by mass) of products going to the oxygen side is $1/(1+s)$, where s is the mass stoichiometric coefficient (s=8 in this case). Mass, momentum and energy balances in the control volumes lead to changes in density, velocity pressure and temperature across them.

The thickness, l, of the combustion layer grows linearly with x. The radial position of the flame, r, is calculated from the cross-sectional area of the two flows. The combusting volume is given by $2\pi r l \delta x$. The average volumetric consumption rate of oxygen is estimated from experimental data.

2.2. Balance equations

The procedure used is similar to an integral method, although all flow variables (ρ, A, u, T and p) are assumed constant in the radial direction. The recirculating zone outside the expanding flame suggests that the axial pressure is constant. We consider small changes $\delta\rho$, δA, δu and δT

Table 38.1. Parameters required in order to calculate the flame shape using the 1-D axisymmetric model. $\dot{w}_{O_2}$ is the oxygen consumption rate.

Variable	Base value	Constant	value
p	10 bar	d_{LOx}	5 mm
U_{H_2}	100 ms^{-1}	T_{LOx}	80 K
U_{LOx}	2 ms^{-1}	T_f	2500 K
$\dot{w}_{O_2}$	100 kgm^{-3}s^{-1}	ρ_{LOx}	1180 kgm^{-3}

over a distance δx. Mass addition from vaporization and combustion is included. The conservation equations for each control volume are:

$$\begin{array}{ll} \text{Mass} & \rho u \delta A + \rho A \delta u + A u \delta \rho = \delta \dot{m}_{in} - \delta \dot{m}_{out} \\ x\text{-momentum} & \rho A u \delta u + \delta p = \delta \dot{m}_{in}(u_{in} - u) \\ \text{Energy} & \rho A u (c_p \delta T + u \delta u) = (c_{p_{in}} T_{in} - c_p T + \frac{u_{in}^2 - u^2}{2}) \delta \dot{m}_{in} \end{array}$$

Equations of state of the form $p = \rho RT$ are substituted into these equations for gaseous oxygen and reaction products. This results in six equations and six unknowns. Given initial upstream conditions and a LOx evaporation profile, the flame shape can be calculated by marching in the x-direction. Some of the parameters of the problem are contained in table 38.1.

2.2.1 Numerical results. The five important parameters are: pressure, hydrogen velocity, LOx mass flowrate, LOx vaporization profile and oxygen consumption rate. An evaporation profile is defined in order to set a base case and the parameters are varied one by one, keeping all the others constant.

Effect of Pressure (figure 38.3a). Increasing the pressure moves the flame closer to the oxygen jet. This is because the density of the evaporated oxygen is higher and takes up less space within the flame. In reality, this effect is compensated by an increase in the rate of oxygen vaporization.

Effect of hydrogen velocity (figure 38.3b). Increasing the hydrogen velocity moves the flame closer to the oxygen core. This is because the products inside the flame have a higher velocity and consequently take up less area. The effect is small between 100ms^{-1} and 1000ms^{-1}, which is the domain in which cryogenic injectors usually operate.

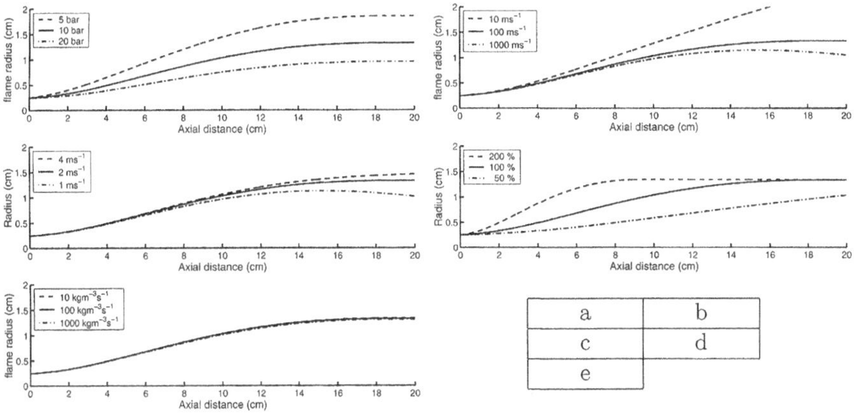

Figure 38.3. Flame profile calculated via the 1-D axisymmetric model. (a) Effect of pressure. (b) Effect of hydrogen velocity. (c) Effect of the oxygen velocity. (d) Effect of gasification rate of oxygen, expressed as a percentage of the base case. (e) Effect of the consumption rate of oxygen per unit volume.

Effect of the oxygen velocity (figure 38.3c). A decrease in the oxygen velocity causes a very slight reduction in the flame expansion angle. Although the mass flow rate reduces, the lower velocity means that the area required by the oxygen flow does not change. The slight reduction arises because less area is taken up by the combustion products, whose area depends also on u_{H2}.

Effect of the gasification rate of oxygen (figure 38.3d). Increasing the rate of dense oxygen gasification greatly increases the expansion angle of the flame.

Effect of the rate of combustion (figure 38.3e). Increasing the rate of combustion has a negligible effect on the expansion angle of the flame. This is only because the combusting oxygen is replaced by hot products, which take up a similar area.

3. Experimental Facility

3.1. General description

The experimental facility is described by [Kendrick, 1999] and [Vilgert, 1997] and optical diagnostics by [Tripathi, 2001] and [Juniper, 2001]. Experimental injection conditions and expansion angles are gathered in Table 38.2. The average heat release rate is estimated by [Tri-

Table 38.2. Operating points of the cryogenic test facility and flame expansion angles.

Pressure	1 bar		5 bar		10 bar		70 bar		
$U_{H2}(\mathrm{ms}^{-1})$	628	893	292	432	207	308	157	192	208
$U_{LOx}(\mathrm{ms}^{-1})$	2.23	2.23	2.23	2.23	2.23	2.23	4.3	4.3	4.2
J	6.3	13.4	6.5	14.4	6.5	14.5	6.0	10.0	12.8
Expansion	3.8°	8.4°	2.6°	6.8°	5.1°	7.2°	3.8°	3.7°	3.6°

pathi, 2001] as 30 $\times 10^8$ Wm^{-3} at 10 bar and 500 $\times 10^8$ Wm^{-3} at 70 bar.

3.1.1 Effect of hydrogen velocity and gasification rate.

Between 1 and 10 bar, increasing the hydrogen velocity increases the expansion angle. This is because jet break up is improved, increasing the surface area and vaporization rate. Jet break-up is characterized well by the momentum flux ratio, J [Gill, 1978]. On the other hand, at 70 bar, increasing the hydrogen velocity tends to decrease the flame expansion angle slightly. This is discussed further in section 4.

3.1.2 Effect of pressure.

In these experiments, the hydrogen velocity is decreased when the pressure is increased in order to maintain a constant momentum flux ratio. These parameters have opposite effects on the flame expansion angle, which is reflected in the experimental results. However, from 10 to 70 bar, where the pressure is multiplied by 7 but the hydrogen velocity divided by only 1.3, the flame expansion angle is much reduced. This confirms that increasing the pressure reduces the flame expansion angle.

4. Discussion

Increased hydrogen velocity causes better jet break-up. This causes increased oxygen gasification only at subcritical pressures. This can be explained by considering the factors affecting gasification rate: (1) Heat transfer from the flame to the dense oxygen concentration, which is mainly governed by turbulent mixing, (2) Heat and mass transfer at small scale across the liquid oxygen surface (vaporization) and its equivalent in the supercritical case. The slowest of these two rates will be dominant. They can be compared via a Damköhler number $D_v = \tau_m/\tau_v$, where τ_m and τ_v are characteristic times of mixing and vaporization. From the analysis below, $\mathrm{D}_v \sim 10^{-1}$ at subcritical pressures, but is $\mathrm{D}_v \sim 10^1$ at supercritical pressures. The Chemical time is given by

$\tau_c = D/S_l^2$. It is typically less than $2\mu s$ and therefore is never a rate-limiting process.

4.1. Characteristic times

4.1.1 Mixing time. Following the analysis by [Broadwell, 1982], a characteristic mixing time is given by:

$$\tau_m = \frac{3}{2}\kappa\frac{L}{\Delta U}\left[1 - \frac{1}{Re^{\frac{1}{2}}}\right] \approx \frac{3}{2}\kappa\frac{L}{\Delta U} \qquad \text{for } Re \gg 1 \qquad (1)$$

Thus the time to reach the smallest scales is proportional to $L/\Delta U$, virtually independently of the Reynolds number and the pressure. κ is between 1 and 10. For typical L and ΔU in the combustion chamber used in the experiments, $\tau_m$is between 2 and 80 μs.

4.1.2 Vaporization time. At subcritical pressures, the vaporization time can be crudely estimated via the model of an isolated droplet in a gas at rest. At 1 to 10 bar, τ_{vap} varies from $10\mu s$ to $600\mu s$. At 50 bar this decreases to between $1\mu s$ and $60\mu s$. However, around and above the critical pressure, surface tension and the latent heat of vaporization diminish. As a result, [Deplanque, 1995] propose that aerodynamic instabilities on the surface of a droplet grow rapidly. The outer layer of the liquid is stripped off and mixes with the surrounding gas in a process analogous to vaporization. According to their numerical simulations, this rate is larger than the vaporization rate at the same pressure by at least an order of magnitude. This brings the new vaporization time to between $0.1\mu s$ and $6\mu s$.

5. Conclusion

A simple model of a cryogenic spray flame demonstrates that the expansion angle is strongly dependent on the spray's gasification rate. Experimental results and a characteristic time argument suggest that the process determining gasification rate changes around the critical pressure. At subcritical pressure it is vaporization-controlled. This rate is dominated by the surface area of the liquid jet, which is determined by jet break-up and hence the momentum flux ratio, J. Above the critical pressure, when there is neither a surface nor a latent heat of vaporization, flame expansion is limited by the rate at which heat can be transferred from the flame to the condensed oxygen, which is determined by turbulent mixing. This rate is no longer affected by the surface area of the condensed jet, which explains why J has little effect at high pressure.

In combustion chambers operating at supercritical pressures good mixing is more important than good atomization.

References

Broadwell, J.E. and Breidenthal, R.E. (1982). A Simple Model of Mixing and Chemical Reaction in a Turbulent Shear Layer. In *Journal of Fluid Mechanics.* 125: 397-410

Delplanque, J-P. and Sirignano, W. (1994). Boundary-layer Stripping Effects on Droplet Transcritical Convective Vaporization. In *Atomization and Sprays.* 4: 325-349

Delplanque, J-P. and Sirignano, W. (1995). Transcritical vaporization and combustion of LOx droplet arrays in a convective environment. In *Combustion Science and Technology.* 105: 327-344

Farago, Z. and Chigier, N. (1991). Morphological classification of disintegration of a round liquid jet in a coaxial air stream. In *ICLASS.* Rouen, France

Gill, G (1978). A Qualitative Technique for Concentric Tube Element Optimization, Utlilizing the Factor (Dynamic Head Ratio - 1). In *AIAA 16th Aerospace Sciences Meeting.* AIAA Paper 78-76

Juniper, M. (2001). Stabilisation des flammes cryotechniques et effets du retrait. PHD thesis, Ecole Centrale de Paris, Chatenay-Malabry, France

Kendrick, D.,Herding, G., Scouflaire, P.,Rolon, C. and Candel, S. (1999). Effects of Recess on Cryogenic Flame Stablilization. In *Combustion and Flame.* 3: 327-339

Tripathi, A. (2001). Structure des flammes cryotechniques à haute pression. PHD thesis, Ecole Centrale de Paris, Chatenay-Malabry, France

L. Vingert, M. Habiballah, P. Hervat, F. Dugué and P. Vuillermoz. (1997). Evolution of the MASCOTTE test bench to high pressure operation and related combustor technology issues. In *3ème Colloque Franco-Allemand sur la Combustion dans les Moteurs-Fusée*

Author Index

Averill, J. 181

Bellan, B. 59
Ben-Dakhlia, R. 347
Bewley, T.R. 45
Blint, R.J. 427
Blossey, P.N. 45
Boulanger, J. 161
Bourlioux, A. 211
Brethouwer, G. 33

Candel, S. 357, 439
Cant, R.S. 259
Caraeni, D.A. 221
Charentenay, J. 287, 347
Charlette, F. 181
Choi, H. 367
Cuenot, B. 427

Darabiha, N. 149, 347
Diep, J. 85
Diers, O. 327
Dimotakis, P.E. 95
Domingo, P. 235
Ducruix, S. 357

Echekki, T. 113
Ewing, D. 73

Ferchichi, M. 201
Fuchs, L. 221
Fujimura, M. 269

Gajan, P. 327
Gamard, S. 23
George, W.K. 23
Gicquel, O. 149, 279
Giovangigli, V. 347
Giuliani, F. 327

Haworth, D.C. 427
Hewson, J.C. 113
Hilbert, R. 149, 251, 279
Hoffmann, S. 405
Holme, T. 377

Hunt, J.C.R. 33

Ito, Y. 269

Janicka, J. 387
Jérôme, H. 169
Jiménez, C. 427
Johnson, M.R. 295
Jung, E. 23

Kang, H.S. 181
Kempf, A. 387
Kerstein, A.R. 113
Khouider, B. 211
Knio, O. 181
Kostiuk, L.W. 295
Krebs, W. 405

Lacas, F. 315
Lartigue, G. 405
Ledoux, M. 327
Légier, J.P. 315
Livescu, D. 125

Madnia, C.K. 125
Majda, A.J. 211
McIlwain, S. 377
Meneveau, C. 181
Mengler, C. 387
Menon, S. 415
Miyauchi, T. 269
Mizobuchi, Y. 337
Moin, P. 305

Nakabe, K. 367
Narayanan, S. 45
Nieuwstadt, F.T.M. 33

Ogawa, S. 337
Okong'o, N. 59

Pantano, C. 137
Poinsot, T. 315, 357, 405, 427
Pollard, A. 377

Sadiki, A. 387

Sankaran, V. 415
Sarkar, S. 137
Schlüter, J. 405
Scouflaire, P. 439
Selle, L. 405
Shinjo, J. 337
Sigurdson, L. 85
Suzuki, K. 367
Szasz, R.Z. 221

Tachibana, S. 337
Takaki, R. 337
Tanahashi, M. 269
Tatsumi, T. xiii
Tavoularis, S. 201
Thévenin, D. 149, 251, 279, 287
Tripathi, A. 439
Trouvé, A. 169
Tullis, S. 259

Varoquié, B. 315
Vervisch, L. 161, 235
Veynante, D. 181, 315
Villermaux, E. 1

Waterman, S. 377
Williams, F. 137
Woodward, S. 23, 73

Zamuner, B. 287

List of Participants

Josette Bellan
Jet Propulsion Lab
California Institute of Technology
4800 Oak Grove Dr. MS 125-109
Pasadena, California, 91109-8099,
USA
josette.bellan@jpl.nasa.gov

Rafik Ben-Dakhlia
CNRS
Ecole Centrale Paris
BP 72- 29 av. de la Division Leclerc
92322 CHATILLON Cedex
Chatenay-Malabry, 92295, France
r.bendakhlia@onera.fr

A. Benaissa
Department of Mechanical Engineering

Royal Military College of Canada
Kingston, Ontario, K7K 7B4, Canada
benaissa@rmc.ca

Jean-Pierre Bertoglio
LMFA-CNRS
Lab. de Mecanique des Fluides
d'Acoustique
36, avenue Guy de Collongue
Ecully, 69130, France
berto@mecaflu.ec-lyon.fr

Peter Blossey
Department of Applied Mathematics
University of Washington
Box 352420
Seattle, Washington, 98195-2420, USA

bloss@amath.washington.edu

Joan Boulanger
Institut National des Sciences
Appliquees de Rouen
UMR-CNRS-6614/CORIA
Saint Etienne du Rouvray
Cedex, Fr-76801, France
Boulanger@coria.fr

Anne Bourlioux
Dept. Math & Statistics
Universite de Montreal
CP 6128 succ Centre-Ville
Montreal, Quebec, H3C 3J7, Canada
Anne.Bourlioux@cerca.umontreal.ca

Mohamed Boutazakhti
Dept. Mechanical Engineering
University of Toronto
c/o 4 Devonshire Place
Toronto, Ontario, M5S 2E1, Canada
mohb@mie.utoronto.ca

Geert Brethouwer
Faculty of Mechanical Eng'g and
Maritime Tech.
TU Delft
Leeghwaterstraat 21
CB Delft, 2628, The Netherlands
g.brethouwer@wbmt.tudelft.nl

Sebastien Caillat
Departement Energetique Industrielle
Ecole des Mines de Douai
941, rue Charles Bourseul B.P. 838
Douai Cedex, 59508, France
Sebastien.Caillat@ensm-douai.fr

Sebastien Candel
Laboratoire EM2C
Ecole Centrale de Paris
Grande Voie des Vignes
Chatenay-Malabry, Cedex, 92295,
France
candel@em2C.ecp.fr

R Stewart Cant
Department of Engineering
University of Cambridge
Trumpington Street
Cambridge, CB2 1PZ, UK
rsc10@eng.cam.ac.uk

Hang-Seok Choi
Department of Mechanical Engineering
Kyoto University
Kyoto, , 606-8501, Japan
t60y0915@ip.media.kyoto-u.ac.jp

Gabby Ciccarelli
Department of Mechanical Eng.
Queen's University
McLaughlin Hall, Stuart Street
Kingston, Ontario, K7L 3N6, Canada
ciccarel@me.queensu.ca

Nasser Darabiha
Laboratoire EM2C
Ecole Centrale de Paris
Grande Voie des Vignes
Chatenay-Malabry, Cedex, , 92295, France
nasser@em2C.ecp.fr

Julien de Charentenay
CNRS
Ecole Centrale Paris
BP 72- 29 av. de la Division Leclerc
92322 CHATILLON Cedex
Chatenay-Malabry, , 92295, France
j.mareschal@onera.fr

Paul Dimotakis
Graduate Aeronautical Laboratories
California Institute if Technology
Mail Code 301-46
Pasadena, California, 91125, USA
dimotakis@caltech.edu

Cecile Drevet
Dept. Mechanical Engineering
Queen's University
Stuart Street
Kingston, Ontario, K7L 3N6, Canada
drevet@me.queensu.ca

Sebastien Ducruix
C.N.R.S.
Ecole Centrale Paris
Grande Voie des Vignes
Chatenay-Malabry, , 92295, France
ducruix@em2c.ecp.fr

Dan Ewing
Dept. of Mechanical Engineering
McMaster University
1280 Main St W.
Hamilton, Ontario, L8S 4L7, Canada
ewingd@mcmaster.ca

Ivan Fedioun
LME/ESEM
rue Leonard de Vinci
Orleans, Cedex 2, , 45072, France
Ivan.Fedioun@univ-orleans.fr

Mohsen Ferchichi
Dept. Mechanical Engineering
Lakehead University
955 Oliver Road
Thunder Bay, Ontario, P7B 5E1, Canada
mohsen.ferchichi@lakeheadu.ca

Pierre Gajan
ONERA
Center de Toulouse
2 avenue Edouard Belin
Toulouse, Cedex 4, , 31055, France
Pierre.Gajan@onecert.fr

Olivier Gicquel
Laboratoire EM2C
Ecole Centrale Paris
Grande Voie des Vignes
Chatenay-Malabry, , 92295, France
gicquel@em2c.ecp.fr

Fabrice Giuliani
ONERA
Center de Toulouse
2 avenue Edouard Belin
Toulouse, Cedex 4, , 31055, France
Fabrice.Giuliani@onecert.fr

Peyman Givi
Dept. Mechanical and Aerospace Engineering
University at Buffalo
SUNY at Buffalo, 334 Jarvis Hall
Buffalo, NY, 14260-4400, USA
givi@eng.buffalo.edu

Mark Glauser
Dept Mechanical, Aerospace and Manufacturing Eng'g
Syracuse University
151 Link Hall
Syracuse, NY, 13244-1240, USA
mglauser@syr.edu

Ted Grandmaison
Dept. of Chem. Eng.
Queen's University
Dupius Hall, Division Street
Kingston, Ontario, K7L 3N6, Canada
grandmai@chee.queensu.ca

Daniel C. Haworth
Dept of Mechanical & Nuclear Engineering
Pennsylvania State University
232 Research Building East
University Park, PA, 16802-2321, USA
dch12@psu.edu

Brian Helenbrooke
MAE Department
Clarkson University
MAE Dept. Box 5725, Clarkson University
Potsdam, NY, 13699, USA
helenbrk@clarkson.edu

Alan Hewson
Combustion Research Facility
Sandia National Laboratories
7011 East Avenue, Bldg. 905, Rm. 121
Livermore, CA, 94551, USA
jchewso@california.sandia.gov

Renan Hilbert
CNRS
Ecole Central Paris
Grande Voie des Vignes
Chatenay-Malabry, , 92295, France
hilbert@em2c.ecp.fr

Trevor Holme
Dept. Mechanical Engineering
Queen's University
Stuart Street
Kingston, Ontario, K7L 3N6, Canada

Johannes Janicka
Energie- und Kraftwerkstechnik
TU Darmstadt
Petersenstr. 30
Darmstadt, , 64287, Germany
janicka@hrz1.hrz.tu-darmstadt.de

Matthew Johnson
Department of Mechanical Engineering
University of Alberta
4-9 Mech. Eng. Bldg
Edmonton, Alberta, T6G 2G8, Canada
Matt.Johnson@ualberta.ca

Daehan Jung
Department of Mechanical Eng.
State University of New York Buffalo

Buffalo, NY, 14260, USA
dj3@eng.buffalo.edu

Matthew Juniper
CNRS.
Ecole Centrale Paris
Grande Voie des Vignes

Chatenay-Malabry, , 92295, France
juniper@em2c.ecp.fr

Nicolas Lardjane
CNRS - LCSR
University d'Orleans
1C, Avenue de la Recherche
Scientifique
Orleans, , 45071, France
lardjane@cnrs-orleans.fr

Philippe Lavoie
Department of Mechanical Engineering
Queen's University
McLaughlin Hall
Kingston, Ontario, K7L 3N6, Canada
lavoie@me.queensu.ca

Diego Lentini
Dipart. di Meccanica Aeronautica
Universita di Roma "La Sapienza"
Via Eudossiana 18
Roma, RM, I - 00184, Italy
d.lentini@dma.ing.uniroma1.it

Jeffrey J. Lienau
Pratt and Whitney
Propulsion Systems Analysis/CFD
Group
400 Main Street M/S 163-17
East Hartford, Connecticut, 06108,
USA
liejeff@pweh.com

Marilyn Lightstone
Dept. Mechanical Engineering
McMaster University
1280 Main Street West
Hamilton, Ontario, L8S 4L7, Canada
lightsm@mcmail.cis.mcmaster.ca

Daniel Livescu
Dept. Mechanical Engineering
State University of New York at Buffalo
318 Jarvis Hall, SUNY
Buffalo, NY, 14260, USA
dlivescu@eng.buffalo.edu

Cyrus K. Madnia
Dept. Mechanical Engineering
State University of New York Buffalo
326 Jarvis Hall
Buffalo, NY, 14260-4400, USA
madnia@eng.buffalo.edu

Darko Matovic
Dept. Mechanical Engineering
Queen's University
McLaughlin Hall
Kingston, Ontario, K7L 3N6, Canada
matovic@me.queensu.ca

Charles Meneveau
Mechanical Engineering
John Hopkins University
3400 N. Charles Street
Baltimore, MD, 21218, USA
meneveau@jhu.edu

Suresh Menon
School of Aerospace Engineering
Georgia Institute of Technology
270 Ferst Drive

Atlanta, Georgia, 30332-0150, USA
suresh.menon@ae.gatech.edu

Berangere Mesnier
Lab. Combustion Systemes Reactifs
CNRS
1C Avenue de la Researche
Scientifique
Orleans Cedex 2, 45971, FRANCE
mesnier@cnrs-orleans.fr

Yasuhiro Mizobuchi
Computational Science Division
National Aerospace Laboratory
7-44-1 Jindaiji-Higash, Chofu
Tokyo, 182-8522, Japan
mizo@nal.go.jp

Parviz Moin
Center for Turbulance Research
Stanford
Building 500, room 500c
Stanford, CA, 94305-3030, USA
moin@stanford.edu

Laurent Mydlarski
Dept. of Mechanical Engineering
McGill University
817 Sherbrooke St. West
Montreal, Quebec, H3A-2K6, Canada
laurentm@mecheng.mcgill.ca

Tore Myhrvold
Dept Applied Mechanics,
Thermodynamics and Fluid D
Norwegian University of Science and Technology
Trondheim, , N-7491, Norway
tore.myhrvold@energy.sintef.no

Kazuyoshi Nakabe
Dept. Mechanical Engineering
Kyoto University
Kyoto, , 606-8501, Japan
nakabe@mech.kyoto-u.ac.jp

Norbert Peters
Department of Mechanical Eng.
Stanford University
Bldg 500, Room 500M
Stanford, CA, 94305, USA
peters@ctr.stanford.edu

Thierry Poinsot
IMF Toulouse & CERFACS Toulouse
Institut de Mecanique des Fluides de Toulouse
42 Avenue G. Coriolis
Toulouse, , 31057, FRANCE
poinsot@imft.fr

Andrew Pollard
Dept. Mechanical Engineering
Queen's University
McLaughlin Hall, Stuart Street
Kingston, Ontario, K7L 3N6, Canada
pollard@me.queensu.ca

Stephen B. Pope
Dept. Mechanical & Aerospace Engineering
Cornell University
Upson Hall, Cornell Universiy
Ithaca, NY, 14853, USA
pope@mae.cornell.edu

James J. Riley
Department of Mechanical Eng.

University of Washingtson
Box 352600
Seattle, Washington, 98195, USA
rileyj@u.washington.edu

Gioivanni Paolo Romano
Dept. Mechanics and Aeronautics
University "la Sapienza"
Via Eudossiana 18
Roma, , 00184, Italy
romano@dma.ing.uniroma1.it

Jimmy Sapede
Departement Energetique Industrielle
Ecole des Mines de Douai
941 Rue Charles Bourseul
Douai, , 59508, France
sapede@ensm-douai.fr

Sutanu Sarkar
Mechanical and Aerospace Engineering
University of California at San Diego
9500 Gilman Drive
La Jolla, CA, 92093, USA
ssarkar@ucsd.edu

Lorenz Sigurdson
Vortex Fluid Dynamics Lab
University of Alberta
4-8D Mech. Eng. Bldg.
Edmonton, Alberta, T6G 2G8,
Canada
lorenz.sigurdson@ualberta.ca

Thierry Souleres
Department of Mechanical and Nuclear Engineering
Pennsylvania State University
232 Research Building East
University Park, PA, 16802-2321, USA
tus5@psu.edu

Eugene Suk
Department of Mechanical Eng.

University of Toronto
5 King's College Road
Toronto, Ontario, M5S 3G8, Canada

Pierre Sullivan
Dept. of Mechanical and Industrial Engineering
University of Toronto
5 King's College Road
Toronto, Ontario, M5S 3G8, Canada
sullivan@mie.utoronto.ca

Robert Szasz
Dept. Heat and Power Engineering
Lund Institute of Technology
P.O. Box 118
Lund, , 22100, Sweden
szrobi@mail.vok.lth.se

Tadao Takeno
Dept. Mechanical Engineering
Meijo University
Tenpaku-ku
Nagoya, 468-8502, Japan
takeno@meijo-u.ac.jp

Mamoru Tanahashi
Dept. Mechanical & Aerospace Eng.
Tokyo Institute of Technology
2-12-1 Ookayama, Meguro-ku
Tokyo, 152-8552, Japan
mtanahas@mes.titech.ac.jp

Qing Tang
Mechanical and Aerospace Eng.
Cornell University
Box 33, Upson Hall
Ithaca, NY, 14853-7501, USA
qtang@eccentric.mae.cornell.edu

Tomomasa Tatsumi
International Institute for Advanced Studies
26-6 Chikuzen-dai, Momoyama, Fushimi-Ku
Kyoto, , 612-8032, Japan
tatsumi@skyblue.ocn.ne.jp

Stavros Tavoularis
Mechanical Engineering
University of Ottawa
P.O. Box 450, Stn. A
Ottawa, Ontario, K1N 6N5, Canada
tav@eng.uottawa.ca

Dominique Thevenin
Laboratoire EM2C
Ecole Centrale de Paris
Grande Voie des Vignes
Chatenay-Malabry Cedex, , 92295, France
thevenin@em2c.ecp.fr

Murray Thomson
Mechanical Engineering
University of Toronto
5 King's College Road
Toronto, Ontario, M5S 3G8, Canada
thomson@mie.utoronto.ca

Arnaud Trouve
Dept. Fire Protection Engineering
University of Maryland
Glenn L. Martin Hall
College Park, Maryland, 20742-3031, USA
atrouve@eng.umd.edu

Stephen Tullis
Department of Engineering
Cambridge University
Trumpington Street
Cambridge, , CB2 1PZ, United Kingdom
swt21@eng.cam.ac.uk

Luc Vervisch
INSA de Rouen
Avenue de l'Universite – BP-8
Saint Etienne-du-Rouvray cedex
76801 France
vervisch@coria.fr

Denis Veynante
Laboratoire EM2C.
CNRS Ecole Centrale Paris
Grande Voie des Vignes
Chatenay-Malabry Cedex, , 92295, France
denis@em2c.ecp.fr

Emmanuel Villermaux
IRHPE
Universite d'Aix-Marseille I & II
49 Avenue F. Joloit Curie
Marseille cedex 13
13384 France
emmanuel.villermaux@irphe.univ-mrs.fr

Stephanie Waterman
Dept. Mechanical Engineering
Queen's University
Stuart Street
Kingston, Ontario, K7L 3N6, Canada

Zhenyu Wen
Mechanical Engineering
University of Toronto
c/o 30 Charles St. W. #622
Toronto, ON M4Y 1R5
Zywen@mie.utoronto.ca

V.I. Zapryagaev
Institute Theoretical & Applied
Mechanics University of California
Russian Academy of Sciences, Siberian Branch
Institutskaya str., 4/1
Novosibirsk, , 630090, Russia
zapr@itam.nsc.ru

Ye Zhou
Lawrence Livermore National Lab

7000 East Avenue, P.O. Box 808, L-023
Livermore, California, 94551, USA
zhou3@llnl.gov

Mechanics

FLUID MECHANICS AND ITS APPLICATIONS

Series Editor: R. Moreau

46. U. Frisch (ed.): *Advances in Turbulence VII.* Proceedings of the Seventh European Turbulence Conference, held in Saint-Jean Cap Ferrat, 30 June–3 July 1998. 1998 ISBN 0-7923-5115-0
47. E.F. Toro and J.F. Clarke: *Numerical Methods for Wave Propagation.* Selected Contributions from the Workshop held in Manchester, UK. 1998 ISBN 0-7923-5125-8
48. A. Yoshizawa: *Hydrodynamic and Magnetohydrodynamic Turbulent Flows.* Modelling and Statistical Theory. 1998 ISBN 0-7923-5225-4
49. T.L. Geers (ed.): *IUTAM Symposium on Computational Methods for Unbounded Domains.* 1998 ISBN 0-7923-5266-1
50. Z. Zapryanov and S. Tabakova: *Dynamics of Bubbles, Drops and Rigid Particles.* 1999 ISBN 0-7923-5347-1
51. A. Alemany, Ph. Marty and J.P. Thibault (eds.): *Transfer Phenomena in Magnetohydrodynamic and Electroconducting Flows.* 1999 ISBN 0-7923-5532-6
52. J.N. Sørensen, E.J. Hopfinger and N. Aubry (eds.): *IUTAM Symposium on Simulation and Identification of Organized Structures in Flows.* 1999 ISBN 0-7923-5603-9
53. G.E.A. Meier and P.R. Viswanath (eds.): *IUTAM Symposium on Mechanics of Passive and Active Flow Control.* 1999 ISBN 0-7923-5928-3
54. D. Knight and L. Sakell (eds.): *Recent Advances in DNS and LES.* 1999 ISBN 0-7923-6004-4
55. P. Orlandi: *Fluid Flow Phenomena.* A Numerical Toolkit. 2000 ISBN 0-7923-6095-8
56. M. Stanislas, J. Kompenhans and J. Westerveel (eds.): *Particle Image Velocimetry.* Progress towards Industrial Application. 2000 ISBN 0-7923-6160-1
57. H.-C. Chang (ed.): *IUTAM Symposium on Nonlinear Waves in Multi-Phase Flow.* 2000 ISBN 0-7923-6454-6
58. R.M. Kerr and Y. Kimura (eds.): *IUTAM Symposium on Developments in Geophysical Turbulence* held at the National Center for Atmospheric Research, (Boulder, CO, June 16–19, 1998) 2000 ISBN 0-7923-6673-5
59. T. Kambe, T. Nakano and T. Miyauchi (eds.): *IUTAM Symposium on Geometry and Statistics of Turbulence.* Proceedings of the IUTAM Symposium held at the Shonan International Village Center, Hayama (Kanagawa-ken, Japan November 2–5, 1999). 2001 ISBN 0-7923-6711-1
60. V.V. Aristov: *Direct Methods for Solving the Boltzmann Equation and Study of Nonequilibrium Flows.* 2001 ISBN 0-7923-6831-2
61. P.F. Hodnett (ed.): *IUTAM Symposium on Advances in Mathematical Modelling of Atmosphere and Ocean Dynamics.* Proceedings of the IUTAM Symposium held in Limerick, Ireland, 2–7 July 2000. 2001 ISBN 0-7923-7075-9
62. A.C. King and Y.D. Shikhmurzaev (eds.): *IUTAM Symposium on Free Surface Flows.* Proceedings of the IUTAM Symposium held in Birmingham, United Kingdom, 10–14 July 2000. 2001 ISBN 0-7923-7085-6
63. A. Tsinober: *An Informal Introduction to Turbulence.* 2001 ISBN 1-4020-0110-X; Pb: 1-4020-0166-5
64. R.Kh. Zeytounian: *Asymptotic Modelling of Fluid Flow Phenomena.* 2002 ISBN 1-4020-0432-X
65. R. Friedrich and W. Rodi (eds.): *Advances in LES of Complex Flows.* Prodeedings of the EUROMECH Colloquium 412, held in Munich, Germany, 4-6 October 2000. 2002 ISBN 1-4020-0486-9
66. D. Drikakis and B.J. Geurts (eds.): *Turbulent Flow Computation.* 2002 ISBN 1-4020-0523-7
67. B.O. Enflo and C.M. Hedberg: *Theory of Nonlinear Acoustics in Fluids.* 2002 ISBN 1-4020-0572-5

Mechanics

FLUID MECHANICS AND ITS APPLICATIONS

Series Editor: R. Moreau

68. I.D. Abrahams, P.A. Martin and M.J. Simon (eds.): *IUTAM Symposium on Diffraction and Scattering in Fluid Mechanics and Elasticity*. Proceedings of the IUTAM Symposium held in Manchester, (UK, 16-20 July 2000). 2002 ISBN 1-4020-0590-3
69. P. Chassaing, R.A. Antonia, F. Anselmet, L. Joly and S. Sarkar: *Variable Density Fluid Turbulence*. 2002 ISBN 1-4020-0671-3

Kluwer Academic Publishers – Dordrecht / Boston / London

GPSR Compliance

The European Union's (EU) General Product Safety Regulation (GPSR) is a set of rules that requires consumer products to be safe and our obligations to ensure this.

If you have any concerns about our products, you can contact us on ProductSafety@springernature.com

In case Publisher is established outside the EU, the EU authorized representative is:

Springer Nature Customer Service Center GmbH
Europaplatz 3
69115 Heidelberg, Germany

Batch number: 10809439

Printed by Printforce, the Netherlands